T0314955

INTRODUCTION TO MODERN POWER ELECTRONICS

INTRODUCTION TO MODERN POWER ELECTRONICS

THIRD EDITION

Andrzej M. Trzynadlowski

Library of Congress Cataloging-in-Publication Data:

Trzynadlowski, Andrzej.
  Introduction to modern power electronics / Andrzej M. Trzynadlowski. – Third edition.
    pages cm
  Includes bibliographical references and index.
  ISBN 978-1-119-00321-2 (cloth)
  1. Power electronics.   I. Title.
  TK7881.15.T79 2016
  621.31'7–dc23

For Dorota, Bart, Nicole, Genie, Gary, and Guy

CONTENTS

PREFACE

This text is primarily intended for a one-semester introductory course in power electronics at the undergraduate level. However, containing a comprehensive overview of modern tools and techniques of electric power conditioning, the book can also be used in more advanced classes. Practicing engineers wishing to refresh their knowledge of power electronics, or interested in branching into that area, are also envisioned as potential readers. Students are assumed to have working knowledge of the electric circuit analysis and basic electronics.

During the five years since the second edition of the book was published, power electronics has enjoyed robust progress. Novel converter topologies, applications, and control techniques have been developed. Utilizing advanced semiconductor switches, power converters reach ratings of several kilovolts and kiloamperes. The threat of unchecked global warming, various geopolitical and environmental issues, and the monetary and ecological costs of fossil fuels represent serious energy challenges, which set off intensive interest in sources of clean power. As a result, power electronic systems become increasingly important and ubiquitous. Changes made to this third edition reflect the dominant trends of modern power electronics. They encompass the growing practical significance of PWM rectifiers, the Z-source dc link, matrix converters, and multilevel inverters, and their application in renewable energy systems and powertrains of electric and hybrid vehicles.

In contrast with most books, which begin with a general introduction devoid of detailed information, Chapter 1 constitutes an important part of the teaching process. Employing a hypothetical generic power converter, basic principles and methods of power electronics are explained. Therefore, whatever content sequence an instructor wants to adopt, Chapter 1 should be covered first.

Chapters 2 and 3 provide description of semiconductor power switches and supplementary components and systems of power electronic converters. The reader should be aware of the existence and function of those auxiliary but important parts although the book is mostly focused on power circuits, operating characteristics, control, and applications of the converters.

The four fundamental types of electrical power conversion—ac to dc, ac to ac, dc to dc, and dc to ac—are covered in Chapters 4 through 7, respectively. Chapters 4 and 7, on rectifiers and inverters, are the longest chapters, reflecting the great importance of those converters in modern power electronics. Chapter 8 is devoted to switching dc power supplies, and Chapter 9 covers applications of power electronics in clean energy systems.

Each chapter begins with an abstract and includes a brief summary that follows the main body. Numerical examples, homework problems, and computer assignments complement most chapters. Several relevant and easily available references are provided after each of them. Three appendices conclude the book.

The textbook is accompanied by a series of forty-six PSpice circuit files constituting a virtual power electronics laboratory, and available at http://www.wiley.com/go/modernpowerelectronics3e. The files contain computer models of most power electronic converters covered in the book. The models are a valuable teaching tool, giving the reader an opportunity to tinker with the converters and visualize their operation. Another teaching tool, a PowerPoint presentation, which contains all figures, tables, and most important formulas, is also available, at http://www.wiley.com/go/modernpowerelectronics3e. It will ease the instructor from drawing the often complex circuit diagrams and waveforms on the classroom board.

Against most of the contemporary engineering textbooks, the book is quite concise. Still, covering the whole material in a single-semester course requires from the students a substantial homework effort. The suggested teaching approach would consist in presenting the basic issues in class and letting the students to broaden their knowledge by reading assigned materials, solving problems, and performing PSpice simulations.

I want to express my gratitude to the reviewers of the book proposal, whose valuable comments and suggestions have been greatly appreciated. My students at the University of Nevada, Reno, who used the first and second editions for so many years, provided very constructive critiques as well. Finally, my wife Dorota and children Bart and Nicole receive apologies for my long preoccupation, and many thanks for their unwavering support.

ANDRZEJ M. TRZYNADLOWSKI

ABOUT THE COMPANION WEBSITE

This book is accompanied by a companion website:

www.wiley.com/go/modernpowerelectronics3e

The website includes:

- PSpice circuit files
- Power Point Presentation
- Solutions Manual available for instructors

1 Principles of Electric Power Conversion

In this introductory chapter, fundamentals of power electronics are outlined, including the scope, tools, and applications of this area of electrical engineering. The concept of generic power converter is introduced to illustrate the operating principles of power electronic converters and the types of power conversion performed. Components of voltage and current waveforms, and the related figures of merit, are defined. Two basic methods of magnitude control, that is, phase control and pulse width modulation (PWM), are presented. Calculation of current waveforms is explained. The single-phase diode rectifier is described as the simplest power electronic converter.

1.1 WHAT IS POWER ELECTRONICS?

Modern society with its conveniences strongly relies on the ubiquitous availability of electric energy. The electricity performs most of the physical labor, provides the heating and lighting, activates electrochemical processes, and facilitates information collecting, processing, storage, and exchange.

Power electronics can be defined as a branch of electrical engineering devoted to conversion and control of electric power, using electronic converters based on semiconductor power switches. The power grid delivers an ac voltage of fixed frequency and magnitude. Typically, homes, offices, stores, and other small facilities are supplied from single-phase, low-voltage power lines, while three-phase supply systems with various voltage levels are available in industrial plants and other large commercial enterprises. The 60-Hz (50-Hz in most other parts of the world) fixed-voltage electric power can be thought of as raw power, which for many applications must be conditioned. The power conditioning involves conversion, from ac to dc or vice-versa, and control of the magnitude and/or frequency of voltages and currents. Using the electric lighting as a simple example, an incandescent bulb can directly be supplied with the raw power. However, a fluorescent lamp requires electronic ballast that starts and stabilizes the electric arc. The ballast is thus a power conditioner, necessary for proper operation of the lamp. If used in a movie theater, the incandescent bulb mentioned before is supplied from an ac voltage controller that allows dimming

Introduction to Modern Power Electronics, Third Edition. Andrzej M. Trzynadlowski.
© 2016 John Wiley & Sons, Inc. Published 2016 by John Wiley & Sons, Inc.
Companion website: www.wiley.com/go/modernpowerelectronics3e

of the light just before the movie begins. Again, this controller constitutes an example of power conditioner, or power converter.

Raw dc power is usually supplied from batteries and, increasingly, from photovoltaic sources and fuel cells. Photovoltaic energy systems are usually connected to the grid, and the necessary power conditioning involves dc-to-ac voltage conversion and control of the ac voltage. If a dc source feeds an electric motor, as in a golf cart or an electric wheelchair, a power electronic converter between the battery and the motor performs voltage control and facilitates reverse power flow during braking or downhill ride.

The birth of power electronics can be traced back to the dawn of twentieth century when the first mercury arc rectifiers were invented. However, for conversion and control of electric power, *rotating electro-machine converters* were mostly used in the past. An electro-machine converter was an electric generator driven by an electric motor. If, for instance, adjustable dc voltage was to be obtained from fixed ac voltage, an ac motor operated a dc generator with controlled output voltage. Conversely, if ac voltage was required and the supply energy came from a battery pack, a speed-controlled dc motor and an ac synchronous generator were employed. Clearly, the convenience, efficiency, and reliability of such systems were inferior in comparison with today's *static power electronic converters* performing motionless energy conversion and control.

Today's power electronics has begun with the development of the *silicon controlled rectifier (SCR)*, also called a *thyristor*, by the General Electric Company in 1958. The SCR is a unidirectional semiconductor power switch that can be turned on ("closed") by a low-power electric pulse applied to its controlling electrode, the gate. The available voltage and current ratings of SCRs are very high, but the SCR is inconvenient for use in dc-input power electronic converters. It is a *semi-controlled* switch, which when conducting current cannot be turned off ("opened") by a gate signal. Within the last few decades, several kinds of *fully controlled* semiconductor power switches that can be turned on and off have been introduced to the market.

Widespread introduction of power electronic converters to most areas of distribution and usage of electric energy is common for all developed countries. The converters condition the electric power for a variety of applications, such as electric motor drives, uninterruptable power supplies, heating and lighting, electrochemical and electro-thermal processes, electric arc welding, high-voltage dc transmission lines, active power filters and reactive power compensators in power systems, and high-quality supply sources for computers and other electronic equipment.

It is estimated that at least half of the electric power generated in the USA flows through power electronic converters, and an increase of this share to close to 100% in the next few decades is expected. In particular, a thorough revamping of the existing US national power grid is envisioned. Introduction of power electronic converters to all stages of the power generation, transmission, and distribution, coupled with extensive information exchange ("smart grid"), allows a dramatic increase of the grid's capabilities without investing in new power plants and transmission lines. The important role of power electronics in renewable energy systems and electric and hybrid vehicles is also worth stressing. It is safe to say that practically every electrical engineer encounters some power electronic converters in his/her professional career.

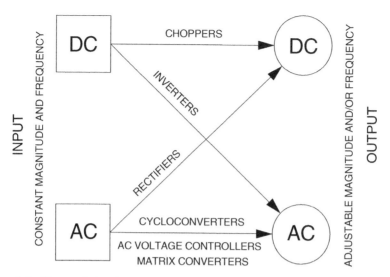

Figure 1.1 Types of electric power conversion and the corresponding power electronic converters.

Types of electric power conversion and the corresponding converters are presented in Figure 1.1. For instance, the ac-to-dc conversion is accomplished using rectifiers, which are supplied from an ac source and whose output voltage contains a fixed or adjustable dc component. Individual kinds of power electronic converters are described and analyzed in Chapters 4 through 8. Basic principles of power conversion and control are explained in the following sections of this chapter.

1.2   GENERIC POWER CONVERTER

Though not a practical apparatus, the hypothetical *generic power converter* shown in Figure 1.2 is a useful tool for illustration of the principles of electric power conversion and control. It is a two-port network of five switches. Switches S1 and

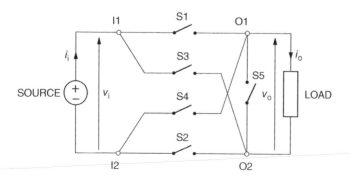

Figure 1.2   Generic power converter.

S2 provide *direct connection* between the input (supply) terminals, I1 and I2, and the output (load) terminals, O1 and O2, respectively, while switches S3 and S4 allow *cross connection* between these pairs of terminals. A voltage source, either dc or ac, supplies the electric power to a load through the converter. Practical loads often contain a significant inductive component, so a resistive–inductive (RL) load is assumed in the subsequent considerations. To ensure a closed path for the load current under any operating conditions, a fifth switch, S5, is connected between the output terminals of the converter and closed when switches S1 through S4 are open. It is assumed that the switches open or close instantaneously.

The supply source is an ideal voltage source and as such it may not be shorted. Also, the load current may not be interrupted. As the voltage across inductance is proportional to the rate of change of current, a rapid drop of that current would cause a high and potentially damaging overvoltage. Therefore, the generic converter can only assume the following three states:

State 0: Switches S1 through S4 are open and switch S5 is closed, shorting the output terminals and closing a path for the lingering load current, if any. The output voltage is zero. The input terminals are cut off from the output terminals so that the input current is also zero.
State 1: Switches S1 and S2 are closed, and the remaining ones are open. The output voltage equals the input voltage and the output current equals the input current.
State 2: Switches S3 and S4 are closed, and the remaining ones are open. Now, the output voltage and current are reversed with respect to their input counterparts.

Let us assume that the generic converter is to perform the ac-to-dc conversion. The sinusoidal input voltage, v_i, whose waveform is shown in Figure 1.3, is given by

$$v_i = V_{i,p} \sin(\omega t), \tag{1.1}$$

where $V_{i,p}$ denotes the peak value of that voltage and ω is the input radian frequency. The output voltage, v_o, of the converter should contain a possibly large dc component.

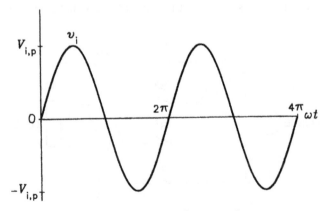

Figure 1.3   Input ac voltage waveform.

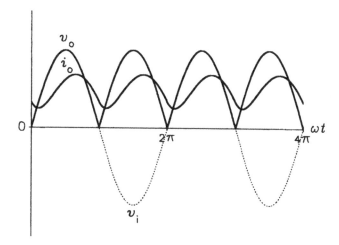

Figure 1.4   Output voltage and current waveforms in the generic rectifier.

Note that the output voltage is not expected to be of ideal dc quality, since such voltage and current waveforms are not feasible in the generic converter, as well as in practical power electronic converters. The same applies to the ideally sinusoidal output voltage and current in ac output converters. If within the first half-cycle of the input voltage, the converter is in state 1, and within the second half-cycle in state 2, the output voltage waveform will be such as depicted in Figure 1.4, that is,

$$v_o = |v_i| = V_{i,p}|\sin(\omega t)|. \tag{1.2}$$

The dc component is the average value of the voltage. Power electronic converters performing the ac-to-dc conversion are called *rectifiers*.

The output current waveform, i_o, can be obtained as a numerical solution of the load equation:

$$L\frac{di_o}{dt} + Ri_o = v_o. \tag{1.3}$$

Techniques for analytical and numerical computation of voltage and current waveforms in power electronic circuits are described at the end of this chapter. Here, only general features of the waveforms are outlined. The output current waveform of the considered generic rectifier is also shown in Figure 1.4. It can be seen that this waveform is closer to an ideal dc waveform than is the output voltage waveform because of the frequency-dependent load impedance. The kth harmonic, $v_{o,k}$, of the output voltage produces the corresponding harmonic, $i_{o,k}$, of the output current such that

$$I_{o,k} = \frac{V_{o,k}}{\sqrt{R^2 + (k\omega_o L)^2}}, \tag{1.4}$$

where $I_{o,k}$ and $V_{o,k}$ denote root mean square (rms) values of the current and voltage harmonics in question, respectively. In the considered rectifier, the fundamental radian frequency, ω_o, of the output voltage is twice as high as the input frequency, ω. The load impedance (represented by the denominator at the right-hand side of Eq. 1.4) for individual current harmonics increases with the harmonic number, k. Clearly, the dc component ($k = 0$) of the output current encounters the lowest impedance, equal to the load resistance only, while the load inductance attenuates only the ac component. In other words, the RL load acts as a low-pass filter.

Interestingly, if an ac output voltage is to be produced and the generic converter is supplied from a dc source, so that the input voltage is $v_i = V_i = $ const., the switches are operated in the same manner as in the previous case. Specifically, for every half period of the desired output frequency, states 1 and 2 are interchanged. In this way, the input terminals are alternately connected and cross-connected with the output terminals, and the output voltage acquires the ac (although not sinusoidal) waveform shown in Figure 1.5. The output current is composed of growth-function and decay-function segments, typical for transient conditions of an RL circuit subjected to dc excitation. Again, thanks to the attenuating effects of the load inductance, the current waveform is closer to the desired sinusoid than is the voltage waveform. In practice, the dc-to-ac power conversion is performed by power electronic *inverters*. In the case described, the generic inverter is said to operate in the *square-wave mode*.

If the input or output voltage is to be a three-phase ac voltage, the topology of the generic power converter portrayed here would have to be expanded, but it still would be a network of switches. Real power electronic converters are *networks of semiconductor power switches*, too. For various purposes, other elements, such as inductors, capacitors, fuses, and auxiliary circuits, are employed besides the switches in power circuits of practical power electronic converters. Yet, in most of these converters, the fundamental operating principle is the same as in the generic converter, that is, the

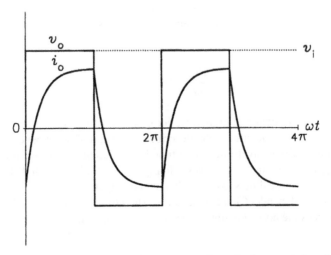

Figure 1.5   Output voltage and current waveforms in the generic inverter.

input and output terminals are being connected, cross-connected, and disconnected in a specific manner and sequence required for the given type of power conversion. Typically, as in the generic rectifier and inverter presented, the load inductance inhibits the switching-related undesirable high-frequency components of the output current.

Although a *voltage source* has been assumed for the generic power converter, some power electronic converters are supplied from *current sources*. In such converters, a large inductor is connected in series with the input terminals to prevent rapid changes of the input current. Analogously, voltage-source converters usually have a large capacitor connected across the input terminals to stabilize the input voltage. Inductors or capacitors are also used at the output of some converters to smooth the output current or voltage, respectively.

According to one of the tenets of circuit theory, two unequal ideal current sources may not be connected in series and two unequal ideal voltage sources may not be connected in parallel. Consequently, the load of a current-source converter may not appear as a current source while that of a voltage-source converter as a voltage source. As illustrated in Figure 1.6, it means that in a current-source power electronic converter a capacitor should be placed in parallel with the load. In addition to smoothing the output voltage, the capacitor prevents the potential hazards of connecting the input inductance conducting certain current with a load inductance conducting different current. In contrast, in voltage-source converters, no capacitor may be connected across the output terminals and it is the load inductance, or an extra inductor between the converter and the load, that is smoothing the output current.

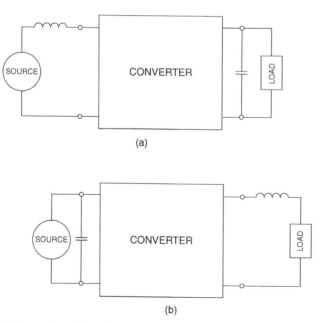

(a)

(b)

Figure 1.6   Basic configurations of power electronic converters: (a) current source, (b) voltage source.

1.3   WAVEFORM COMPONENTS AND FIGURES OF MERIT

Terms such as the "dc component," "ac component," and "harmonics" mentioned in the preceding section deserve closer examination. Knowledge of the basic components of voltage and current waveforms allows evaluation of performance of a converter. Certain relations of these components are commonly used as performance indicators, or *figures of merit*.

A time function, $\psi(t)$, here a waveform of voltage or current, is said to be *periodic* with a period T if

$$\psi(t) = \psi(t + T), \tag{1.5}$$

that is, if the pattern (shape) of the waveform is repeated every T seconds. In the realm of power electronics, it is often convenient to analyze voltages and currents in the *angle domain* instead of the *time domain*. The so-called *fundamental frequency*, f_1, in Hz, is defined as

$$f_1 = \frac{1}{T}, \tag{1.6}$$

and the corresponding *fundamental radian frequency*, ω, in rad/s, as

$$\omega = 2\pi f_1 = \frac{2\pi}{T}. \tag{1.7}$$

Now, a periodic function $\psi(\omega t)$ can be defined as such that

$$\psi(\omega t) = \psi(\omega t + 2\pi). \tag{1.8}$$

The *rms value*, Ψ, of waveform $\psi(\omega t)$ is defined as

$$\Psi \equiv \sqrt{\frac{1}{2\pi} \int_0^{2\pi} \psi^2(\omega t)\, d\omega t}, \tag{1.9}$$

and the *average value*, or *dc component*, Ψ_{dc}, of the waveform as

$$\Psi_{dc} \equiv \frac{1}{2\pi} \int_0^{2\pi} \psi(\omega t)\, d\omega t. \tag{1.10}$$

When the dc component is subtracted from the waveform, the remaining waveform, $\psi_{ac}(\omega t)$, is called the *ac component*, or *ripple*, that is,

$$\psi_{ac}(\omega t) = \psi(\omega t) - \Psi_{dc}. \tag{1.11}$$

The ac component has an average value of zero and the fundamental frequency of f_1.

The rms value, Ψ_{ac}, of $\psi_{ac}(\omega t)$ is defined as

$$\Psi_{ac} \equiv \sqrt{\frac{1}{2\pi} \int_0^{2\pi} \psi_{ac}^2(\omega t)d\omega t}, \tag{1.12}$$

and it can be shown that

$$\Psi^2 = \Psi_{dc}^2 + \Psi_{ac}^2. \tag{1.13}$$

For waveforms of the desirable ideal dc quality, such as the load current of a rectifier, a figure of merit called a *ripple factor*, *RF*, is defined as

$$RF = \frac{\Psi_{ac}}{\Psi_{dc}}. \tag{1.14}$$

A low value of the ripple factor indicates high quality of a waveform.

Before proceeding to other waveform components and figures of merit, the terms and formulas introduced so far will be illustrated using the waveform of output voltage, v_o, of the generic rectifier, shown in Figure 1.4. The waveform pattern is repeating itself every π radians and, within the 0 to π interval, $v_o = v_i$. Therefore, the average value, $V_{o,dc}$, of the output voltage can most conveniently be determined by calculating the area under the waveform from $\omega t = 0$ to $\omega t = \pi$ and dividing it by the length, π, of the considered interval. Thus,

$$V_{o,dc} = \frac{1}{\pi} \int_0^{\pi} V_{i,p}\sin(\omega t)d\omega t = \frac{2}{\pi}V_{i,p} = 0.64V_{i,p}. \tag{1.15}$$

Note that the formula above differs from Eq. (1.10). Since $\omega_1 = \omega_o = 2\omega$, the integration is performed in the 0 to π interval of ωt instead of the 0 to 2π interval of $\omega_1 t$.

Similarly, the rms value, V_o, of the output voltage can be calculated as

$$V_o = \sqrt{\frac{1}{\pi} \int_0^{\pi} [V_{i,p}\sin(\omega t)]^2 d\omega t} = \frac{V_{i,p}}{\sqrt{2}} = 0.71V_{i,p}. \tag{1.16}$$

The result in Eq. (1.16) agrees with the well-known relation for a sine wave as $v_o^2 = v_i^2$.

Based on Eqs. (1.13) and (1.14), the rms value, $V_{o,ac}$, of ac component of the voltage in question can be calculated as

$$V_{o,ac} = \sqrt{V_o^2 - V_{o,dc}^2} = \sqrt{\left(\frac{V_{i,p}}{\sqrt{2}}\right)^2 - \left(\frac{2}{\pi}V_{i,p}\right)^2} = 0.31V_{i,p}, \qquad (1.17)$$

and the ripple factor, RF_V, of the voltage as

$$RF_V = \frac{V_{o,ac}}{V_{o,dc}} = \frac{0.31V_{i,p}}{0.64V_{i,p}} = 0.48. \qquad (1.18)$$

Decomposition of the analyzed waveform into the dc and ac components is shown in Figure 1.7.

To analytically determine the ripple factor, RF_I, of the output current, the output current waveform, $i_o(\omega t)$, would have to be expressed in a closed form. Instead, numerical computations were performed on the waveform in Figure 1.4, and RF_I was found to be 0.31. This value is 36% lower than that of the output voltage. This is an example only, but output currents in power electronic converters routinely have higher quality than the output voltages. It is worth mentioning that the obtained value of RF_I is poor. Practical high-quality dc current waveforms have the ripple factor in the order of few percentage points, and below the 5% level the current is considered as ideal. The current ripple factor depends on the type of converter, and it decreases with an increase in the inductive component of the load. Components of the current waveform evaluated are shown in Figure 1.8.

The ripple factor is of no use for quality evaluation of ac waveforms, such as the output current of an inverter, which ideally should be pure sinusoids. However, as

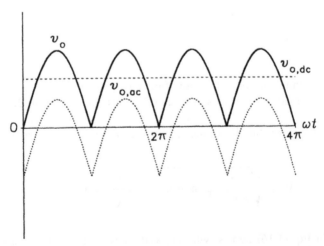

Figure 1.7 Decomposition of the output voltage waveform in the generic rectifier.

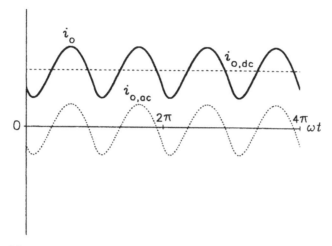

Figure 1.8   Decomposition of the output current waveform in the generic rectifier.

already mentioned and exemplified by the waveforms in Figure 1.5, purely sinusoidal voltages and currents cannot be produced by switching power converters. Therefore, an appropriate figure of merit must be defined as a measure of deviation of a practical ac waveform from its ideal counterpart.

Following the theory of Fourier series (see Appendix B), the ac component, $\psi_{ac}(t)$, of a periodic function, $\psi(t)$, can be expressed as an infinite sum of harmonics, that is, sine waves whose frequencies are multiples of the fundamental frequency, f_1, of $\psi(t)$. In the angle domain,

$$\psi_{ac}(\omega t) = \sum_{k=1}^{\infty} \psi_k(k\omega t) = \sum_{k=1}^{\infty} \Psi_{k,p} \cos(k\omega t + \varphi_k), \tag{1.19}$$

where k is the *harmonic number*, and $\Psi_{k,p}$ and φ_k denote the peak value and phase angle of the kth harmonic, respectively. The first harmonic, $\psi_1(\omega t)$, is called a *fundamental*. Terms "fundamental voltage" and "fundamental current" are used throughout the book to denote the fundamental of a given voltage or current.

The peak value, $\Psi_{1,p}$, of fundamental of a periodic function, $\psi(\omega t)$, is calculated as

$$\Psi_{1,p} = \sqrt{\Psi_{1,c}^2 + \Psi_{1,s}^2}, \tag{1.20}$$

where

$$\Psi_{1,c} = \frac{1}{\pi} \int_0^{2\pi} \psi(\omega t) \cos(\omega t) d\omega t, \tag{1.21}$$

$$\Psi_{1,s} = \frac{1}{\pi} \int_0^{2\pi} \psi(\omega t) \sin(\omega t) d\omega t, \tag{1.22}$$

and the rms value, Ψ_1, of the fundamental is

$$\Psi_1 = \frac{\Psi_{1,p}}{\sqrt{2}}. \tag{1.23}$$

Since the fundamental of a function does not depend on the dc component of the function, the ac component, $\psi_{ac}(\omega t)$, can be used in Eqs. (1.21) and (1.22) in place of $\psi(\omega t)$.

When the fundamental is subtracted from the ac component, the so-called *harmonic component*, $\psi_h(\omega t)$, is obtained as

$$\psi_h(\omega t) = \psi_{ac}(\omega t) - \psi_1(\omega t). \tag{1.24}$$

The rms value, Ψ_h, of $\psi_h(\omega t)$, called a *harmonic content* of function $\psi(\omega t)$, can be calculated as

$$\Psi_h = \sqrt{\Psi_{ac}^2 - \Psi_1^2} = \sqrt{\Psi^2 - \Psi_{dc}^2 - \Psi_1^2} \tag{1.25}$$

and used for calculation of the so-called *total harmonic distortion, THD*, defined as

$$\text{THD} \equiv \frac{\Psi_h}{\Psi_1}. \tag{1.26}$$

The concept of total harmonic distortion is also widely employed outside power electronics as, for instance, in the characterization of quality of audio equipment. Conceptually, the total harmonic distortion constitutes an ac counterpart of the ripple factor.

In the generic inverter, whose output waveforms have been shown in Figure 1.6, the rms value, V_o, of output voltage equals the dc input voltage, V_i. Since v_o is either V_i or $-V_i$, $v_o^2 = V_i^2$. The peak value, $V_{o,1,p}$, of fundamental output voltage is

$$V_{o,1,p} = V_{o,1,s}, \tag{1.27}$$

because the waveform in question has the odd symmetry (see Appendix B). Consequently,

$$V_{o,1,p} = \frac{2}{\pi} \int_0^\pi V_i \sin(\omega t) d\omega t = \frac{4}{\pi} V_i = 1.27 V_i. \tag{1.28}$$

Now, the fundamental output voltage, $v_{o,1}(\omega t)$, can be expressed as

$$v_{o,1}(\omega t) = V_{o,1,p} \sin(\omega t) = \frac{4}{\pi} V_i \sin(\omega t). \tag{1.29}$$

The rms value, $V_{o,1}$, of fundamental output voltage is

$$V_{o,1} = \frac{V_{o,1,p}}{\sqrt{2}} = \frac{2\sqrt{2}}{\pi} V_i = 0.9V_i, \qquad (1.30)$$

and the harmonic content, $V_{o,h}$, is

$$V_{o,h} = \sqrt{V_o^2 - V_{o,1}^2} = \sqrt{V_i^2 - \left(\frac{2\sqrt{2}}{\pi} V_i\right)^2} = 0.44V_i. \qquad (1.31)$$

Thus, the total harmonic distortion of the output voltage, THD_V, is

$$THD_V = \frac{V_{o,h}}{V_{o,1}} = \frac{0.44V_i}{0.9V_i} = 0.49. \qquad (1.32)$$

The high value of THD_V is not surprising since the output voltage waveform of the generic inverter operating in the square-wave mode significantly differs from a sine wave. Decomposition of the waveform analyzed is illustrated in Figure 1.9.

The numerically determined total harmonic distortion, THD_I, of the output current, i_o, in this example is 0.216, that is, less than that of the output voltage by as much as 55%. Indeed, as seen in Figure 1.10 that shows decomposition of the current waveform, the harmonic component is quite small in comparison with the fundamental. As in the generic rectifier, it shows the attenuating influence of the load inductance on the output current. In practical inverters, the output current is considered to be of high quality if THD_I does not exceed 0.05 (5%).

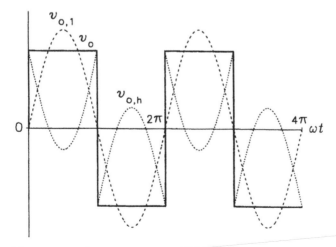

Figure 1.9   Decomposition of the output voltage waveform in the generic inverter.

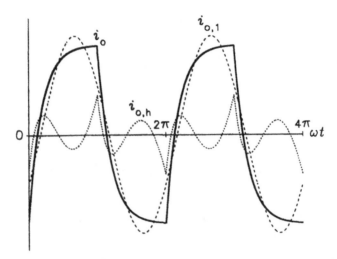

Figure 1.10 Decomposition of the output current waveform in the generic inverter.

Other figures of merit often employed for performance evaluation of power electronic converters are

(1) *Power efficiency*, η, of the converter is defined as

$$\eta \equiv \frac{P_o}{P_i}, \tag{1.33}$$

where P_o and P_i denote the output and input powers of the converter, respectively.

(2) *Conversion efficiency*, η_c, of the converter is defined as

$$\eta_c \equiv \frac{P_{o,dc}}{P_i} \tag{1.34}$$

for dc output converters, and

$$\eta_c \equiv \frac{P_{o,1}}{P_i} \tag{1.35}$$

for ac output converters. Symbol $P_{o,dc}$ denotes the dc output power, that is, the product of the dc components of the output voltage and current, while $P_{o,1}$ is the ac output power carried by the fundamental components of the output voltage and current.

(3) *Input power factor*, PF, of the converter is defined as

$$PF = \frac{P_i}{S_i}, \tag{1.36}$$

where S_i is the apparent input power. The power factor can also be expressed as

$$PF = K_d K_\Theta. \tag{1.37}$$

Here, K_d denotes the so-called *distortion factor* (not to be confused with the total harmonic distortion, THD), defined as the ratio of the rms fundamental input current, $I_{i,1}$, to the rms input current, I_i, and K_Θ is the *displacement factor*, that is, cosine of the phase shift, Θ, between the fundamentals of input voltage and current.

The power efficiency, η, of a converter simply indicates what portion of the power supplied to the converter reaches the load. In contrast, the conversion efficiency, η_c, expresses the relative amount of *useful* output power and, therefore, constitutes a more valuable figure of merit than the power efficiency. Since the input voltage to a converter is usually constant, the power factor serves mainly as a measure of utilization of the input current, drawn from the source that supplies the converter. With a constant power consumed by the converter, a high power factor implies a low current and, consequently, low power losses in the source. Most likely, the reader is familiar with the term "power factor" as the cosine of the phase shift between the voltage and current waveforms, as used in the theory of ac circuits. However, it must be stressed that it is true for purely sinusoidal waveforms only, and the general definition of the power factor is given by Eq. (1.36).

In an ideal power converter, all the three figures of merit defined above would be in equal unity. To illustrate the relevant calculations, the generic rectifier will again be employed. Since ideal switches have been assumed, no losses are incurred in the generic converter, so that the input power, P_i, and output power, P_o, are equal and the power efficiency, η, of the converter is always unity.

For the RL-load assumed for the generic rectifier, it can be shown that the conversion efficiency, η_c, is a function of the current ripple factor, RF_I. Specifically,

$$\eta_c = \frac{P_{o,dc}}{P_i} = \frac{P_{o,dc}}{\frac{P_o}{\eta}} = \eta \frac{RI_{o,dc}^2}{RI_o^2} = \eta \frac{I_{o,dc}^2}{I_{o,dc}^2 + I_{o,ac}^2}$$

$$= \frac{\eta}{1 + \left(\frac{I_{o,ac}}{I_{o,dc}}\right)^2} = \frac{\eta}{1 + (RF_I)^2}, \tag{1.38}$$

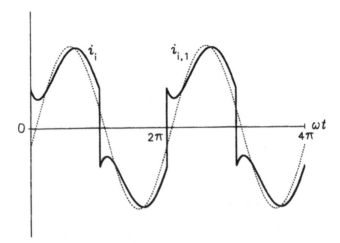

Figure 1.11 Decomposition of the input current waveform in the generic rectifier.

where R is the load resistance. The ripple factor for the example current in Figure 1.4 has already been found to be 0.307. Hence, with $\eta = 1$, the conversion efficiency is $1/(1 + 0.307^2) = 0.914$.

The input current waveform, $i_i(\omega t)$, of the rectifier is shown in Figure 1.11 with the numerically found fundamental, $i_{i,1}(\omega t)$. The converter either directly passes the input voltage and current to the output or inverts them. Therefore, the rms values of input voltage, V_i, and current, I_i, are equal to those of the output voltage, V_o, and current, I_o. The apparent input power, S_i, is a product of the rms values of input voltage and current. Hence,

$$ \text{PF} = \frac{P_i}{S_i} = \frac{\frac{P_o}{\eta}}{V_i I_i} = \frac{RI_o^2}{\eta V_o I_o} = \frac{RI_o}{\eta V_o}, \tag{1.39} $$

and specific values of R, I_o, and V_o are needed to determine the power factor. If, for example, the peak input voltage, $V_{i,p}$, in the example rectifier described is 100 V and the load resistance, R, is 1.3 Ω, then, according to Eq. (1.16), the dc output voltage, V_o, is 70.7 V. The numerically computed rms output current, I_o, is 51.3 A. Consequently, $\text{PF} = (1.3 \times 51.3)/(1 \times 70.7) = 0.943$ (lag), the "lag" indicating that the fundamental input current lags the fundamental input voltage (compare Figures 1.1 and 1.11).

1.4 PHASE CONTROL AND SQUARE-WAVE MODE

Based on the idea of generic converter, whose switches connect, cross-connect, or disconnect the input and output terminals, and short the output terminals in the last case, the principles of the ac-to-dc and dc-to-ac power conversion were explained in

Section 1.2. The question how to control the magnitude of the output voltage and, consequently, that of the output current has not yet been answered though.

The reader is likely familiar with electric transformers and autotransformers that allow magnitude regulation of ac voltage and current. These are heavy and bulky apparatus designed for a fixed frequency and impractical for wide-range magnitude control. Moreover, their principle of operation inherently excludes transformation of dc quantities. In the early days of electrical engineering, adjustable resistors were predominantly employed for voltage and current control. Today, the *resistive control* can still be encountered in relay-based starters for electric motors and obsolete adjustable-speed drive systems. On the other hand, small rheostats and potentiometers are still widely used in low-power electric and electronic circuits, in which the power efficiency is not of major importance.

Resistive control does not have to involve real resistors. Actually, any of the existing transistor-type power switches could serve this purpose. Between the state of saturation, in which a transistor offers minimum resistance in the collector–emitter path, and the blocking state resulting in practically zero collector and emitter currents, a wide range of intermediate states is available. Therefore, such a switch can be viewed as a controlled resistor and one may wonder if, for instance, the transistor switches used in power electronic converters could be operated in the same way as are the transistors in low-power analog electronic circuits.

To show why the resistive control *should not* be used in high-power applications, two basic schemes, depicted in Figure 1.12, will be considered. For simplicity, it is assumed that the circuits shown are to provide control of a dc voltage supplied to a resistive load. The dc input voltage, V_i, is constant, while the output voltage, V_o, is to be adjustable within the zero to V_i range. Generally, for power converters with a controlled output quantity (voltage or current), the so-called *magnitude control ratio*, M, can be defined as

$$M \equiv \frac{\Psi_{o,adj}}{\Psi_{o,adj(max)}}, \tag{1.40}$$

where $\Psi_{o,adj}$ denotes the value of the adjustable component of the output quantity, for example, the dc component of the output voltage in a controlled rectifier, while $\Psi_{o,adj(max)}$ is the maximum available value of this component. Usually, it is required

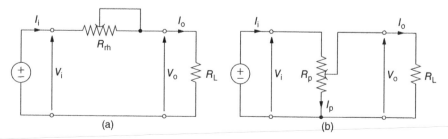

Figure 1.12 Resistive control schemes: (a) rheostatic control, (b) potentiometric control.

that M be adjustable from certain minimum value to unity. In certain converters, the minimum value can be negative, down to -1, implying polarity reversal of the controlled variable.

The magnitude control ratio should not be confused with the so-called *voltage gain*, K_V, which, generally, represents the ratio of the output voltage to the input voltage. Specifically, the average value is used for dc voltages and the peak value of the fundamental for ac voltages. Hence, for instance, the voltage gain of a rectifier is defined as the ratio of the dc output voltage, $V_{o,dc}$, to the peak input voltage, $V_{i,p}$. In the resistive control schemes considered, $K_V = V_o/V_i$ and, since the maximum available value of the output voltage equals V_i, the voltage gain equals the magnitude control ratio, M.

Figure 1.12a illustrates the *rheostatic control*. The active part, R_{rh}, of the controlling rheostat forms a voltage divider with the load resistance, R_L. Here,

$$M = \frac{V_o}{V_i} = \frac{R_L}{R_{rh} + R_L}, \tag{1.41}$$

and since the input current, I_i, equals the output current, I_o, the efficiency, η, of the power transfer from the source to the load is

$$\eta = \frac{R_L I_o^2}{(R_{rh} + R_L)I_i^2} = \frac{R_L I_o^2}{(R_{rh} + R_L)I_o^2} = \frac{R_L}{R_{rh} + R_L} = M. \tag{1.42}$$

The identity relation between η and M is a serious drawback of the rheostatic control as decreasing the output voltage causes an equal reduction of the efficiency.

The *potentiometric control*, shown in Figure 1.12b and based on the principle of current division, fares even worse. Note that the input current, I_i, is greater than the output current, I_o, by the amount of the potentiometer current, I_p. The power efficiency, η, is

$$\eta = \frac{V_o I_o}{V_i I_i} = M\frac{I_o}{I_i}, \tag{1.43}$$

that is, less than M.

It can be seen that the main trouble with the resistive control is that the load current flows through the controlling resistance. As a result, power is lost in that resistance, and the power efficiency is reduced to a value equal to, or less than, the magnitude control ratio. In practical power electronic systems, this is unacceptable. Imagine, for example, a 120-kVA converter (not excessively large against today's standards) that at $M = 0.5$ loses so much power in the form of heat as do 40 typical, 1.5-kW domestic heaters! Efficiencies of power electronic converters are seldom lower than 90% in low-power converters and exceed 95% in high-power ones. Apart from the economic considerations, large power losses in a converter would require an extensive cooling system. Even in the contemporary high-efficiency power conversion schemes, the cooling is often quite a problem since the semiconductor power switches are of relatively small size and, consequently, of limited thermal capacity. Therefore, they tend to overheat quickly if cooling is inadequate.

The resistive control allows adjustment of the *instantaneous values* of voltage and current, which is important in many applications, for example, those requiring amplification of analog signals, such as radio, TV, and tape recorder. There, transistors and operational amplifiers operate on the principle of resistive control, and because of the low levels of power involved, the low efficiency is of a minor concern. In power electronic converters, as illustrated later, it is sufficient to control the *average value* of dc waveforms and *rms value* of ac waveforms. This can be accomplished by periodic application of state 0 of the converter (see Section 1.2), in which the connection between the input and output is broken and the output terminals are shorted. In this way, the output voltage is made zero within specified intervals of time and, depending on the length of zero intervals, its average or rms value is more or less reduced compared to that of the full waveform.

Clearly, the mode of operation described can be implemented by appropriate use of switches of the converter. Note that there are no power losses in *ideal* switches, because when a switch is on (closed), there is no voltage across it, while when it is off (open), there is no current through it. For this reason, both the power conversion and the control in power electronic converters are accomplished by means of switching. Analogously to the switches in the generic power converter, the semiconductor devices used in practical converters are allowed to assume two states only. The device is either fully conducting, with a minimum voltage drop between its main electrodes (on-state), or fully blocking, with a minimum current passing between these electrodes (off-state). That is why the term "semiconductor power switches" is used for the devices employed in power electronic converters.

The only major difference between the ideal switches in the hypothetical generic converter in Figure 1.2 and practical semiconductor switches is in the unidirectionality of the latter devices. In the on-state, a current in the switch can only flow in one direction, for instance, from the anode to the cathode in an SCR. Therefore, an idealized semiconductor power switch can be thought of as a series connection of an ideal switch and an ideal diode.

Historically, for a major part of twentieth century, only semicontrolled power switches, such as mercury arc rectifiers, gas tubes (thyratrons), and SCRs, had been available for power-conditioning purposes. As mentioned in Section 1.1, a semicontrolled switch once turned on ("fired") cannot be turned off ("extinguished") as long as the conducted current drops below certain minimum level for a sufficient amount of time. This condition is required for extinguishing the arc in the thyratrons and mercury arc rectifiers, or for the SCRs to recover the blocking capability. If a switch operates in an ac circuit, the turn-off occurs naturally when the current changes polarity from positive to negative. After a turn-off, a semicontrolled switch must be re-fired in every cycle when the anode–cathode voltage becomes positive, that is, when the switch becomes *forward biased*.

As the forward bias of a switch in an ac input power electronic converter lasts a half-cycle of the input voltage, the firing can be delayed by up to a half-cycle from the instant when the bias changes from reverse to forward. This creates an opportunity for

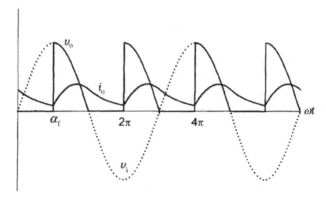

Figure 1.13   Output voltage and current waveforms in the generic rectifier with the firing angle of 90°.

controlling the average or rms value of output voltage of the converter. To demonstrate this method, the generic power converter will again be employed. For simplicity, the firing delay of the converter switches, previously assumed zero, is now set to a quarter of the period of the ac input voltage, that is, 90° in the angle domain.

Controlled ac-to-dc power conversion is illustrated in Figure 1.13. As explained in Section 1.2, when switches S1 through S4 of the generic converter are open, switch S5 must close to provide a path for the load current which, because of the inductance of the load, may not be interrupted. Thus, states 1 and 2 of the converter are separated by state 0. It can be seen that the sinusoidal half-waves of the output voltage in Figure 1.4 have been replaced with quarter-waves. As a result, the dc component of the output voltage has been reduced by 50%. Clearly, a longer delay in closing switches S1–S2 and S3–S4 would further reduce this component until, with the delay of 180°, it would drop to zero. The generic power converter operates now as the *controlled rectifier*. Practical controlled rectifiers based on SCRs do not need to employ an equivalent of switch S5. As already explained, an SCR cannot be turned off when conducting a current. Therefore, state 1 can only be terminated by switching to state 2, and the other way round, so that one pair of switches takes over the current from another pair. Both these states provide for the closed path for the output current. State 0, if any, occurs only when the output current has died out.

In the angle domain, the firing delay is referred to as a *delay angle*, or *firing angle*, and the method of output voltage control described is called the *phase control*, since the firing occurs at a specified phase of the input voltage waveform. In practice, the phase control is limited to power electronic converters based on SCRs. Fully controlled semiconductor switches allow more effective control by means of the so-called *pulse width modulation* (*PWM*), described in the next section.

The voltage control characteristic, $V_{o,dc}(\alpha_f)$, of the generic controlled rectifier can be determined as

$$V_{o,dc}(\alpha_f) = \frac{1}{\pi} \int_0^\pi v_o(\omega t) d\omega t = \frac{1}{\pi} \int_{\alpha_f}^\pi V_{i,p} \sin(\omega t) d\omega t = \frac{V_{i,p}}{\pi}[1 + \cos(\alpha_f)]. \qquad (1.44)$$

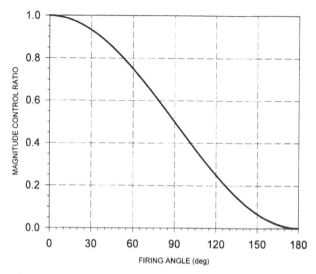

Figure 1.14   Control characteristic of the generic phase-controlled rectifier.

If $\alpha_f = 0$, Eq. (1.44) becomes identical with Eq. (1.15). The control characteristic, which is nonlinear, is shown in Figure 1.14.

The ac-to-ac conversion performed by the generic converter for adjusting the rms value of an ac output voltage is illustrated in Figure 1.15. Power electronic converters used for this type of power conditioning are called *ac voltage controllers*. Practical ac voltage controllers are mostly based on the so-called *triacs*, whose internal structure is equivalent to two SCRs connected in antiparallel. Phase-controlled ac voltage controllers do not require a counterpart of switch S5.

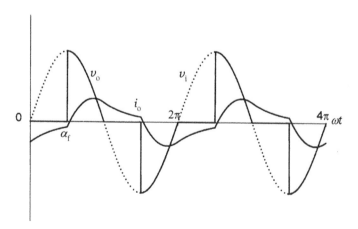

Figure 1.15   Output voltage and current waveforms in the generic ac voltage controller with the firing angle of 90°.

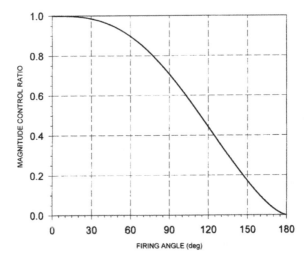

Figure 1.16   Control characteristic of the phase-controlled generic ac voltage controller.

The voltage control characteristic, $V_o(\alpha_f)$, of a generic ac voltage controller, given by

$$V_o(\alpha_f) = \sqrt{\frac{1}{\pi} \int_0^\pi v_o^2(\omega t)d\omega t} = \sqrt{\frac{1}{\pi} \int_{\alpha_f}^\pi [V_{i,p} \sin(\omega t)]^2 d\omega t}$$

$$= V_{i,p} \sqrt{\frac{1}{2\pi} \left[\pi - \alpha_f + \frac{\sin 2\alpha_f}{2} \right]}$$ (1.45)

is shown in Figure 1.16. Again, the characteristic is nonlinear. The fundamental output voltage, $V_{o,1}$, can be shown to depend nonlinearly on the firing angle, too.

The concept of using the zero-output state 0 to control the magnitude of output voltage can be extended to the generic inverter. Analogously to the firing angle α_f, the active states 1 and 2 can be delayed by the *delay angle* α_d (the same name is also used as an alternative to "firing angle" in ac input converters). The resultant *square-wave mode* is illustrated in Figure 1.17 for the delay angle of 30°. Using the Fourier series, the rms value, $V_{o,1}$, of the fundamental output voltage is found to be

$$V_{o,1} = \frac{2\sqrt{2}}{\pi} V_i \cos(\alpha_d) \approx 0.9 V_i \cos(\alpha_d).$$ (1.46)

1.5   PULSE WIDTH MODULATION

As explained in the preceding section, control of the output voltage of a power electronic converter by means of an adjustable firing delay has been primarily dictated

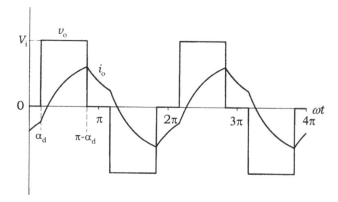

Figure 1.17   Output voltage and current waveforms in the generic inverter with the delay angle of 30°.

by the operating properties of semicontrolled power switches. The phase control, although conceptually simple, results in serious distortion of the output current of a converter. The distortion increases with the length of delay. Clearly, the distorted current is a consequence of the distorted voltage. However, the shape of the output voltage waveforms in power electronic converters is of much less practical concern than that of the output current waveforms. Precisely, *it is the current that does the job*, whether it is the emission of a light bulb, torque production in an electric motor, or electrolytic process in an electrochemical plant.

As mentioned before, most practical loads contain inductance. As such a load constitutes a low-pass filter, high-order harmonics of the output voltage of a converter have weaker impact on the output current waveform than do the low-order ones. Therefore, the quality of the current strongly depends on the amplitudes of low-order harmonics in the frequency spectrum of output voltage. Such spectra for the phase-controlled generic rectifier and ac voltage controller are illustrated in Figure 1.18. Amplitudes of the harmonics are expressed in the per-unit format, with

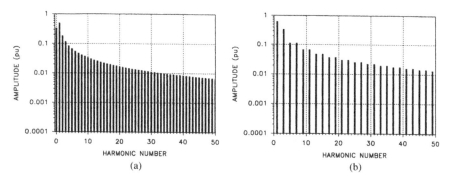

Figure 1.18   Harmonic spectra of output voltage with the firing angle of 90° in (a) phase-controlled generic rectifier, (b) phase-controlled generic ac voltage controller.

the peak value, $V_{i,p}$, of input voltage taken as the base voltage. Both spectra display several high-amplitude low-order harmonics.

In ac input converters, the distortion of the *input* current is of equal importance. Distorted currents drawn from the power system cause the so-called *harmonic pollution* of the system resulting in faulty operation of system protection relays and electromagnetic interference (EMI) with communication systems. To alleviate these problems, utility companies require that input filters be installed, which raises the total cost of power conversion. The lower is the frequency of harmonics to be attenuated, the larger and more expensive filters are needed. The already mentioned alternate method of voltage and current control by *pulse width modulation (PWM)* results in better spectral characteristics of converters and smaller filters. Therefore, PWM schemes are increasingly adopted in modern power electronic converters.

The principle of PWM can best be explained considering dc-to-dc power conversion performed by the generic converter supplied with a fixed dc voltage. The converter controls the dc component of the output voltage. As shown in Figure 1.19, this is accomplished by using the converter switches in such a way that the output voltage consists of a train of pulses (state 1 of the generic converter) interspersed with notches (state 0). Fittingly, the respective practical power electronic converters are called *choppers*. Low-power converters used in power supplies for electronic equipment are usually referred to as *dc voltage regulators* or, simply, *dc-to-dc converters*.

In the case illustrated, the pulses and notches are of equal duration, that is, switches S1 and S2 operate with the so-called *duty ratio* of 0.5. A duty ratio, d, of a switch is defined as

$$d \equiv \frac{t_{ON}}{t_{ON} + t_{OFF}}, \qquad (1.47)$$

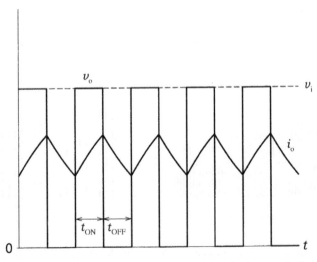

Figure 1.19   Output voltage and current waveforms in the generic chopper.

where t_{ON} denotes the on-time, that is, the interval within which the switch is closed, and t_{OFF} is the off-time, that is, the interval within which the switch is open. Here, switch S5 also operates with the duty ratio of 0.5. However, if the duty ratio of switches S1 and S2 were, for example, 0.6, the duty ratio of switch S5 would have to be 0.4. Switches S3 and S4 of the generic converter are not utilized (unless a reversal of polarity of the output voltage is required), so their duty ratio is zero.

It is easy to see that the average value (dc component), $V_{o,dc}$, of the output voltage is proportional to the fixed value, V_i, of the input voltage and to the duty ratio, d_{12}, of switches S1 and S2, that is,

$$V_{o,dc} = d_{12}V_i. \tag{1.48}$$

Since the possible range of a duty ratio is zero (switch open all the time) to unity (switch closed all the time), adjusting the duty ratio of appropriate switches allows setting $V_{o,dc}$ at any level between zero and V_i. As follows from Eq. (1.48), the voltage control characteristic, $V_{o,dc} = f(d_{12})$, of the generic chopper is linear.

The switching frequency, f_{sw}, defined as

$$f_{sw} \equiv \frac{1}{t_{ON} + t_{OFF}} \tag{1.49}$$

does not affect the dc component of the output voltage. However, the quality of output current strongly depends on f_{sw}. As illustrated in Figure 1.20, if the number of pulses per second is doubled compared to that in Figure 1.19, the increased switching frequency results in the reduction of the current ripple by about 50%. The time

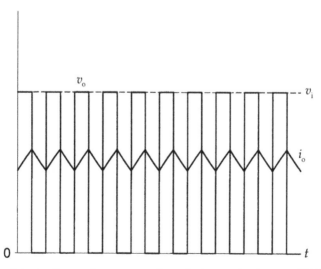

Figure 1.20   Output voltage and current waveforms in the generic chopper with the switching frequency twice as high as that in Figure 1.19.

between consecutive state changes of the converter is simply short enough to prevent significant current changes between consecutive "jumps" of the output voltage.

The reduction of current ripple can also be explained by the harmonic analysis of the output voltage. Note that the fundamental output frequency equals the switching frequency. As a result, harmonics of the ac component of the voltage appear around frequencies that are integer multiples of f_{sw}. The corresponding inductive reactances of the load are proportional to these frequencies. Therefore, if the switching frequency is sufficiently high, the ac component of the output current is so strongly attenuated that the current is practically of ideal dc quality.

Voltage control using PWM can be employed in all the other types of electric power conversion described in this chapter. Instead of taking out solid chunks of the output voltage waveforms by the firing delay as shown in Figures 1.13 and 1.15, a number of narrow segments can be removed in the PWM operating mode illustrated in Figure 1.21. Here, the ratio, N, of the switching to input frequency is 12. This is also the number of pulses of output voltage per cycle. The output current waveforms have significantly higher quality than those in Figures 1.13 and 1.15, exemplifying the operational superiority of pulse width modulated converters over phase-controlled ones.

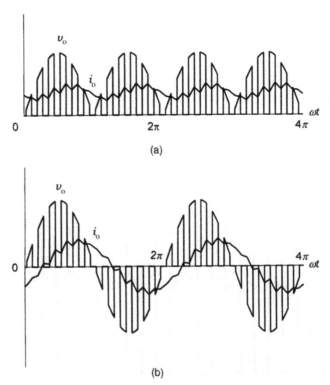

Figure 1.21   Output voltage and current waveforms in (a) generic PWM rectifier, (b) generic PWM ac voltage controller ($N = 12$).

It should be mentioned that, for clarity, in most examples of PWM converters throughout the book, the switching frequencies employed in preparing the figures are lower than the practical ones. Typically, depending on the type of power switches, switching frequencies are of the order of few kHz, seldom less than 1 kHz and, in the so-called *supersonic converters*, they are higher than 20 kHz. Therefore, if, for example, a switching frequency in a 60-Hz PWM ac voltage controller is 3.6 kHz, the output voltage is "sliced" into 60 segments per cycle, instead of the 12 segments shown in Figure 1.21b. Generally, a switching frequency should be several times higher than the reciprocal of the dominant (the longest) time constant of the load.

It can be shown that the linear relation (1.47) pertaining to the chopper can be extended on other types of PWM converters operating with fixed duty ratios of switches, such as rectifiers and ac voltage controllers. In all these converters,

$$V_{o,adj} = dV_{o,adj(max)}, \tag{1.50}$$

that is, the magnitude control ratio, M, equals the duty ratio, d, of switches connecting the input and output terminals. Depending on the type of converter, symbol $V_{o,adj}$ represents the adjustable dc component or fundamental ac component of the output voltage, while $V_{o,adj(max)}$ is the maximum available value of this component. However, concerning the *rms value*, V_o, of output voltage, the dependence on the duty ratio is radical, that is,

$$V_o = \sqrt{d}V_{o,max}, \tag{1.51}$$

where $V_{o(max)}$ is the maximum available rms value of output voltage. Specific values of $V_{o,adj(max)}$ and $V_{o(max)}$ depend on the type of converter and the magnitude of the input voltage. Control characteristics of the generic PWM rectifier (Eq. (1.48)) and an ac voltage controller (Eq. (1.51)) are shown in Figure 1.22 for comparison with those of phase-controlled converters depicted in Figures 1.14 and 1.16.

Harmonic spectra of the output voltage of the generic PWM rectifier and ac voltage controller are shown in Figure 1.23, again in the per-unit format. Here, the switching frequency is 24 times higher than the input frequency, that is, the output voltage has 24 pulses per cycle. The duty ratio of converter switches is 0.5. Comparing these spectra with those in Figure 1.18, essential differences can be observed. Amplitudes of the low-order harmonics have been suppressed, especially in the spectrum for the ac voltage controller. Higher harmonics appear in clusters centered about integer multiples of 12 for the rectifier and of 24 for the ac voltage controller. The reduced amplitudes of low-order harmonics of the output voltage are translated into enhanced quality of the output current waveforms.

It should be pointed out that the simple PWM described, with a constant duty ratio of switches, has only been used for illustration of the basic principles of PWM converters. In practice, constant duty ratios are typical for choppers and common for ac voltage controllers. PWM rectifiers and inverters employ more sophisticated PWM

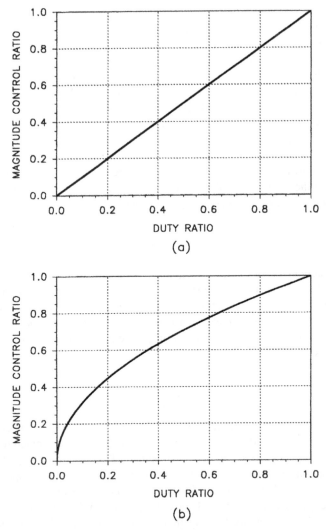

Figure 1.22   Control characteristics of (a) generic PWM rectifier, (b) generic PWM ac voltage controller.

techniques, in which the duty ratios of switches change throughout the cycle of the output voltage. Such techniques can also be applied in PWM ac voltage controllers.

PWM techniques characterized by variable duty ratios of converter switches will be explained in detail in Chapters 4 and 7. Example waveforms of the output voltage and current of a generic inverter with $N = 10$ voltage pulses per cycle and with variable duty ratios are shown in Figure 1.24. The magnitude control ratio, M, pertaining to the amplitude of the fundamental output voltage, is 1 in Figure 1.24a and 0.5 in Figure 1.24b. The varying widths of the voltage pulses and proportionality of these widths to the magnitude control ratio can easily be observed. The output current

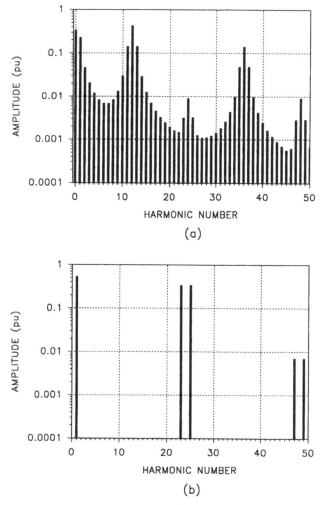

Figure 1.23 Harmonic spectra of output voltage in (a) generic PWM rectifier, (b) generic PWM ac voltage controller ($N = 24$).

waveforms, although rippled, are much closer to ideal sine waves than those in the square-wave operating mode of the inverter (see Figure 1.5). The harmonic content of the current would decrease further with an increase in the switching frequency.

All the examples of PWM converters presented in this section show that *the higher the switching frequency, the better quality of the output current is obtained*. However, the allowable switching frequency in practical power electronic converters is limited by several factors. First, any power switch requires certain amount of time for the transitions from the on-state to off-state and vice-versa. Therefore, the operating frequency of a switch is restricted, the maximum value depending on the type and ratings of the switch. Second, the control system of a converter has a limited operating speed. Finally, as explained in Chapter 2, the so-called *switching losses* in practical switches

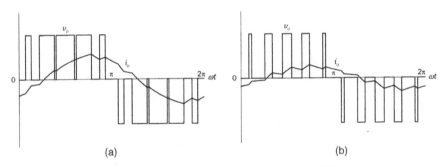

Figure 1.24   Output voltage and current waveforms in the generic PWM inverter: (a) $M = 1$, (b) $M = 0.5$ ($N = 10$).

increase with the switching frequency, reducing the efficiency of power conversion. Therefore, the switching frequency employed in a PWM converter should represent a sensible tradeoff between the quality and efficiency of operation of the converter.

1.6   COMPUTATION OF CURRENT WAVEFORMS

Sources of raw electric power, such as generators or batteries, are predominantly voltage sources. Since, as already demonstrated, a static power converter is a network of switches, waveforms of output voltages are easy to determine from the converter's principle of operation. It is not so with output currents, which depend on both the voltages and load.

As exemplified by the generic converter, power electronic converters operate most of the time in the so-called *quasi-steady state*, which is a sequence of transient states. The converters are variable-topology circuits, and each change of topology initiates a new transient state. Final values of currents and voltages in one topology become initial values in the next topology. Since linear electric circuits can be described by ordinary linear differential equations, the task of finding a current waveform becomes a classic initial value problem. As shown later, in place of differential equations, difference equations can conveniently be applied to converters operating in the PWM mode.

If the load of a converter is linear and the output voltage can be expressed in a closed form, the output current waveform corresponding to individual states of the converter can also be expressed in a closed form. However, when a computer is used for converter analysis, numerical algorithms can be employed to compute consecutive points of the current waveform. Both the analytical and numerical approaches will be illustrated in the subsequent sections.

1.6.1   Analytical Solution

To show the typical way of finding the closed-form expressions for the output current in a power electronic converter, the generic rectifier, generic inverter, and

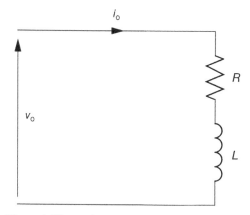

Figure 1.25   Resistive–inductive (RL) load circuit.

generic PWM ac voltage controller will be considered. In all three cases, an RL-load
is assumed.

The RL-load circuit of the generic converter is shown in Figure 1.25. If the converter operates as a rectifier, the input and output terminals are cross-connected when
the ac input voltage, given by Eq. (1.1), is negative. Consequently, the output voltage,
$v_o(t)$, and current, $i_o(t)$, are periodic with the period of $T/2$, where T is the period of
the input voltage, equal to $2\pi/\omega$. Therefore, all the subsequent considerations of the
generic rectifier are limited to the interval 0 to $T/2$.

The Kirchhoff Voltage Law for the circuit in Figure 1.25 can be written as

$$Ri_o(t) + L\frac{di_o(t)}{dt} = V_{i,p}\sin(\omega t). \tag{1.52}$$

Equation (1.52) can be solved for $i_o(t)$ using the Laplace transformation or, less
tediously, taking advantage of the known property of linear differential equations
allowing $i_o(t)$ to be expressed as

$$i_o(t) = i_{o,F}(t) + i_{o,N}(t), \tag{1.53}$$

where $i_{o,F}(t)$ and $i_{o,N}(t)$ denote the so-called *forced* and *natural* components of $i_o(t)$,
respectively. The convenience of this approach lies in the fact that the forced component constitutes the steady-state solution of Eq. (1.52), that is, the steady-state current
excited in the circuit in Figure 1.26 by the sinusoidal voltage $v_i(t)$. As well known to
every electrical engineer,

$$i_{o,F}(t) = \frac{V_{i,p}}{Z}\sin(\omega t - \varphi), \tag{1.54}$$

where

$$Z = \sqrt{R^2 + (\omega L)^2} \tag{1.55}$$

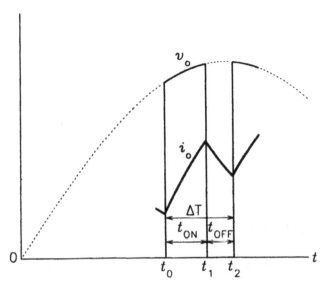

Figure 1.26 Fragments of the output voltage and current waveforms in the generic PWM ac voltage controller.

and

$$\varphi = \tan^{-1}\left(\frac{\omega L}{R}\right) \tag{1.56}$$

are the impedance and phase angle of the RL-load, respectively. The shortcut from Eq. (1.52) to the forced solution (1.54) saves a lot of work.

The natural component, $i_{o,N}(t)$, represents a solution of the homogenous equation

$$Ri_{o,N}(t) + L\frac{di_{o,N}(t)}{dt} = 0 \tag{1.57}$$

obtained from Eq. (1.52) by equating the right-hand side (the excitation) to zero. It can be seen that $i_{o,N}(t)$ must be such that the linear combination of it and its derivative is zero. Clearly, an exponential function

$$i_{o,N}(t) = Ae^{BT} \tag{1.58}$$

is the best candidate for the solution. Substituting $i_{o,N}(t)$ in Eq. (1.57) yields

$$RAe^{BT} + LABe^{BT} = 0 \tag{1.59}$$

from which, $B = -R/L$, and

$$i_{o,N}(t) = Ae^{-\frac{R}{L}t}. \tag{1.60}$$

Thus, based on Eqs. (1.53) and (1.54),

$$i_o(t) = \frac{V_{i,p}}{Z} \sin(\omega t - \varphi) + Ae^{-\frac{R}{L}t}. \tag{1.61}$$

To determine the constant A, note that in the quasi-steady state of the rectifier, $i_o(0) = i_o(T/2) = i_o(\pi/\omega)$ (see Figure 1.4). Hence,

$$\frac{V_{i,p}}{Z} \sin(-\varphi) + A = \frac{V_{i,p}}{Z} \sin(\pi - \varphi) + Ae^{-\frac{R}{L}\frac{\pi}{\omega}}, \tag{1.62}$$

where

$$-\frac{R\pi}{L\omega} = -\frac{\pi}{\tan(\varphi)}. \tag{1.63}$$

Equation (1.62) can now be solved for A, yielding

$$A = \frac{2V_{i,p}\sin(\varphi)}{Z\left[1 - e^{-\frac{\pi}{\tan(\varphi)}}\right]} \tag{1.64}$$

and

$$i_o(t) = \frac{V_{i,p}}{Z}\left[\sin(\omega t - \varphi) + \frac{2\sin(\varphi)}{1 - e^{-\frac{\pi}{\tan(\varphi)}}}e^{-\frac{R}{L}t}\right]. \tag{1.65}$$

The shown approach to derivation of the rather complex relation (1.65) takes much less time and effort than those required when the Laplace transformation is employed. The reader is encouraged to confirm this observation by his/her own computations.

When the generic power converter works as an inverter, the output voltage equals V_i when the converter is in state 1 and $-V_i$ in state 2. If T now denotes the period of the output voltage, it can be assumed that the state 1 lasts from 0 to $T/2$ and state 2 from $T/2$ to T. As it is only the polarity of output voltage that changes every half-cycle, the output current, $i_{o(2)}(t)$, in the second half-cycle is equal to $-i_{o(1)}(t-T/2)$, where $i_{o(1)}(t)$ is the output current in the first half-cycle. Consequently, it is sufficient to determine $i_{o(1)}(t)$ only.

The forced component, $i_{o,F(1)}(t)$, of current $i_{o(1)}(t)$, excited in the RL-load by the dc voltage $v_o = V_i$ is given by the Ohm's Law as

$$i_{o,F(1)} = \frac{V_i}{R}. \tag{1.66}$$

The natural component, which depends solely on the load, is the same as that of the output current of the generic rectifier. Thus,

$$i_{o(1)}(t) = \frac{V_i}{R} + Ae^{-\frac{R}{L}t}. \tag{1.67}$$

Constant A can be found from the condition $i_{o(1)}(0) = -i_{o(1)}(T/2)$ (see Figure 1.5), that is,

$$\frac{V_i}{R} + A = -\left[\frac{V_i}{R} + Ae^{-\frac{\pi}{\tan(\varphi)}}\right], \tag{1.68}$$

which yields

$$A = \frac{2}{1 + e^{-\frac{\pi}{\tan(\varphi)}}} \frac{V_i}{R}. \tag{1.69}$$

This gives

$$i_o(t) = \begin{cases} \dfrac{V_i}{R}\left[1 - \dfrac{2}{1 + e^{-\frac{\pi}{\tan(\varphi)}}}e^{-\frac{R}{L}t}\right] & \text{for } 0 < t \leq \dfrac{T}{2} \\[4mm] -\dfrac{V_i}{R}\left[1 - \dfrac{2}{1 + e^{-\frac{\pi}{\tan(\varphi)}}}e^{-\frac{R}{L}(t-\frac{T}{2})}\right] & \text{for } \dfrac{T}{2} < t \leq T \end{cases}. \tag{1.70}$$

In the final example, a PWM converter, specifically the PWM ac voltage controller, is dealt with. Within a single cycle of output voltage, the converter undergoes a large number of state changes, and within each corresponding interval of time, the waveform of output current is described by a different equation. Therefore, to express the waveform in a useful format, iterative formulas are employed. The general form of such a formula is $i_o(t + \Delta t) = f[i_o(t), t]$, where Δt is a small increment of time, meaning that consecutive values of the output current are calculated from the previous values.

In developing the formulas, two simplifying assumptions were made: first, that the current waveform is piecewise linear; second, that the initial value, $i_o(0)$, of the output current is equal to that of a hypothetical sinusoidal current that would be generated in the same load by the fundamental of output voltage of the converter. Thus, if the input voltage of the generic ac voltage controller is given by Eq. (1.1), the initial value of the output current is

$$i_o(0) = M\frac{V_{i,p}}{Z}\sin(\omega t - \varphi)|_{t=0} = -M\frac{V_{i,p}}{Z}\sin(\varphi). \tag{1.71}$$

The output voltage waveform of the controller, shown in Figure 1.21b, is a chopped sinusoid. A single switching cycle of that waveform is depicted in Figure 1.26, which also shows the coinciding fragment of the output current waveform. The time interval,

t_0 to t_2, corresponding to the switching cycle is called a *switching interval*. From t_0 to t_1, the controller is in state 1 and the output voltage equals the input voltage, while from t_1 to t_2, the output voltage is zero, the controller assuming state 0. Denoting the length of the switching interval, $t_2 - t_0$, by ΔT, the on-time, t_{ON}, is $M\Delta T$, and the off-time, t_{OFF}, is $(1 - M)\Delta T$.

The differential equation of the load circuit is

$$Ri_o(t) + L\frac{di_o(t)}{dt} = v_o(t), \tag{1.72}$$

which, for the instant $t = t_0$, can be rewritten as

$$Ri_o(t_0) + L\frac{\Delta i_o}{\Delta t} = V_{i,p} \sin(\omega t_0), \tag{1.73}$$

where $\Delta i_o = i_o(t_1) - i_o(t_0)$ and $\Delta t = t_1 - t_0$, that is, as a *difference equation*. Taking into account that $\Delta t = M\Delta T$, Eq. (1.73) can be solved for $i_o(t_0)$ to yield

$$i_o(t_1) = i_o(t_0) + \frac{M}{L}[V_{i,p} \sin(\omega t_0) - Ri_o(t_0)]\Delta T. \tag{1.74}$$

Similarly, at $t = t_1$, the load circuit equation

$$Ri_o(t_1) + L\frac{\Delta i_o}{\Delta t} = 0, \tag{1.75}$$

where $\Delta i_o = i_o(t_2) - i_o(t_1)$ and $\Delta t = t_2 - t_1$, can be rearranged to

$$i_o(t_2) = i_o(t_1)\left[1 - \frac{R}{L}(1 - M)\Delta T\right]. \tag{1.76}$$

Formulas (1.71), (1.74), and (1.76) allow computation of consecutive segments of the piecewise linear waveform of the output current. If, as typical in practical PWM converters, the switching frequency, $f_{sw} = 1/\Delta T$, is at least one order of magnitude higher than the input or output frequency, the accuracy of those approximate relations is sufficient for all practical purposes.

1.6.2 Numerical Solution

When a computer program is used to simulate a converter, the output voltage is calculated as a series of values, $v_{0,0}, v_{0,1}, v_{0,2}, \dots$, for the consecutive instants, $t_0, t_1, t_2, \dots$ These instants should be close apart so that $t_{n+1} - t_n \ll \tau$, where τ denotes the shortest time constant of the simulated system. Then, the respective values of the output current, $i_{0,1}, i_{0,2}, i_{0,3}, \dots$, can be computed as responses of the load circuit to step excitations $v_{0,0}u(t - t_0), v_{0,1}u(t - t_1), v_{0,2}u(t - t_2), \dots$, where $u(t - t_n)$ denotes a unit step function at $t = t_n$.

For instance, for the RL-load considered in the previous section, the differential equation of the load circuit for $t \geq t_n$ can be written as

$$Ri_o(t) + L\frac{di_o(t)}{dt} = v_{o,n}u(t - t_n), \qquad (1.77)$$

where $u(t - t_n) = 1$. The forced component, $i_{o,F}(t)$, of the solution, $i_o(t)$, of Eq. (1.77) is given by

$$i_{o,F}(t) = \frac{v_{o,n}}{R}, \qquad (1.78)$$

and the natural component, $i_{o,N}(t)$, by

$$i_{o,N}(t) = Ae^{-\frac{R}{L}t}. \qquad (1.79)$$

Hence,

$$i_o(t) = \frac{v_{o,n}}{R} + Ae^{-\frac{R}{L}t}, \qquad (1.80)$$

where constant A can be determined from equation

$$i_o(t_n) = i_{o,n} = \frac{v_{o,n}}{R} + Ae^{-\frac{R}{L}t_n} \qquad (1.81)$$

as

$$A = \left(i_{o,n} - \frac{v_{o,n}}{R}\right)e^{-\frac{R}{L}t_n}. \qquad (1.82)$$

Substituting $t = t_{n+1}$ and Eq. (1.82) in Eq. (1.80), and denoting $i_o(t_n + 1)$ by $i_{o,n+1}$ yields

$$i_{o,n+1} = \frac{v_{o,n}}{R} + \left(i_{o,n} - \frac{v_{o,n}}{R}\right)e^{-\frac{R}{L}(t_{n+1}-t_n)}, \qquad (1.83)$$

which is an iterative formula, $i_{o,n+1} = f(i_{o,n}, t_{n+1} - t_n)$, that allows easy computation of consecutive points of the output current waveform.

The other common loads and corresponding numerical formulas for the output current are

Resistive load (R-load):

$$i_{o,n} = \frac{v_{o,n}}{R}. \qquad (1.84)$$

Inductive load (L-load):

$$i_{o,n+1} = i_{o,n} + \frac{v_{o,n}}{L}(t_{n+1} - t_n). \tag{1.85}$$

Resistive-EMF load (RE-load):

$$i_{o,n} = \frac{v_{o,n} - E_n}{R}, \tag{1.86}$$

where E_n denotes the value of load EMF at $t = t_n$. The EMF, in series with resistance R, is assumed to oppose the output current.

Inductive-EMF load (LE-load):

$$i_{o,n+1} = i_{o,n} + \frac{v_{o,n} - E_n}{L}(t_{n+1} - t_n), \tag{1.87}$$

where the load EMF is connected in series with inductance L.

Resistive-inductive-EMF load (RLE-load):

$$i_{o,n+1} = \frac{v_{o,n} - E_n}{R} + \left(i_{o,n} - \frac{v_{o,n} - E_n}{R}\right) e^{-\frac{R}{L}(t_{n+1}-t_n)}, \tag{1.88}$$

where the load is represented by a series connection of resistance R, inductance L, and EMF E.

The presented considerations can be adapted for analysis of the less common current-source converters. There, those are currents that are easily determinable from the operating principles of a converter, while voltage waveforms require analytical or numerical solution of appropriate differential equations.

In the engineering practice, specialized computer programs are used for modeling and analysis of power electronic converters. The free-download software package LTspice for simulation of electronic circuits is called for in most of the computer assignments in this book (see Appendix A). Other commercial versions of the original UC Berkeley's SPICE are also available on the software market. Many advanced simulation programs, of which Saber is the best known example, have been developed in several countries specifically for modeling of switching power converters. Program EMTP, used primarily by utilities for the power system analysis, allows simulation of power electronic converters, which is particularly useful for studies of impact of the converters on the power system operation. General-purpose dynamic simulators, such as Simulink, Simplorer, or ACSL, can successfully be used for analysis of not only the converters themselves, but also whole systems that include converters, such as electric motor drives.

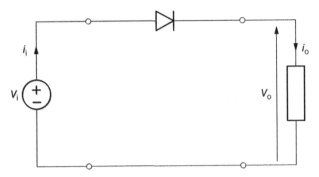

Figure 1.27   Single-pulse diode rectifier.

1.6.3   Practical Example: Single-Phase Diode Rectifiers

To illustrate the considerations presented in the preceding two sections and to start introduction of practical power electronic converters, single-phase diode rectifiers will be described. They are the simplest static converters of electrical power. The power diodes employed in the rectifiers can be thought of as uncontrolled semiconductor power switches. A diode in a closed circuit begins conducting (turns on) when its anode voltage becomes higher than the cathode voltage, that is, when the diode is forward biased. The diode ceases to conduct (turns off) when its current changes polarity.

The *single-pulse* (single-phase half-wave) diode rectifier is shown in Figure 1.27. Functionally, it is analogous to a reduced generic power converter having only switches S1 and S2, which stay closed as long as they conduct the load current. If the load is purely resistive (R-load), and the sinusoidal input voltage is given by Eq. (1.1), the output current waveform resembles that of the output voltage, as shown in Figure 1.28.

Comparing Figures 1.28 and 1.4 it can easily be seen that the average output voltage, $V_{o,dc}$, in the single-pulse rectifier is only half as large as that in the generic rectifier described in Section 1.2, that is, $V_{o,dc} = V_{i,p}/\pi \approx 0.32\ V_{i,p}$ (see Eq. 1.15). The single "pulse" of output current per cycle of the input voltage explains the name of the rectifier.

The voltage gain is even lower when the load contains an inductance (RL-load). Equation (1.60) can be used to describe the output current, but only until it reaches zero at $\omega t = \alpha_e$, where α_e is called an *extinction angle*. Thus,

$$i_o(0) = \frac{V_{i,p}}{Z} \sin(-\varphi) + A = 0, \tag{1.89}$$

from which

$$A = \frac{V_{i,p}}{Z} \sin(\varphi) \tag{1.90}$$

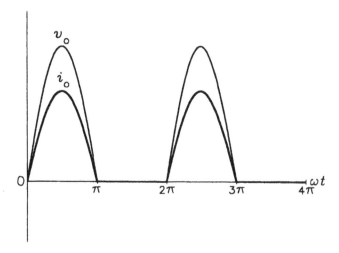

Figure 1.28   Output voltage and current waveforms in the single-pulse diode rectifier with an R-load.

and

$$i_o(t) = \frac{V_{i,p}}{Z}\left[\sin(\omega t - \varphi) + e^{-\frac{R}{L}t}\sin(\varphi)\right],\qquad(1.91)$$

when $i_o > 0$, that is, when $0 < \omega t \le \alpha_e$. Substituting α_e/ω for t and zero for i_o, the extinction angle can be found from equation

$$\frac{V_{i,p}}{Z}\left[\sin(\alpha_e - \varphi) + e^{-\frac{\alpha_e}{\tan(\varphi)}}\sin(\varphi)\right] = 0.\qquad(1.92)$$

Clearly, no closed-form expression for α_e exists, and the extinction angle, which is a function of the load angle φ, can only be found numerically. This problem will be analyzed in greater detail in Chapter 4.

The output voltage and current waveforms of a single-pulse rectifier with an RL-load are shown in Figure 1.29. As the extinction angle is greater than 180°, the output voltage is negative in the $\omega t = \pi$ to $\omega t = \alpha_e$ interval, and its overall average value is lower than that with the resistive load. This can be remedied by connecting the so-called *freewheeling diode*, DF, across the load, as shown in Figure 1.30.

The freewheeling diode, which corresponds to switch S5 in the generic power converter, shorts the output terminals when the output voltage reaches zero and provides a path for the output current in the π to α_e interval. Until the voltage reaches zero, the waveform of output voltage is the same as that in Figure 1.29 and that of output current is described by Eq. (1.91). Later, the current dies out while freewheeled by diode DF. As now $v_o = 0$, the equation of the current is simply

$$i_o(t) = Ae^{-\frac{R}{L}t},\qquad(1.93)$$

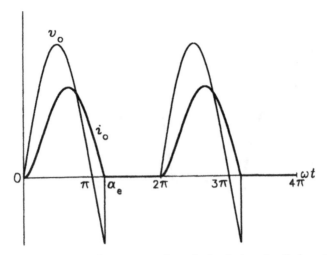

Figure 1.29 Output voltage and current waveforms in the single-pulse diode rectifier with an RL-load.

where constant A can be found by equaling currents given by Eqs. (1.91) and (1.93) at $\omega t = \pi$, that is,

$$\frac{V_{i,p}}{Z}\left[1 - e^{-\frac{\pi}{\tan(\varphi)}}\right]\sin(\varphi) = Ae^{-\frac{\pi}{\tan(\varphi)}}. \tag{1.94}$$

From Eq. (1.94),

$$A = \frac{V_{i,p}}{Z}\left[e^{\frac{\pi}{\tan(\varphi)}} - 1\right]\sin(\varphi) \tag{1.95}$$

and

$$i_o(t) = \frac{V_{i,p}}{Z}\left[e^{\frac{\pi}{\tan(\varphi)}} - 1\right]e^{-\frac{R}{L}t}\sin(\varphi). \tag{1.96}$$

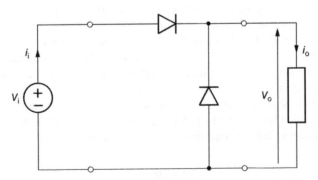

Figure 1.30 Single-pulse diode rectifier with a freewheeling diode.

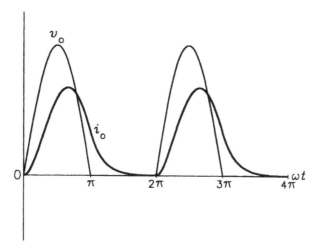

Figure 1.31   Output voltage and current waveforms in the single-pulse diode rectifier with a freewheeling diode and an RL-load.

Waveforms of the output voltage and current of a single-pulse rectifier with the free-wheeling diode are shown in Figure 1.31.

More radical enhancement of the single-pulse rectifier, shown in Figure 1.32, consists in connecting a capacitor across the output terminals. The capacitor charges up when the input voltage is high and it discharges through the load when the input voltage drops below a specific level that depends on the capacitor and load. Typical waveforms of the output voltage, v_o, and capacitor current, i_C, with a resistive load are shown in Figure 1.33. Advanced readers are encouraged to try and obtain analytical expressions for these waveforms using the approach sketched in this chapter.

Practical single-pulse rectifiers do not belong in the realm of true power electronics, as the output capacitors would have to be excessively large. The average output voltage, $V_{o,dc}$, can be increased to $2V_{i,p}/\pi \approx 0.64V_{i,p}$ in a two-pulse (single-phase full-wave) diode rectifier in Figure 1.34. Diodes D1 through D4 correspond to the respective switches in the generic converter, directly connecting and cross-connecting the input terminals with the output terminals in dependence on polarity of the input

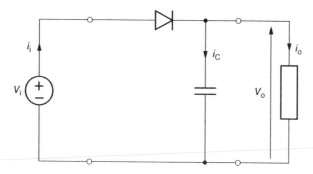

Figure 1.32   Single-pulse diode rectifier with an output capacitor.

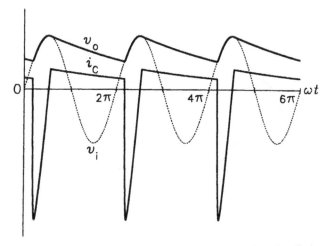

Figure 1.33   Output voltage and current waveforms in the single-pulse diode rectifier with output capacitor and RL-load.

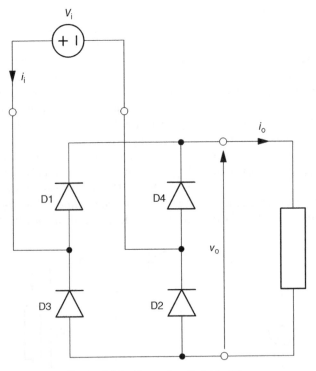

Figure 1.34   Two-pulse diode rectifier.

voltage. As a result, the output voltage and current waveforms are the same as in the generic converter (see Figure 1.4). In low-power rectifiers, a capacitor can be placed across the load, similarly as in the single-pulse rectifier in Figure 1.32, to smooth out the output voltage and increase its dc component. However, the operating characteristics of single-phase rectifiers are distinctly inferior to those of three-phase rectifiers, to be presented in Chapter 4.

SUMMARY

Power conversion and control are performed in power electronic converters, which are networks of semiconductor power switches. Three basic states of a voltage-source converter can be distinguished (although in some converters only two states suffice): the input and output terminals are directly connected, cross-connected, or separated. In the last case, the output terminals must be shorted to maintain a closed path for the output current. An appropriate sequence of states results in conversion of the given input (supply) voltage to the desired output (load) voltage.

Current-source converters are also feasible although less common than the voltage-source ones. A load of a current-source converter must appear as a voltage source, and that of a voltage-source converter as a current source. In practice, the current-source and voltage-source requirements imply a series-connected inductor and a parallel-connected capacitor, respectively.

Control of the output voltage in voltage-source power electronic converters is realized by periodic use of state 0 of the converter. This results in the removal of certain portions of the output voltage waveform. Two approaches are employed: phase control and PWM. The latter technique, promoted by the availability of fast fully controlled semiconductor power switches, is increasingly employed in modern power electronics. The PWM results in higher quality of power conversion and control than does the phase control.

The output voltage waveforms in voltage-source converters are usually easy to determine. Current waveforms, however, require analytical or numerical solution of differential or, especially for PWM converters, difference equations describing the quasi-steady state operation of the converters. A dual approach is applicable to current-source converters. Simulation software packages are extensively used in the engineering practice.

Single-phase diode rectifiers are the simplest ac-to-dc power converters. However, their voltage gain, especially that of the single-pulse rectifier, is low, and ripple factors of the output voltage and current are high. An improvement can be accomplished by installing an output capacitor, but it is feasible in low-power rectifiers only.

EXAMPLES

Example 1.1   The generic power converter is supplied from a 120-V, 50-Hz ac voltage source and its output voltage waveform is shown in Figure 1.35. It can be seen that the converter performs ac-to-ac conversion resulting in the output fundamental

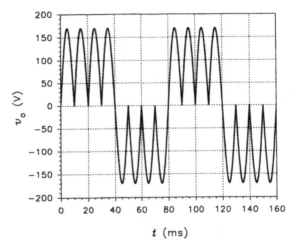

Figure 1.35   Output voltage waveform in the generic cycloconverter in Example 1.1.

frequency reduced with respect to the input frequency. This kind of power conversion is characteristic for *cycloconverters*. The generic cycloconverter in this example operates in a simple *trapezoidal mode*, in which the ratio of the input frequency to output fundamental frequency is an integer. What is the fundamental frequency of the output voltage? Sketch the timing diagram of switches of the converter.

Solution: The output fundamental frequency in question is four times lower than the input frequency of 50 Hz, that is, 12.5 Hz. The timing diagram of switches of the cycloconverter is shown in Figure 1.36. Switch S5 is open all the time as either switches S1 and S2 or S3 and S4 are closed.

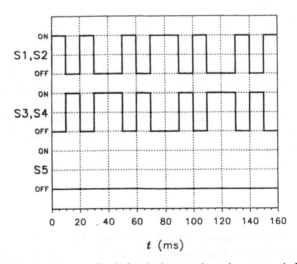

Figure 1.36   Timing diagram of switches in the generic cycloconverter in Example 1.1.

Example 1.2   For the output voltage waveform in Figure 1.35, find:

(a) rms value, V_o

(b) peak value, $V_{o,1,p}$, and rms value, $V_{o,1}$, of fundamental

(c) harmonic content, $V_{o,h}$

(d) total harmonic distortion, THD_V

Solution: The peak value of input voltage is $\sqrt{2} \times 120 = 170$ V. Denoting the fundamental output radian frequency by ω, the output voltage waveform can be expressed as

$$v_o(\omega t) = \begin{cases} |170| \sin(4\omega t) & \text{for} \quad 2n\frac{\pi}{4} \leq \omega t < (2n+1)\frac{\pi}{4} \\ -|170| \sin(4\omega t) & \text{for} \quad (2n+1)\frac{\pi}{4} \leq \omega t < (2n+2)\frac{\pi}{4} \end{cases},$$

where $n = 0, 1, 2, \ldots$, and the rms value, V_o, of the voltage is the same as that of the input voltage, that is, 120 V. The waveform has the odd and half-wave symmetries, so the peak value, $V_{o,1,p}$, of the fundamental is

$$V_{o,1,p} = \frac{4}{\pi} \int_0^{\frac{\pi}{2}} v_o(\omega t) \sin(\omega t) d\omega t = \frac{4}{\pi} \int_0^{\frac{\pi}{4}} 170 \sin(4\omega t) d\omega t - \frac{4}{\pi} \int_{\frac{\pi}{4}}^{\frac{\pi}{2}} 170 \sin(4\omega t) d\omega t$$

$$= \frac{4 \times 170}{\pi} \left\{ \left[\frac{\sin(3\omega t)}{6} - \frac{\sin(5\omega t)}{10} \right]_0^{\frac{\pi}{4}} - \left[\frac{\sin(3\omega t)}{6} - \frac{\sin(5\omega t)}{10} \right]_{\frac{\pi}{4}}^{\frac{\pi}{2}} \right\} = 139 \text{ V}.$$

The rms value of the fundamental is

$$V_{o,1} = \frac{V_{o,1,p}}{\sqrt{2}} = \frac{139}{\sqrt{2}} = 98 \text{ V}$$

and, since there is no dc component, the harmonic content is

$$V_{o,k} = \sqrt{V_o^2 - V_{o,1}^2} = \sqrt{120^2 - 98^2} = 69 \text{ V}.$$

Thus, the total harmonic distortion is

$$THD_V = \frac{V_{o,h}}{V_{o,1}} = \frac{69}{98} = 0.7.$$

Example 1.3   For the generic phase-controlled rectifier, find the relation between the ripple factor of the output voltage, RF_V, and the firing angle, α_f.

Solution: Eq. (1.44) gives the relation between the dc component, $V_{o,dc}$, of the output voltage in question and the firing angle, while Eq. (1.45), concerning the rms value, V_o, of output voltage of a generic ac voltage controller, can directly be applied to the generic rectifier (why?). Consequently,

$$
\begin{aligned}
RF_V(\alpha_f) &= \frac{V_{o,ac}(\alpha_f)}{V_{o,dc}(\alpha_f)} = \frac{\sqrt{V_o^2(\alpha_f) - V_{o,dc}^2(\alpha_f)}}{V_{o,dc}^2(\alpha_f)} = \sqrt{\frac{V_o^2(\alpha_f)}{V_{o,dc}^2(\alpha_f)} - 1} \\
&= \sqrt{\frac{\frac{1}{2\pi}\left[\pi - \alpha_f + \frac{\sin(2\alpha_f)}{2}\right]}{\frac{1}{\pi^2}[1 + \cos(\alpha_f)]^2} - 1} = \sqrt{\frac{\pi}{2} \frac{\pi - \alpha_f + \frac{\sin(2\alpha_f)}{2}}{[1 + \cos(\alpha_f)]^2} - 1}.
\end{aligned}
$$

Graphical representation of the derived relation is shown in Figure 1.37. It can be seen that when the firing angle exceeds 150°, the voltage ripple rapidly increases because the dc component approaches zero.

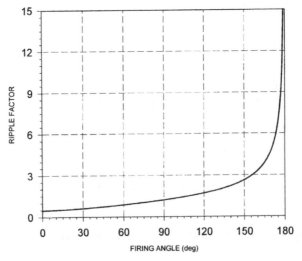

Figure 1.37   Voltage ripple factor versus firing angle in the generic rectifier in Example 1.3.

Example 1.4   The generic converter is supplied from a 100-V dc source and operates as a chopper with the switching frequency, f_{sw}, of 2 kHz. The average output voltage, $V_{o,dc}$, is −60 V. Determine duty ratios of all switches of the converter and the corresponding on- and off-times, t_{ON} and t_{OFF}.

Solution: The negative polarity of the output voltage implies that switches S3, S4, and S5 perform the modulation, while switches S1 and S2 are permanently open (turned off). Therefore, the duty ratio, d_{12}, of the latter switches is zero. Adapting Eq. (1.48) to switches S3 and S4, the duty ratio, d_{34}, of these switches is

$$d_{34} = \frac{V_{o,dc}}{V_1} = \frac{60}{100} = 0.6.$$

The negative sign at 60 V is omitted since the very use of switches S3 and S4 causes reversal of the output voltage (obviously, a duty ratio can only assume values between zero and unity). Switch S5 is turned on when the other switches are off. Hence, its duty ratio, d_5, is

$$d_5 = 1 - d_{34} = 1 - 0.6 = 0.4.$$

From Eq. (1.48),

$$t_{ON} + t_{OFF} = \frac{1}{f_{sw}} = \frac{1}{2 \times 10^3} = 0.0005 \text{ s} = 0.5 \text{ ms}$$

and from Eq. (1.45),

$$t_{ON,34} = d_{34}(t_{ON} + t_{OFF}) = 0.6 \times 0.5 = 0.3 \text{ ms},$$

while

$$t_{ON,5} = d_5(t_{ON} + t_{OFF}) = 0.4 \times 0.5 = 0.2 \text{ ms}.$$

Consequently,

$$t_{OFF,34} = 0.5 - 0.3 = 0.2 \text{ ms}$$

and

$$t_{OFF,5} = 0.5 - 0.2 = 0.3 \text{ ms}.$$

Example 1.5   The generic converter is supplied from a 120-V, 60-Hz ac voltage source and operates as a PWM rectifier with an RL-load, where $R = 2 \ \Omega$ and $L = 5$ mH. The switching frequency is 720 Hz and the magnitude control ratio is 0.6. Develop iterative formulas for the output current and calculate values of that current at the switching instants for one cycle of the output voltage.

Solution: Inspecting similar formulas, (1.74) and (1.76), for the generic PWM ac voltage controller, it can be seen that they are easy to adapt to the PWM rectifier by replacing the $\sin(\omega t_0)$ term in Eq. (1.74) with $|\sin(\omega t_0)|$. Then, the output current of the rectifier can be computed from equations

$$i_o(t_1) = i_o(t_0) + \frac{M}{L}[V_{i,p}| \sin(\omega t_0)| - Ri_o(t_0)]\Delta T$$

and

$$i_o(t_2) = i_o(t_1)\left[1 - \frac{R}{L}(1 - M)\Delta T\right].$$

The input frequency, ω, is $2\pi \times 60 = 377$ rad/s, and the length, ΔT, of a switching interval is 1/720 s. Thus, the general iterative formulas above give

$$i_o(t_1) = i_o(t_0) + \frac{0.6}{5 \times 10^{-3}}\left[\sqrt{2} \times 120| \sin(377t_0)| - 2i_o(t_0)\right] \times \frac{1}{720}$$
$$= 0.667i_o(t_0) + 28.3| \sin(377t_0)|$$

and

$$i_o(t_2) = i_o(t_1)\left[1 - \frac{2}{5 \times 10^{-3}}(1 - 0.6)\frac{1}{720}\right] = 0.778i_o(t_1).$$

As seen in Figure 1.20a, the period of the output voltage of the rectifier equals to the half of that of the input voltage, that is, 1/120 s $= 8.333$ ms. Comparing it with the switching frequency, the number of switching intervals per cycle of the output voltage turns out to be six. With $M = 0.6$, the on-time, t_{ON}, equals $t_1 - t_0$, is 0.6/720 s $= 0.833$ ms, and the off-time, t_{OFF}, equals $t_2 - t_1$, is 0.4/720 s $= 0.556$ ms.

To start the computations, the initial value, $i_o(0)$, is required as $i_o(t_0)$ in the first switching interval. It can be estimated as the dc component, $I_{o,dc}$, of the output current, given by

$$I_{o,dc} = \frac{M\frac{2}{\pi}V_{i,p}}{R} = \frac{0.6 \times \frac{2}{\pi} \times \sqrt{2} \times 120}{2} = 32.4 \text{ A}.$$

Now, the computations of $i_o(t)$ can be performed for the consecutive switching intervals; the t_2 instant for a given interval being the t_0 instant for the next interval.
First interval ($t_0 = 0$, $t_1 = 0.833$ ms, $t_2 = 1.389$ ms):

$$i_o(0.833 \text{ ms}) = 0.667 \times 32.4 + 28.3| \sin(377 \times 0)| = 21.6 \text{ A}$$

and

$$i_o(1.389 \text{ ms}) = 0.778 \times 21.6 = 16.8 \text{ A}.$$

Second interval ($t_0 = 1.389$ ms, $t_1 = 2.222$ ms, $t_2 = 2.778$ ms):

$i_o(2.222 \text{ ms}) = 0.667 \times 16.8 + 28.3| \sin(377 \times 1.389 \times 10^{-3})| = 25.4 \text{ A}$

and

$$i_o(2.778 \text{ ms}) = 0.778 \times 25.4 = 19.8 \text{ A.}$$

Third interval ($t_0 = 2.778$ ms, $t_1 = 3.611$ ms, $t_2 = 4.167$ ms):

$i_o(3.611 \text{ ms}) = 0.667 \times 19.8 + 28.3| \sin(377 \times 2.778 \times 10^{-3})| = 37.7 \text{ A}$

and

$$i_o(4.167 \text{ ms}) = 0.778 \times 37.7 = 29.3 \text{ A.}$$

Fourth interval ($t_0 = 4.167$ ms, $t_1 = 5$ ms, $t_2 = 5.556$ ms):

$i_o(4.167 \text{ ms}) = 0.667 \times 29.3 + 28.3| \sin(377 \times 4.167 \times 10^{-3})| = 47.8 \text{ A}$

and

$$i_o(5.556 \text{ ms}) = 0.778 \times 47.8 = 37.2 \text{ A.}$$

Fifth interval ($t_0 = 2.778$ ms, $t_1 = 3.611$ ms, $t_2 = 4.167$ ms):

$i_o(6.389 \text{ ms}) = 0.667 \times 37.2 + 28.3| \sin(377 \times 5.556 \times 10^{-3})| = 49.3 \text{ A}$

and

$$i_o(6.944 \text{ ms}) = 0.778 \times 49.3 = 38.4 \text{ A.}$$

Sixth interval ($t_0 = 6.944$ ms, $t_1 = 7.778$ ms, $t_2 = 8.333$ ms):

$i_o(7.778 \text{ ms}) = 0.667 \times 38.4 + 28.3| \sin(377 \times 6.944 \times 10^{-3})| = 39.8 \text{ A}$

and

$$i_o(8.333 \text{ ms}) = 0.778 \times 39.8 = 31.0 \text{ A.}$$

As the rectifier is assumed to operate in the quasi-steady state, the last, final value, $i_o(8.333 \text{ ms})$, should be equal to the initial value, $i_o(0)$. It is not exactly so, which implies an incorrect assumption of the initial value of 32.4 A. Note that the impact of the initial value on the final value is minimal, since at each step of the computations the previous value of the output current is multiplied by either 0.667 or 0.778.

Thus, after the 12 steps resulting from the 6 switching intervals, the initial value error, $\Delta i_o(0)$, is translated into the respective final value error, $\Delta i_o(8.333 \text{ ms})$, of $(0.667 \times 0.778)^6 \Delta i_o(0) \approx 0.02 \Delta i_o(0)$ only. Therefore, it is a safe guess that the obtained final value of 31 A is only slightly greater than the actual initial value. Assuming now that $i_o(0)$ is 30.9 A, and repeating the iterative calculations, yields the following points of the output current waveform:

0	30.9 A
0.833 ms	20.6 A
1.389 ms	16.0 A
...	...
6.944 ms	38.3 A
7.778 ms	39.7 A
8.333 ms	30.9 A.

Waveforms of the output voltage and current are shown in Figure 1.38. Clearly, the calculations presented can easily be computerized, greatly reducing the amount of work involved.

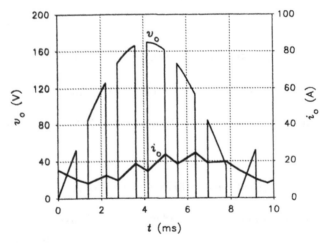

Figure 1.38 Output voltage and current waveforms in the generic PWM rectifier in Example 1.5.

PROBLEMS

P1.1   Refer to Figure 1.13 and sketch the waveforms of the output voltage of the generic rectifier with the firing angle of:

(a) 36°

(b) 72°

(c) 108°

(d) 144°

P1.2 Refer to Figure 1.15 and sketch the waveforms of the output voltage of the generic ac voltage controller with the firing angle of:

(a) 36°

(b) 72°

(c) 108°

(d) 144°

P1.3 From the harmonic spectrum in Figure 1.18a, find:

(a) per-unit value of the dc component of output voltage of the generic rectifier with the firing angle of 90° (use Eq. 1.44 to verify the result)

(b) peak and rms per-unit values of the fundamental output voltage with the same firing angle as in (a)

P1.4 From the harmonic spectrum in Figure 1.18b, read the peak per-unit value of fundamental output voltage of the generic ac voltage controller with the firing angle of 90°, and find:

(a) rms per-unit value of the output voltage (use Eq. 1.44)

(b) per-unit harmonic content of the output voltage

(c) total harmonic distortion of the output voltage

P1.5 For the generic inverter in the square-wave operating mode, derive an equation for the peak per-unit value of kth harmonic of the output voltage (take the dc input voltage, V_i, as the base voltage). Calculate peak per-unit values of the first 10 harmonics.

P1.6 A generic rectifier is supplied from a 230-V, 60-Hz ac source and operates with the firing angle of 30°. For the output voltage of the rectifier, find:

(a) dc component

(b) rms value of the ac component

(c) ripple factor

P1.7 For the generic rectifier in P1.6, find the fundamental frequency of the output voltage.

P1.8 For the generic ac voltage controller, derive an equation for the peak per-unit value of kth harmonic of the output voltage as a function of the firing angle (take the peak value, $V_{i,p}$, of input voltage as the base voltage). Use the spectrum in Figure 1.18b to verify the equation for the firing angle of 90° and the five lowest-order harmonics.

P1.9 Review the output voltage waveforms in Figures 1.13 and 1.15 and the corresponding harmonic spectra in Figure 1.18. Which harmonics present in the spectrum for the generic phase-controlled rectifier are absent in the spectrum for the generic ac voltage controller? Why?

P1.10 The generic converter operates as a chopper and is supplied from a 200-V dc source. What is the magnitude control ratio of the chopper and what are the duty ratios of individual switches if the average output voltage is 80 V?

P1.11 Repeat Problem 1.10 for the output voltage of –95 V.

P1.12 The generic converter operates as a chopper with the magnitude control ratio of 0.4. Duration of a pulse of the output voltage is 125 μs. Find the switching frequency.

P1.13 The generic converter operates as a chopper with 2500 pulses of output voltage per second and with the magnitude control ratio of 0.7. Find the durations of a pulse and a notch of the output voltage.

P1.14 The generic PWM rectifier operates with the average output voltage reduced by 30% with respect to the maximum available value of this voltage. Find the magnitude control ratio and the ratio of the width of pulse of the output voltage to the notch width.

P1.15 The generic PWM rectifier operates as in Problem 1.14 with 100 pulses of the output voltage per period of the 60-Hz input voltage. Find the width of a pulse and the switching frequency.

P1.16 The generic PWM ac voltage controller is supplied from a 60-Hz ac voltage source. The fundamental output voltage is reduced by two-thirds with respect to the supply voltage, and the switching frequency is 5 kHz. Find:

(a) number of pulses of the output voltage per cycle of this voltage

(b) duty ratios of switches S1, S2, and S5

(c) pulse width

P1.17 Refer to Figure 1.23a and determine the states (on or off) of all the switches of the generic PWM inverter at:

(a) $\omega t = \pi/2$ rad

(b) $\omega t = \pi$ rad

(c) $\omega t = 3\pi/2$ rad

P1.18 Waveforms in Figure 1.24 for the generic PWM inverter, with 10 pulses of the output voltage per cycle, were obtained in the following way:

(1) The 360° period of the voltage was divided into 10 equal switching intervals, 36° each

(2) Denoting by α_n the central angle of nth switching interval ($\alpha_1 = 18°$, $\alpha_2 = 54°$, etc.), the duty ratio, d_n, of operating switches S1–S2 or S3–S4 for this interval was calculated as

$$d_n = M|\sin(\alpha_n)|$$

where M denotes the magnitude control ratio.

For eight pulses per cycle and $M = 0.6$, find widths (in degrees) of individual pulses of the output voltage and sketch to scale the resultant voltage waveform, $v_o(\omega t)$. The voltage pulses should be located in the middle of switching intervals.

P1.19 Refer to Figure 1.35 and sketch the output voltage waveform of a generic *phase-controlled* cycloconverter with the input/output frequency ratio of 3 and firing angle of 45°. Mark the states of the converter.

P1.20 Refer to Figure 1.35 and sketch the output voltage waveform of a generic PWM cycloconverter with the input/output frequency ratio of 2 and the magnitude control ratio of 0.5 (for convenience, assume a low number of pulses of the output voltage).

P1.21 The generic PWM rectifier is supplied from a 460-V, 60-Hz ac line and operates with the magnitude control ratio of 0.6 and switching frequency of 840 Hz. The rectifier feeds a dc motor, which, under the given operating conditions, can be represented as an RLE-load with $R = 0.5\ \Omega$, $L = 10$ mH, and $E = 200$ V. Determine the piecewise linear waveform of the output current for one cycle of output voltage.

P1.22 The generic PWM ac voltage controller with an RL-load is supplied from a 220-V, 50-Hz source and operates with 10 switching intervals per cycle and the magnitude control ratio of 0.75. The resistance and inductance of the load are 22 Ω and 55 mH, respectively. Determine the piecewise linear waveform of the output current for one cycle of the output voltage.

COMPUTER ASSIGNMENTS

The generic power converter, which represents an idealized theoretical concept, cannot be modeled precisely using Spice, which is designed for the simulation of practical circuits. In contrast to the ideal, infinitely fast switches assumed for the generic converter, Spice switch models have finite time of transition from one state to another. To avoid interruptions of the output current, the output terminals of the converter must, therefore, be shorted by switch S5 just before the output is separated from the input by opening switches S1–S2 or S3–S4. As a result, the supply source is temporarily shorted too, albeit very briefly, producing large impulses ("spikes") of the input current. Such short-circuit currents are not present in practical, correctly designed and controlled power electronic converters. However, the output voltage and current in the generic converter can be simulated quite accurately.

To calculate the figures of merit, make use of the avg(x) (average value of x) and rms(x) (rms value of x) functions. Also, employ the *Fourier* option for the X-axis to obtain harmonic spectra of voltages and currents. Refer to Appendix A and Reference 3 for instructions on Spice simulations.

Assignments with the asterisk "*" denote circuit files available at the Publisher's website listed in the Preface and Appendix A.

CA1.1* Run Spice file *Gen_Ph-Contr_Rect.cir* for the generic phase-controlled rectifier. Perform simulations for firing angles of 0° and 90° and find for the output voltage of the converter:

(a) dc component

(b) rms value

(c) rms value of the ac component

(d) ripple factor

Observe oscillograms of the input and output voltages and currents. Explain what causes spikes in the oscillograms of the input quantities, and what limits the amplitude of the current spikes.

CA1.2 Develop a Spice circuit file for the generic inverter operating in the square-wave mode with an adjustable fundamental output frequency. Perform a simulation for the output frequency of 50 Hz and find for the output voltage of the inverter:

(a) rms value

(b) rms value of the fundamental (from harmonic spectrum)

(c) harmonic content

(d) total harmonic distortion

Observe oscillograms of the input and output voltages and currents.

CA1.3 Develop a Spice circuit file for the generic ac voltage controller. Perform simulations for firing angles of 0° and 90° and find for the output voltage of the converter:

(a) rms value

(b) rms value of the fundamental (from harmonic spectrum)

(c) harmonic content

(d) total harmonic distortion

Observe oscillograms of the input and output voltages and currents.

CA1.4 Refer to Example 1.1 and Problem P1.19 and develop a Spice circuit file for a generic phase-controlled cycloconverter with the input frequency of 50 Hz and fundamental output frequency of 25 Hz. Perform simulations for firing angles of 0° and 90° and find for the output voltage of the converter:

(a) rms value

(b) rms value of the fundamental (from harmonic spectrum)

(c) harmonic content

(d) total harmonic distortion

Observe oscillograms of the input and output voltages and currents.

CA1.5 Develop a Spice circuit file for the generic chopper. Perform simulations for the switching frequency of 1 kHz and magnitude control ratios of 0.6 and –0.3 and find for the output voltage of the converter:

(a) dc component

(b) rms value

(c) rms value of the ac component

(d) ripple factor

Observe oscillograms of the input and output voltages and currents.

CA1.6* Run Spice program *Gen_PWM_Rect.cir* for the generic PWM rectifier. Perform simulations for 12 pulses of the output voltage per cycle of the input voltage and magnitude control ratio of 0.5, and find for the output voltage of the converter:

(a) dc component

(b) rms value

(c) rms value of the ac component

(d) ripple factor

Observe oscillograms of the input and output voltages and currents. Explain what causes the spikes in the oscillograms of the input quantities, and what limits the amplitude of the current spikes.

CA1.7 Develop a Spice circuit file for the generic PWM ac voltage controller. Perform a simulation for 12 pulses of output voltage per cycle and the magnitude control ratio of 0.5, and find for the output voltage of the converter:

(a) rms value

(b) rms value of the fundamental (from harmonic spectrum)

(c) harmonic content

(d) total harmonic distortion

Observe oscillograms of the input and output voltages and currents.

CA1.8 Refer to Example 1.1 and Problem P1.20 and develop a Spice circuit file for a generic PWM cycloconverter. Perform a simulation for the input frequency of 50 Hz, fundamental output frequency of 25 Hz, magnitude control ratio of 0.5, and 10 pulses of output voltage per cycle of the input voltage. Find for the output voltage of the converter:

(a) rms value

(b) rms value of the fundamental (from harmonic spectrum)

(c) harmonic content

(d) total harmonic distortion

Observe oscillograms of the input and output voltages and currents.

CA1.9 Develop a computer program for calculation of harmonics of a given periodic waveform, $\psi(\omega t)$. Data points that represent one cycle of the waveform are stored in an ASCII file in the ωt, $\psi(\omega t)$ format. Generate and store the voltage waveform in Figure 1.13 (generic phase-controlled rectifier) and use your program to obtain the harmonic spectrum of the waveform. Compare the results with those in Figure 1.18a.

CA1.10 Develop a computer program for calculation of the following parameters of a given periodic waveform, $\psi(\omega t)$:

(a) rms value

(b) dc component

(c) rms value of the ac component

(d) rms value of the fundamental

(e) harmonic content

(f) total harmonic distortion

Data points that represent one cycle of the waveform are stored in an ASCII file in the ωt, $\psi(\omega t)$ format. Generate and store the voltage waveform in Figure 1.15 (generic phase-controlled ac voltage controller) and apply the program to compute parameters (a) through (f).

CA1.11 Develop a computer program for determination of a current waveform generated in a given load by a given voltage. The load can be of the R, RL, RE, LE, or RLE type, and the voltage waveform, $v(t)$, is given as either a closed-form function of time, $v = f(t)$, or an ASCII file of (t, v) pairs.

CA1.12* Run Spice program *Diode_Rect_1P.cir* for a single-pulse diode rectifier. Repeat the simulation for the rectifier with a freewheeling diode and with an output capacitor ("comment out" the unused components). In each case, determine the average output voltage and ripple factor of that voltage.

FURTHER READING

[1] Rashid, M. H., *Power Electronics Handbook*, 2nd ed., Academic Press, Boston, MA, 2010, Chapter 1.

[2] Rashid, M. H., *Power Electronics: Circuits, Devices, and Applications*, 4th ed., Prentice Hall, Upper Saddle River, NJ, 2013, Chapter 1.

[3] Rashid, M. H., *SPICE for Power Electronics and Electric Power*, 3rd ed., CRC Press, Boca Raton, FL, 2012.

2 Semiconductor Power Switches

In this chapter: semiconductor power switches used in power electronic converters are presented; behavioral classification of switches as uncontrolled, semi-controlled, and fully controlled is introduced; parameters and characteristics of power diodes, silicon-controlled rectifiers (SCRs), triacs, gate turn-off thyristors (GTOs), integrated gate commutated thyristors (IGCTs), power bipolar junction transistors (BJTs), metal oxide semiconductor field effect transistors (MOSFETs), and insulated-gate bipolar-junction transistors (IGBTs) are described and compared; and example power modules, which integrate several semiconductor devices in a single package, are shown.

2.1 GENERAL PROPERTIES OF SEMICONDUCTOR POWER SWITCHES

The hypothetical generic power converter in Chapter 1 was a simple two-port network of five switches capable of various types of electric power conversion and control. Practical power electronic converters are based on *semiconductor power switches*, which differ from the switches of the generic converter in two aspects. The chief difference consists in the fact that semiconductor devices can conduct substantial currents in one direction only. The other difference lies in control properties of the switches. An ideal switch is assumed to instantly react to the turn-on (close) or turn-off (open) signals. In contrast, certain semiconductor power switches are *uncontrolled*, that is, they lack the control electrode, or *semi-controlled*, that is, they lack the controlled turn-off capability. In all switches, including the *fully controlled* ones, the speed of response to control signals is limited, depending on the type and size of the switch. As a result, the maximum allowable switching frequencies of practical power switches vary from about 1 kHz for large SCRs and GTOs to more than 1 MHz for small power MOSFETs.

Also in contrast to the lossless switches of the generic converter, practical semiconductor power switches dissipate certain amount of energy when conducting and switching. Voltage drop across a conducting semiconductor device is at least on the order of 1 V, often higher, up to few volts. The *conduction power loss*, P_c, is given by

$$P_c = \frac{1}{T} \int_0^T p \, dt, \tag{2.1}$$

Introduction to Modern Power Electronics, Third Edition. Andrzej M. Trzynadlowski.
© 2016 John Wiley & Sons, Inc. Published 2016 by John Wiley & Sons, Inc.
Companion website: www.wiley.com/go/modernpowerelectronics3e

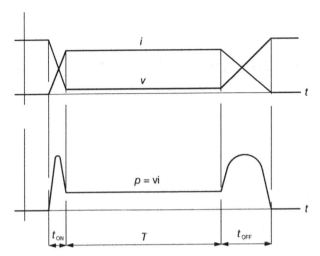

Figure 2.1   Waveforms of voltage, current, and power loss in a semiconductor power switch at turn-on, conduction, and turn-off.

where T denotes the conduction period and p is the instantaneous power loss, that is, the product of the voltage drop across the switch and the conducted current. Power losses increase during turn-on and turn-off of the switch, because during the transition from one conduction state to another both the voltage and the current are temporarily large. The resultant *switching power loss*, P_{sw}, is given by

$$P_{sw} = \left(\int_0^{t_{ON}} p\,dt + \int_0^{t_{OFF}} p\,dt \right) f_{sw}, \qquad (2.2)$$

where f_{sw} denotes the switching frequency, that is, the number of switching cycles (on-off) per second, and t_{ON} and t_{OFF} are the turn-on and turn-off times, whose precise meaning is explained later. Losses in a typical semiconductor switch during the turn-on, conduction, and turn-off are illustrated in Figure 2.1 that shows waveforms of voltage, current, and instantaneous power loss.

An *intrinsic semiconductor* is defined as a material whose resistivity is too low for an insulator and too high for a conductor. Semiconductor power switches are based on silicon, a member of Group IV of the periodic table of elements, which has four electrons in its outer orbit. If pure silicon is doped with a small amount of a Group V element, such as phosphorus, arsenic, or antimony, each atom of the dopant forms a covalent bond within the silicon lattice, leaving a loose electron. These loose electrons greatly increase the conductivity of the material, which is referred to as *n-type* (for "negative") semiconductor.

Vice-versa, if the dopant is a Group III element, such as boron, gallium, or indium, a vacant location called a *hole* is introduced into the silicon lattice. Analogously to an electron, a hole can be considered a mobile charge carrier as it can be filled by an adjacent electron, which in this way leaves a hole behind. Such material, also more conductive than pure silicon, is known as *p-type* (for "positive") semiconductor. The doping involves a single atom of the added impurity per over a million of silicon atoms. Semiconductor power switches are based on various structures of n-type and p-type semiconductor layers.

Because of the introductory nature of this book, no detailed solid-state physics considerations are presented in the subsequent coverage of semiconductor power switches. Instead, major parameters and characteristics are described to give the reader a general idea about capabilities of individual types of those devices. It should be pointed out that the specific parameter values provided in the coverage of individual switches, particularly the maximum voltage and current ratings, are those for the typical, easily available devices. The fast pace of technology and a large number of manufacturers make it difficult to keep precise track of the most recent developments and extreme ratings.

2.2 POWER DIODES

Power diodes, which are uncontrolled semiconductor power switches, are widely used in power electronic converters. The semiconductor structure and circuit symbol of a diode are shown in Figure 2.2. A high current in a diode can only flow from *anode* (A) to *cathode* (C). Voltage-current characteristic of the power diode is illustrated in Figure 2.3. Note that different scales are used for positive and negative half-axes. The *maximum reverse leakage current*, I_{RM}, is lower by many orders of magnitude than the current that can safely be conducted in the forward direction. Similarly, the *reverse breakdown voltage*, V_{RB}, which causes a dangerous *avalanche breakdown*, is much higher than the *maximum forward voltage drop*, V_{FM}, across the diode. Typically, V_{FM}, specified for a given forward current, I_{FM}, is on the order of 1–2 V, and it can be neglected in most practical considerations of power converters. However, the

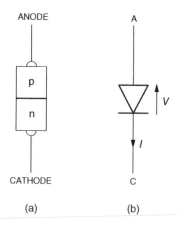

Figure 2.2 Power diode: (a) semiconductor structure, (b) circuit symbol.

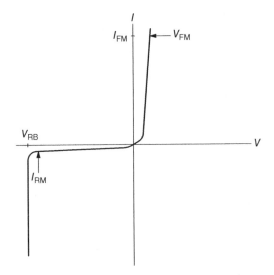

Figure 2.3   Voltage–current characteristic of the power diode.

voltage drop across a diode must be taken into account in the calculations of converter losses and the design of a cooling system. Otherwise, a diode can be considered as an almost ideal switch. It turns on when *forward biased*, that is, with a positive anode–cathode voltage, and turns off when a forward current reaches zero, usually on its way to negative polarity.

Main parameters of power diodes, which are listed in catalogs of semiconductor devices and form a base for proper selection of diodes for a given application, are described below. The symbols used are the most common ones, but they may somewhat vary with individual manufacturers. The same observation applies to the number and choice of parameters provided in various catalogs.

(1) Maximum allowable *reverse repetitive peak voltage*, V_{RRM} (V). Note the peak value employed since it is the instantaneous voltage that, if equal to or greater than the reverse breakdown voltage, V_{BR}, may damage a semiconductor device. High voltage across a diode can only occur when the diode is *reverse biased*, and $V_{RRM} < V_{RB}$. This parameter can simply be called a *rated voltage* since it appears in price lists of diodes as one of only two parameters specified, the other one being the rated current. Therefore, besides V_{RRM}, symbol V_{rat} will also be used in this book.

(2) Maximum allowable *full-cycle average forward current*, $I_{F(av)}$ (A). This parameter will subsequently be called a *rated current* and, alternatively, denoted by I_{rat}. Note that overcurrent damages a semiconductor device by excessive heat. Therefore, the average value is more appropriate here than the peak one. Because of the typical use of diodes as rectifiers, the average current for standard diodes is assumed as that in a single-pulse rectifier operating with

an R-load and frequency of 60 Hz. For *fast recovery* diodes, the frequency is 1 kHz. The rated current is less than the value, I_{FM}, of forward current for which the maximum forward voltage drop is defined.

(3) Maximum allowable *full-cycle rms forward current*, $I_{F(rms)}$ (A). This is the rms value of current in a diode operating as a single-pulse rectifier with an R-load and conducting an average current of $I_{F(av)}$. Thus, if the diode selected on the basis of $I_{F(av)}$ actually works as a rectifier, then there is no danger for the rms forward current to be excessive. Only in applications characterized by high-ripple currents, the rms forward current in a diode must be checked against the $I_{F(rms)}$ value. This, if not listed in a catalog, can easily be determined as

$$I_{F(rms)} = \frac{\pi}{2} I_{F(av)}. \tag{2.3}$$

(4) Maximum allowable *nonrepetitive surge current*, I_{FSM} (A). It is a maximum one-half cycle (60 Hz) peak surge current in excess of the rated current that the diode can survive. In practice, it pertains to possible short circuits in a converter. Duration of the surge current must be limited to a short period of time, specifically 8.3 ms (the 60-Hz half-period), within which the overcurrent protection system is expected to interrupt the current.

(5) Maximum allowable junction and case temperatures, Θ_{JM} and Θ_{CM} (°C). They usually vary from 150°C to 200°C. The range of allowable storage temperatures can also be specified, e.g., from −65°C to 175°C.

(6) Junction-to-case and case-to-sink *thermal resistances*, $R_{\Theta JC}$ and $R_{\Theta CS}$ (°C/W). These are used in designing the cooling system, proper selection of the *heat sink* (radiator) in particular. It must be stressed that current-related parameters of semiconductor power switches pertain to the switches with heat sinks. Because of the small size of contemporary switches, the thermal capacity of a bare device is too low to safely dissipate the heat generated by internal losses.

(7) *Fuse coordination*, I^2t (A²s). This parameter is used for fuse selection for a given diode. If I^2t of the fuse is less than I^2t of the diode, an overcurrent will cause the fuse to burn before the diode gets damaged. The I^2t value for a diode is obtained by squaring the rms value of current that can be withstood for a period of 8.3 ms and multiplying the result by the same 8.3 ms. Taking into account the definition of maximum allowable nonrepetitive surge current, I^2t is given by

$$I^2t = \frac{I_{FSM}^2}{240} \tag{2.4}$$

(in 50-Hz countries, the 240 should be replaced with 200). Vice-versa, if I_{FSM} is not given in a catalog, it can be calculated as

$$I_{FSM} = \sqrt{240 I^2 t}. \tag{2.5}$$

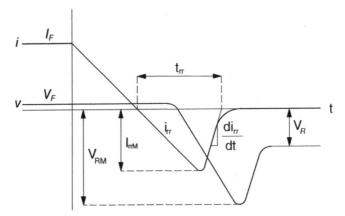

Figure 2.4   Voltage and current waveforms during reverse recovery in the power diode.

Parameters (1) through (5) can be called *restrictive*, since their values may not be exceeded without endangering the diode's integrity. In contrast, parameters (6) and (7), as well as the previously introduced maximum leakage current, I_{RM}, and maximum forward voltage drop, V_{FM}, are *descriptive* in nature, providing information about certain properties of the device. Another important descriptive parameter called a *reverse recovery time*, t_{rr}, requires a more detailed explanation.

When a conducting diode becomes abruptly reverse biased, the device does not regain its reverse-blocking capability until of the so-called *reverse recovery charge*, Q_{rr}, is removed from the junction capacitance. To discharge that capacitance, for a short time the diode passes a high current in the reverse direction. This phenomenon is illustrated in Figure 2.4 which shows the voltage and current waveforms when a reverse voltage, V_R, is applied to a conducting diode at $t = 0$. It can be seen that the reverse recovery time, t_{rr}, can loosely be defined as a duration of the period when a diode recovers its blocking capability. Within this period, a negative current overshoot, I_{rrM}, occurs, closely followed by a voltage overshoot, V_{RM}. The latter overshoot is proportional to the slope, di_{rr}/dt, of the *current tail*, that is, the decaying portion of the current waveform.

The value of di_{rr}/dt, often listed in catalogs as another descriptive parameter, affects the reverse recovery time, which for standard diodes varies from several to over 20 μs. In comparison with standard diodes, t_{rr} is shorter by one order of magnitude in the already mentioned fast recovery diodes which, however, produce higher voltage overshoots. Small so-called ultra-fast recovery diodes have the reverse recovery times on the order of a hundred of nanoseconds only. Generally, for a given class of diodes, the length of reverse recovery time depends mostly on the size of the device.

The area of triangle outlined by the negative current represents the reverse recovery charge, Q_{rr}. Consequently,

$$Q_{rr} \cong 0.5 t_{rr} I_{rrM}. \tag{2.6}$$

Table 2.1   Example High-Power Diodes

Symbol:	5SDD 31H6000	R6012625	5SDF 10H6004	R6031435
Maker:	ABB	Powerex	ABB	Powerex
Type:	General Purpose	General Purpose	Fast Recovery	Fast Recovery
Case:	Disc	Stud	Disc	Stud
V_{RRM}:	6 kV	2.6 kV	6 kV	1.4 kV
$I_{F(av)}$:	3.25 kA	0.25 kA	1.1 kA	0.35 kA
$I_{F(rms)}$:	5.1 kA	0.4 kA	1.7 kA	0.55 kA
I_{FSM}:	42.7 kA	6 kA	18 kA	6 kA
I^2t:	7.6×10^6 A^{2}s	1.5×10^5 A^{2}s	1.6×10^6 A^{2}s	1.5×10^5 A^{2}s
V_{FM}:	1.55 V	1.5 V	3 V	1.5 V
I_{RRM}:	120 mA	50 mA	50 mA	50 mA
t_{rr}:	25 μs	11 μs	6 μs	2 μs
Diameter:	102 mm	27 mm	95 mm	27 mm
Thickness:	27 mm	59 mm[a]	27 mm	59 mm[a]

[a]Excluding the stud and terminal wire.

Some manufacturers list only values of Q_{rr} and I_{rrM} in the data sheets of their diodes. Eq. (2.6) allows then to estimate the reverse recovery time, t_{rr}.

It must be noted that the descriptive parameters are not fixed but dependent on operating conditions of a diode. The anode current and junction temperature are particularly influential. This observation applies not only to power diodes but also to all semiconductor devices.

Power diodes are among the largest semiconductor devices. Their voltage and current ratings extend to 8.5 kV and 12 kA, respectively. However, the highest voltage diodes are not the highest current ones, and vice-versa. The same is true for other semiconductor power switches. Parameters of example high-power diodes are listed in Table 2.1. It can be seen that the disc ("hockey puck") case, in which the device constitutes a flat cylinder to be sandwiched between two heat sinks (radiators), allows for higher ratings than those of stud-type diodes, whose only one end is screwed into the heat sink. The general purpose diodes are mostly used in uncontrolled rectifiers, while the fast recovery ones in other converters.

Less powerful uncontrolled semiconductor switches, the *Shottky diodes*, based on a metal–semiconductor junction, deserve mentioning because of their superior characteristics. With a practically zero reverse recovery time, they are very fast. They are also characterized by low forward voltage drop (down to 0.2 V for small diodes) and capable of conducting currents of up to 0.6 kA, with the reverse repetitive peak voltage exceeding 1.5 kV. For example, the silicon-carbide CPW5-1700-Z050B Shottky diode by Cree Inc. has the rated voltage and current of 1.7 kV and 50 A, respectively.

2.3   SEMI-CONTROLLED SWITCHES

Members of the *thyristor family* of semiconductor devices can be considered semi-controlled power switches. They can be turned on by an appropriate gate signal, but

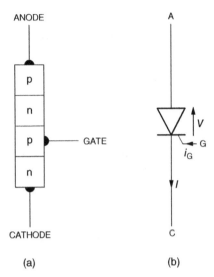

Figure 2.5   SCR: (a) semiconductor structure, (b) circuit symbol.

they turn off by themselves similarly to a diode, that is, when the conducted current decreases below the so-called *holding* level. Two subsequently presented prominent thyristor-type devices are the SCR, already mentioned in Chapter 1, and the *triac*, an electric equivalent of two SCRs connected in antiparallel.

2.3.1   SCRs

The SCR (termed "thyristor" in many parts of the world), a four-layer, three-electrode semiconductor device shown in Figure 2.5, can be thought of as a controlled diode which, when off, blocks currents of either polarity. When forward biased and turned on ("fired"), the SCR operates as an ordinary diode as long as the conducted current remains above the *holding current*, I_H, level. In converters employing a number of SCRs, such as the phase-controlled rectifiers described in Chapter 4, the so-called *commutation* occurs, that is, one SCR turns off when another SCR takes over conduction of the current.

In power electronic converters, an SCR is turned on by a gate current, i_G, supplied by an external source connected between the *gate* (G) and cathode. A high forward voltage between the anode and cathode, and even a rapid change, *dv/dt*, of such voltage can also cause turn-on, albeit unwanted one. The voltage–current characteristic of the SCR is shown in Figure 2.6 (for clarity, the forward and reverse leakage currents are exaggerated). With no gate current, when a forward voltage applied to the SCR exceeds the *forward breakover voltage*, V_{BF}, the forward leakage current increases to a *latching current*, I_L, level and the SCR starts conducting. Injecting a current through the gate into the central p-type layer reduces the forward breakover voltage to a value lower than that of the actually applied voltage, providing controlled turn-on.

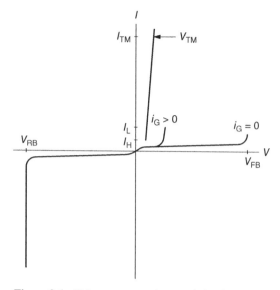

Figure 2.6   Voltage–current characteristic of the SCR.

Most of the diode parameters listed in the preceding section are shared by SCRs although somewhat different symbols are usually used to account for the fact that SCRs can block voltages of either polarity. Consequently, for the on-state, subscript "T" is used in place of an "F." For instance, as seen in Figure 2.6, the maximum forward voltage drop is denoted by V_{TM} instead of V_{FM}. Quantities pertaining to the forward-blocking state carry subscripts beginning with a "D." For example, V_{DRM} represents the maximum allowable *forward* repetitive peak voltage across the blocking SCR that will not cause turn-on without the firing signal. Typically, $V_{DRM} = V_{RRM}$.

In addition to parameters related to the cathode–anode path, catalogs of SCRs provide information on the dc gate current and voltage (precisely, gate-cathode voltage) required for successful firing and designated I_{GT} and V_{GT}, respectively. Data sheets of SCRs, which are more detailed than general catalogs, contain diagrams of firing areas with the instantaneous gate current and voltage as coordinates. Depending on the application, the gate current in an SCR to be fired is produced by employing a single pulse of the gate voltage, v_G, or a *multipulse*, both shown in Figure 2.7. A multipulse is obtained by rectifying and clipping a high-frequency sinusoidal voltage. Typically, frequencies on the order of several kilohertz are used, so that a multipulse may contain tens of individual half-wave sine pulses. The multipulse firing is employed in situations when it is not certain that a single pulse would accomplish the required turn-on. Typical gate current for large SCRs is of the order of 0.1–0.3 A, resulting in current gains, that is, the ratio of the anode and gate currents, of several thousands.

The *critical dv/dt* parameter expresses the minimum rate of change of anode voltage that causes turn-on without a gate current. Also listed in catalogs and data sheets is the *maximum allowable repetitive di/dt* value. The limit on the allowable rate of

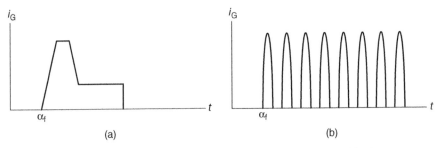

Figure 2.7 SCR gate voltage signals: (a) single pulse, (b) multipulse.

rise of the anode current is required, in order to spread the conduction area over the whole cross-section of the SCR before the current reaches a high level. Otherwise, an excessive current density in the small initial area of conduction would cause spot overheating and permanent damage to the device.

Two time parameters are usually listed in SCR catalogs. The turn-on time, t_{ON}, is counted from the instant of application of the gate signal to the instant when the anode voltage drops to 10% of its initial, full value. The turn-on time period consists of two sub-periods: the *delay time*, from the firing instant to the instant when the anode voltage has decreased from 100% to 90% of the initial value, and the *voltage fall time*, from 90% to 10%. Alternatively, the turn-on time can be defined in terms of the anode current. Then, it is a sum of the delay time, counted from the firing instant to the instant when the current has risen to 10% of its final value, and the *current rise time*, during which the current increases to 90% of the final value. Typical turn-on times are on the order of few microseconds, and the firing pulses should be an order of magnitude longer.

The meaning of the other time parameter, the turn-off time, t_{OFF}, is illustrated in Figure 2.8, which shows the anode voltage and current waveforms during the so-called *forced commutation* (turn-off) followed by reapplication of the forward voltage. The forced commutation is realized by applying a reverse-biasing voltage across a conducting SCR to reverse the anode current and extinguish the SCR. Even when

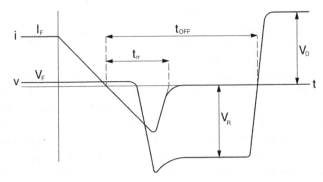

Figure 2.8 Anode voltage and current waveforms during forced commutation of the SCR.

Table 2.2   Example High-Power SCRs

Symbol:	5STP 20Q8500	TDS4453302	5STF 15F2040	T7071230
Maker:	ABB	Powerex	ABB	Powerex
Type:	Phase Control	Phase Control	Fast Switching	Fast Switching
Case:	Disc	Disc	Disc	Stud
V_{RRM}/V_{DRM}:	8 kV	4.5 kV	2 kV/1.8 kV	1.4 kV
$I_{T(av)}$:	2.15 kA	3.32 kA	1.49 kA	0.3 kA
$I_{T(rms)}$:	3.38 kA	5.22 kA	2.34 kA	0.47 kA
I_{TSM}:	47.5 kA	56 kA	18.2 kA	8 kA
I^2t:	11.3×10^6 A^{2}s	1.31×10^7 A^{2}s	1.37×10^6 A^{2}s	2.65×10^5 A^{2}s
V_{TM}:	2 V	1.8 V	1.6 V	1.45 V
I_{RRM}/I_{DRM}:	1 A	300 mA	150 mA	30 mA
t_{ON}:	3 μs	3 μs	2 μs	3 μs
t_{OFF}:	1080 μs	600 μs	40 μs	60 μs
I_{GT}:	400 mA	300 mA	300 mA	150 mA
V_{GT}:	2.6 V	4 V	3 V	3 V
Diameter:	150 mm	144 mm	75 mm	38 mm
Thickness:	27 mm	27 mm	27 mm	102 mm[a]

[a]Excluding the stud and terminal wire.

the reverse recovery process has been finished, a negative anode voltage must be maintained for some time, so that the SCR recovers its forward-blocking capability. SCRs are usually classified as either phase-control or inverter-grade ones. The inverter-grade SCRs, to be used in power inverters, are significantly faster than the phase-control SCRs, which are designed for 60-Hz applications such as rectifiers and ac voltage controllers.

Besides power diodes, SCRs are the largest semiconductor power switches. Standard, phase-control SCRs are available with ratings as high as 8 kV and 6 kA. The fast switching, inverter-grade SCRs are rated up to 2.5 kV and 3 kA. Special, light-activated thyristors used in HVDC transmission have voltage and current ratings similar to those of regular phase-control SCRs. Table 2.2 shows the parameters of example high-power SCRs.

2.3.2   Triacs

The triac is a semiconductor device that is electrically equivalent to two SCRs connected in antiparallel, although, as seen in Figure 2.9, its internal structure is not exactly the same as that of two SCRs. Because of the resultant capability of bidirectional current conduction, the power electrodes are simply called *main terminal 1* (T1) and *main terminal 2* (T2), instead of anode and cathode. The gate signal is applied between the gate and terminal 1. The triac can be turned on by a positive or negative gate current, and the direction of conducted current depends on the polarity of supply voltage.

In comparison with a pair of equivalent SCRs, a triac has a longer turn-off time, lower critical *dv/dt*, and lower current gain. However, the compact construction is

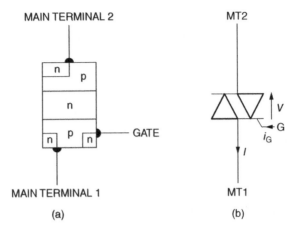

Figure 2.9 Triac: (a) semiconductor structure, (b) symbol.

advantageous in certain applications, such as the lighting and heating control, solid-state relays, and control of small motors. The available voltage and current ratings are limited to 1.4 kV and 0.1 kA, respectively.

The so-called *bi-directionally controlled thyristors* (BCTs) are worth mentioning in this context, as they are functionally similar to triacs. The BCT consists of two integrated antiparallel SCR-like devices on one silicon wafer. Unlike in the triac, the constituent thyristors are individually triggered. BCTs are much larger devices than triacs, with ratings of up to 6.5 kV and 2.6 kA.

2.4 FULLY CONTROLLED SWITCHES

Although SCRs initiated the era of semiconductor power electronics, several types of fully controlled switches have gradually phased them out from many applications, particularly the dc-input converters. The feasibility of controlled turn-on and turn-off makes fully controlled semiconductor power switches very attractive for use in modern power electronic converters, especially those with pulse width modulation. Common representatives of this class of semiconductor devices are described in the subsequent sections.

2.4.1 GTOs

The GTO is an acronym for the *gate turn-off thyristor*, whose structure and circuit symbol are shown in Figure 2.10. It is a thyristor-type semiconductor switch that is turned on similarly to an SCR by a low-positive gate current. However, in contrast to the SCR, the GTO can also be turned off, using a large negative pulse of the gate current, often exceeding the value of rated current. Thus, the turn-off current gain is poor, but the turn-off gate pulse is only on the order of tens of microseconds, so that

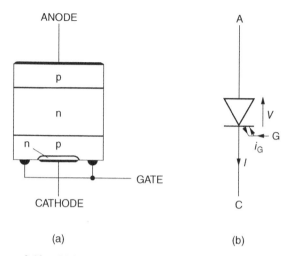

Figure 2.10   GTO: (a) semiconductor structure, (b) circuit symbol.

the associated energy of the gate signal is low. Ratings of GTOs are comparable with those of SCRs, reaching 6 kV and 6 kA.

GTOs have been the first fully controlled high-power semiconductor switches, but they are slow and their switching and conduction losses are high. They require snubbers, which suppress ("snub") voltage transients during turn-off.

2.4.2   IGCTs

The *integrated gate commutated thyristor*, IGCT, whose circuit symbol is shown in Figure 2.11, is similar to the GTO as it can be turned on and off by the gate signal. The gate turn-off current is greater than the anode current, which results in short turn-off times. The *gate driver*, that is, the circuitry generating the gate current, is integrated

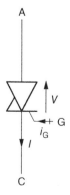

Figure 2.11   Circuit symbol of IGCT.

Table 2.3   Example IGCT and GTOs

Symbol:	5SHY 42L6500	5SHX 19L6020	5SHX19L6010	FG6000AU
Maker:	ABB	ABB	ABB	Mitsubishi
Type:	Asymmetric IGCT	Rev. Cond. IGCT	GTO (disc)	GTO (disc)
V_{DRM}:	6.5 kV	5.5 kV	4.5 kV	6 kV
$I_{T(av)}$:	1.29 kA	0.84 kA	1 kA	2 kA
$I_{T(rms)}$:	2.03 kA	1.32 kA	1.57 kA	3.1 kA
I_{TSM}:	40 kA	25.5 kA	25 kA	40 kA
I^2t:	2.4×10^6 A²s	1.6×10^6 A²s	3.1×10^6 A²s	6.7×10^6 A²s
V_{TM}:	3.7 V	2.9 V	4.4 V	6 V
I_{DRM}:	50 mA	50 mA	100 mA	320 mA
t_{ON}:	4 µs	3.5 µs	100 µs	10 µs
t_{OFF}:	8 µs	7 µs	100 µs	30 µs
I_{GQM}[a]:	3.8 kA	1.8 kA	1.1 kA	2.4 kA
E_{off}[b]:	44 J	11 J	14 J	N/A
Size:	429 × 173 × 41 mm	429 × 173 × 41 mm	85 × 85 × 26 mm	190 × 190 × 36 mm

[a]I_{GQM}, peak turn-off gate current.
[b]E_{off}, turn-off energy per pulse of gate current.

into the package of the device and surrounds the IGCT. As a result, IGCTs differ from the typical stud or disc packages of other high-power switches by their boxy shape. The impedance of the connection between the driver and the device is low, thanks to the large contact area and short length of the connection. This is a crucial feature considering the high- and fast-changing gate current, which precludes the use of wire leads. No snubbers are required.

The IGCTs cannot block reverse voltages and are referred to as *asymmetrical*. Asymmetrical IGCTs integrated in a single package with a freewheeling diode are called *reverse conducting* IGCTs.

IGCTs can switch much faster than GTOs, so that they can operate at high switching frequencies. However, high switching losses limit the sustained switching frequencies to less than 1 kHz. The voltage and current ratings reach 6.5 kV and 6 kA. Parameters of example IGCTs and GTOs are listed in Table 2.3.

2.4.3   Power BJTs

The n-p-n *bipolar junction transistor*, BJT in short, is shown in Figure 2.12. The collector (C) to emitter (E) path serves as the switch, conducting or interrupting the main current, while the base (B) is the control electrode. In contrast to thyristors, the collector current, I_C, can be continuously controlled by the base current, I_B, as

$$I_C = \beta I_B, \tag{2.7}$$

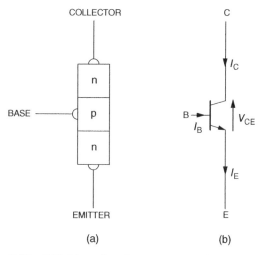

Figure 2.12 BJT: (a) semiconductor structure, (b) circuit symbol.

where β denotes a *dc current gain* of the transistor. In high-power BJTs, the current gain is low, on the order of 10. The emitter current, I_E, is a sum of the collector and base currents.

The voltage–current characteristics of the BJT, specifically the collector current, I_C, versus collector–emitter voltage, V_{CE}, relations for various values of the base current, I_B, are shown in Figure 2.13. The conduction power loss, P_c, is given by

$$P_c = V_{CE} I_C. \tag{2.8}$$

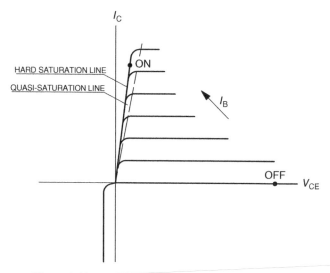

Figure 2.13 Voltage–current characteristics of the BJT.

Therefore, for the BJT to emulate an ideal, lossless switch, the base current in the on-state must be high enough for the operating point to lie on, or close to, the *hard saturation line* associated with the lowest voltage drop across the device. In the off-state, the base current is zero, and the collector current is reduced to the leakage level (assumed zero in Figure 2.13).

The rated current of a BJT represents the maximum allowable dc collector current, usually denoted simply as I_C. The rated voltage, denoted by V_{CEO}, is the maximum collector–emitter voltage that the transistor can safely block with a zero base current. As indicated in Figure 2.13, BJTs cannot block negative collector–emitter voltages. Thus, when used in an ac-input converter, a BJT must be protected from the reverse breakdown with a diode connected in series with the collector. The same applies to other types of asymmetric blocking semiconductor switches.

BJTs are susceptible to the so-called *second breakdown*, named so to distinguish them from the reverse, avalanche breakdown (first breakdown). In contrast to the reverse breakdown caused by an excessive voltage across a blocking device, the second breakdown occurs when both the collector–emitter voltage and the collector current are high, that is, during the turn-on or turn-off. As a result of crystal faults or doping fluctuations and the high power loss, local hot spots appear in the semiconductor. The positive temperature coefficient of the collector current causes a positive feedback effect, manifested in an increased current density in the hot spot region. Given sufficient time, this *thermal runaway* can cause an irreparable damage. Limiting the power dissipation in the transistor is the best means to prevent the second breakdown. To counterbalance the positive temperature coefficient, the so-called *emitter ballast resistance* can be incorporated in the transistor structure. However, it increases the on-state voltage drop.

To accelerate the turn-off process, it is recommended that the base current be temporarily changed from positive to negative. The base current and collector current waveforms for turn-on and turn-off are shown in Figure 2.14. The turn-off time can further be reduced by avoiding full saturation in the on-state. Instead, by reducing the on-state base current, the operating point of the transistor is shifted to the so-called *quasi-saturation zone* in the vicinity of the hard saturation line.

Single BJTs are seldom used in power electronic converters because of their low current gain. Instead, the *Darlington connection* (cascade) of two or three transistors, as illustrated in Figure 2.15, is employed. The current gain of a two-transistor Darlington connection is on the order of 100, and that of a three-transistor connection, of 1000. Power BJTs can operate with switching frequencies on the order of 10 kHz, and their maximum available voltage and current ratings have reached 1.5 kV and 1.2 kA.

In recent years, power BJTs, especially the high-rating ones, have been losing their market share to other switches, particularly the IGBTs, described in Section 2.4.5. The voltage-controlled IGBTs possess all the advantages of BJTs, without the weaknesses of the latter devices, such as the second breakdown or the current-controlled switching. Thus, the IGBTs are perfectly suited for most of the applications where power BJTs have been employed. Several manufacturers have already discontinued the production of BJTs.

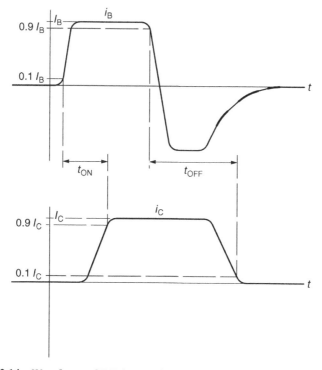

Figure 2.14   Waveforms of BJT base and collector currents at turn-on and turn-off.

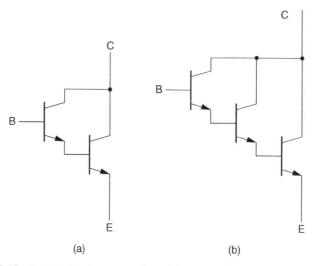

Figure 2.15   BJT Darlington connections: (a) two transistors, (b) three transistors.

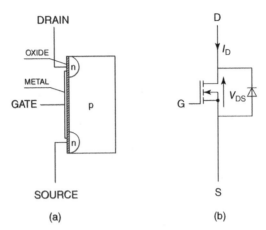

Figure 2.16 Power MOSFET: (a) semiconductor structure, (b) circuit symbol.

2.4.4 Power MOSFETs

The power MOSFET, whose simplified structure and circuit symbol are shown in Figure 2.16, is a semiconductor power switch characterized by the highest switching speed. Three electrodes of the MOSFET, the *drain* (D), *source* (S), and *gate* (G) correspond to the collector, emitter, and base of the BJT, respectively. However, in contrast to that transistor, the power MOSFET is voltage controlled, and the dc impedance of the gate-source path is practically infinite ($10^9 \ \Omega$–$10^{11} \ \Omega$). Only during fast turn-on and turn-off, the gate circuit does carry a short current pulse, which is associated with the respective charging and discharging of the gate-source capacitance.

Voltage–current characteristics of the power MOSFET are illustrated in Figure 2.17. The characteristics show the drain current, I_D, as a function of the

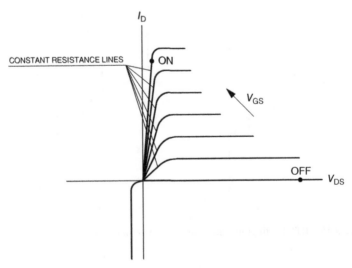

Figure 2.17 Voltage–current characteristics of the power MOSFET.

drain-source voltage, V_{DS}, for various values of the gate-source voltage, V_{GS}. Although the characteristics bear certain resemblance to those of the BJT (see Figure 2.13), there is no hard saturation line. Instead, each characteristic has a constant-resistance portion in the area of low V_{DS} values, and the control voltage, V_{GS}, should be made sufficiently high for the on-state operating point to lie on that portion. The rated voltage and current represent the maximum allowable values, V_{DSS} and I_{DM}, of the drain-source voltage and drain current, respectively. Similarly to BJTs, power MOSFETs may not be exposed to negative drain-source voltages, unless protected by a diode connected in series with the transistor.

Technical advantages of power MOSFETs are not limited to high switching speeds. Little power is required to control them, and the control circuitry is simpler than that for BJTs. The typical turn-on gate-source voltage is 20 V, while 0 V is used to turn the device off. MOSFETs have a negative temperature coefficient on drain current, which facilitates paralleling several transistors for an increased current-handling capability. If the temperature of one of the component MOSFETs increases, the conducted current drops, restoring the thermal balance among the connected devices. This characteristic also makes for uniform current density within the MOSFET, preventing the second breakdown. On-state resistance of high-voltage power MOSFETs used to be quite high, but the recent technological advances have resulted in significant reduction of that parameter. The switching losses are low, even with high switching frequencies, thanks to short turn-on and turn-off times. These, usually less than 100 ns, are defined similarly as for the BJTs (see Figure 2.14).

The diode in parallel with the MOSFET in Figure 2.16b, called a *body diode*, is a byproduct of the technological process. It can serve as a freewheeling diode, but being relatively slow it should be bypassed in fast switching converters with an external fast recovery diode. Switching frequencies can be as high as hundreds of kilohertz in medium-power converters and on the order of 1 MHz in low-power switching power supplies. Power MOSFETs are available with the voltage and current ratings of up to 1.5 kV and 1.8 kA.

2.4.5 IGBTs

IGBTs, the *insulated gate bipolar transistors*, are hybrid semiconductor devices, combining advantages of MOSFETs and BJTs. They are voltage controlled, like MOSFETs, but have lower conduction losses and higher voltage and current ratings. The equivalent circuit and circuit symbol of the IGBT are shown in Figure 2.18. To stress the hybrid nature of the IGBT, the control electrode is called a gate (G), while the main current is conducted or interrupted in the collector (C) to emitter (E) path.

Voltage–current characteristics of the IGBT are shown in Figure 2.19 as relations between the collector current, I_C, and collector–emitter voltage, V_{CE}, for various values of the gate-emitter voltage, V_{GE}. Majority of the IGBTs available on the market are of the *asymmetric* (punch-through) type. They do not have the reverse voltage-blocking capability and, with low conduction losses, are destined for application in dc-input converters, such as choppers and inverters. Usually, they are integrated with an antiparallel freewheeling diode. The *symmetrical* (nonpunch-through) IGBTs can block reverse voltages as high as the rated, forward-blocked voltage. It is an

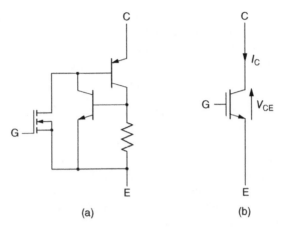

Figure 2.18 IGBT: (a) equivalent circuit, (b) circuit symbol.

additional advantage of these devices, whose conduction losses, however, are higher than those in asymmetrical IGBTs. The symmetrical IGBTs are primarily used in ac-input PWM converters, such as rectifiers and ac voltage controllers.

Regarding the on-state voltage drop, IGBTs are comparable to BJTs, but superior to power MOSFETs. Similarly to MOSFETs, IGBTs are turned on by the gate-emitter voltage on the order of 20 V and turned off by 0 V. IGBTs can be switched with "supersonic" (exceeding 20 kHz) frequencies. The maximum available voltage and current ratings are 6.5 kV and 2.4 kA. In addition to their operational superiority, IGBTs have wider ranges of the voltage and current ratings than power BJTs. Therefore, IGBTs are now the most popular semiconductor power switches. Table 2.4 shows the parameters of example high-power power transistors.

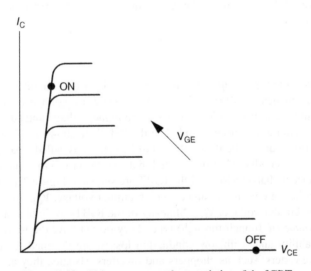

Figure 2.19 Voltage–current characteristics of the IGBT.

Table 2.4   Example High-Power Transistors

Symbol:	IXFK160N30T	ESM3030DV	5SNA 2000K45	CM750HG-130R
Maker:	IXYS	STMicroelectronics	ABB	Mitsubishi
Type:	Power MOSFET	Darlington BJT	IGBT	IGBT
V_{DSS}/V_{CE}:	0.3 kV	0.4 kV	4.5 kV	6.5 kV
I_D/I_C:	0.16 kA	0.1 kA	2 kA	0.75 kA
I_{DM}/I_{CM}:	0.44 kA	0.15 kA	4 kA	1.5 kA
$V_{GS}/V_{BE}/V_{GE}$:	20 V	7 V	20 V	20 V
I_G/I_B:	0.2 μA	5 A	0.5 μA	0.5 μA
t_{ON}:	34 ns	4.1 μs	1.35 μs	2.3 μs
t_{OFF}:	90 ns	1.2 μs	5.2 μs	10.2 μs
Size:	$42^a \times 16 \times$ 5 mm	$38 \times 25 \times$ 12 mm	$247 \times 237 \times$ 32 mm	$190 \times 140 \times$ 41 mm

[a]Including leads.

2.5   COMPARISON OF SEMICONDUCTOR POWER SWITCHES

Designers of modern power electronic converters have the choice of many types of semiconductor power switches. The variety of available devices allows optimal selection of switches for a given converter. As each switch type has its advantages and disadvantages, it would be difficult to unequivocally judge one of them as generally better than the others.

A hypothetical perfect switch that can be used as a reference in evaluating real devices would have the following characteristics:

(1) High-rated voltage and current, allowing application of single switches in high-power converters.

(2) Low, possibly zero, leakage current in off-state and voltage drop across the switch in on-state. These would result in the minimum conduction and off-state power losses in the switch.

(3) Short turn-on and turn-off times, allowing the device to be switched with high frequencies and minimum switching losses.

(4) Low power requirements to turn the switch on and off. This would simplify the design of control circuits and improve the efficiency and reliability of the whole converter.

(5) Negative temperature coefficient of the conducted current, to result in the equal current sharing by paralleled devices.

(6) Large allowable dv/dt and di/dt values, limiting the need for snubbers for protection of the switch from failures and structural damage.

(7) Low price—an important consideration in today's highly competitive field of commercial power electronics.

Finally, it is desirable for semiconductor power switches to have large *safe operating areas* (*SOA*), both the forward-bias and reverse-bias ones. The concept of SOAs, omitted for conciseness from the description of individual types of switches, requires some elaboration. Notice that the voltage–current characteristics in Figures 2.13, 2.17, and 2.19 do not extend uniformly over the whole range of the voltage. The hyperbolical envelope of these characteristics results from the fact that the amount of heat generated in the device is proportional to the power loss, that is, the voltage–current product. For this product to be constant and equal to the maximum allowable value, the current must be inversely proportional to the voltage. This yields a hyperbolical limit on the allowable steady-state operating points in the voltage–current coordinates. When a switch turns on or off, the instantaneous operating points can fall far beyond that limit. Barring a significant overvoltage, this is not necessarily dangerous for the switch since the turn-on and turn-off may be too short for an excessive temperature buildup.

A typical SOA for a power MOSFET is shown in Figure 2.20. It is drawn in the logarithmic scale to replace hyperbolical borderlines with linear ones. The sloped line on the left-hand side results from the on-state resistance of the MOSFET, which determines the voltage drop across the device for a given drain current. The three sloped lines on the right-hand side are the SOA limits for current pulses of the indicated durations. A forward-bias SOA corresponds to the situation when the gate-source voltage is positive, as during turn-on. If that voltage is negative, as for turn-off, the respective SOA is called a reverse-bias SOA. Both of them are identical for the power MOSFET, but it is not necessarily so for the other fully controlled switches. Similarly to high dv/dt and di/dt ratings, a robust SOA reduces the need for external protection circuits.

The most powerful semiconductor switches tend to be slow, while the high-frequency switches have lower power handling capabilities. Progressing from the slow high-power SCRs and GTOs through IGCTs, BJTs, and IGBTs, to the fast but relatively low-power MOSFETs, the best balance between the voltage–current ratings

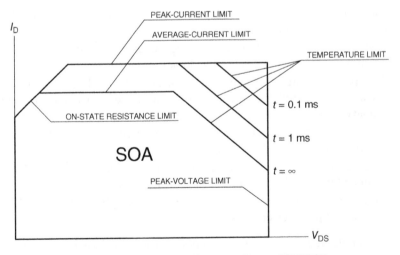

Figure 2.20   Safe operating area of power MOSFET.

Table 2.5   Properties and Maximum Ratings of Semiconductor Power Switches

Type	Switching Signal	Switching Characteristic	Switching Frequency	Forward Voltage	Rated Voltage	Rated Current
Diode			20 kHz[a]	1.2–1.7 V	6.5 kV	12 kA
SCR	Current	Trigger	0.5 kHz	1.5–2.5 V	8.5 kV	9 kA
Triac	Current	Trigger	0.5 kHz	1.5–2 V	1.4 kV[b]	0.1 kA[b]
GTO	Current	Trigger	1 kHz	3–4 V	6 kV	6 kA
IGCT	Current	Trigger	5 kHz	3–4 V	6.5 kV	6 kA
BJT	Current	Linear	20 kHz	1.5–3 V	1.7 kV	1.2 kA
IGBT	Voltage	Linear	20 kHz	3–4 V	6.5 kV	3.6 kA
MOSFET	Voltage	Linear	1 MHz	3–4 V	1.7 kV	1.8 kA

[a] Fast recovery diodes. General purpose diodes operate at 50 or 60 Hz.
[b] BCTs, whose operating principle is similar to that of the triac, reach 6.5 kV of rated voltage and 5.5 kA of rated current.

and high-frequency switching ability is stricken by the IGBT. The IGBT, although not free from certain weaknesses, such as relatively high conduction losses, seems to be closest to the perfection expressed by the presented wish list of properties.

The future of inverter-grade SCRs looks bleak, although the slower, phase-control SCRs still constitute the best choice for phase-controlled rectifiers and ac voltage controllers. BJTs are gradually phased out by the hybrid-technology devices. In conclusion, it is worth mentioning that the variety of the existing semiconductor power switches are by no means limited to those presented in this chapter. Besides the popular switches described, there is another batch of other, less common devices, such as, for instance, *MOS-controlled thyristors* (MCTs), *static induction transistors* (SITs), or *static induction thyristors* (SITHs).

Basic properties and maximum ratings of most common semiconductor power switches are summarized in Table 2.5. Most often, switches with the highest rated voltage offered by the manufacturers have only a medium rated current, and vice-versa. The values provided are typical, but is possible to find devices, usually special-purpose ones, whose ratings are even higher. For example, ABB lists on its website welding diodes, whose rated current exceeds 13.5 kA.

Figures 2.21 and 2.22 show an assortment of semiconductor power switches in their typical housings. It can be seen that the switches with highest voltage and current ratings, such as power diodes, SCRs, and GTOs use exclusively the disc case, epoxy or ceramic, whose shape allows sandwiching the switch between two heat sinks. The distinct form of the IGCT allows fast injection of the large turn-off current to the gate area. Smaller switches are packaged in a variety of cases.

2.6 POWER MODULES

To facilitate the design and simplify the physical layout of power electronic converters, manufacturers of semiconductor devices offer a variety of *power modules*. A power module is a set of semiconductor power switches interconnected into a

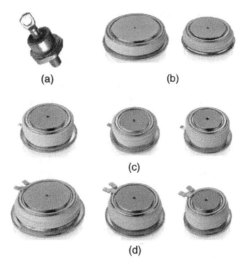

Figure 2.21 Semiconductor power switches I: (a) stud diode, (b) disc diodes, (c) disc SCRs, (d) disc GTOs. (Courtesy of ABB)

specific topology and enclosed in a single case, as those in Figure 2.22d. Most popular topologies are the single- and three-phase bridges, or their sub-circuits. Power modules may also contain several switches of the same type connected in series, parallel, or series-parallel, to increase the overall voltage and/or current ratings.

Example configurations of the available power modules are illustrated in Figures 2.23, 2.24, 2.25, and 2.26. Six assemblies of power diodes and SCRs are shown in Figure 2.23. The series connections of the devices in Figures 2.23a–2.23c are used in rectifiers, while the antiparallel connection of two SCRs in Figure 2.23d, which has

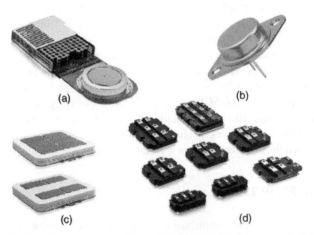

Figure 2.22 Semiconductor power switches II: (a) IGCT, (b) BJT or MOSFET, (c) IGBTs, (d) IGBTs, MOSFETs, or power modules. (Courtesy of ABB)

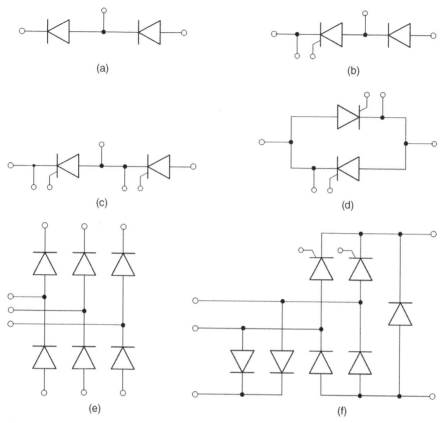

Figure 2.23   Diode and SCR modules: (a) two power diodes, (b) power diode and SCR, (c) two SCRs, (d) antiparallel connection of two SCRs (BCT), (e) diode rectifier, (f) controlled rectifier for dc motor control.

already been introduced in Section 2.3.2 as the BCT, can form a part of an ac voltage controller or a static ac switch. A six-pulse (or two-pulse) diode rectifier module is illustrated in Figure 2.23e, and Figure 2.23f depicts a single-phase rectifier bridge for dc motor control. The two diodes on the left-hand side form an additional rectifier, which allow producing a dc current for the field winding of the motor.

The two-transistor Darlington cascade is used in modules in Figure 2.24, showing a dual-switch module, quad-switch module, and six-switch module. Each BJT is equipped with a parallel freewheeling diode, while the resistor-diode circuits connected between the emitter and base of each composite Darlington transistor help to reduce the leakage current and speed up the turn-off. The quad-switch bridge topology can serve, depending on the control algorithm, as a four-quadrant chopper or a single-phase voltage-source inverter. The six-switch bridge configuration constitutes a three-phase voltage-source inverter. However, the more devices are crammed together in a single module, the less heat per device can safely be dissipated, and

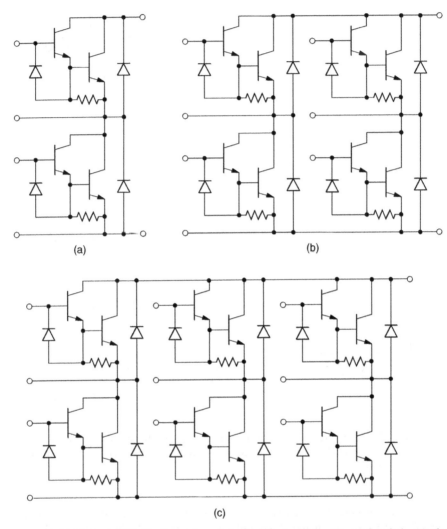

(a)

(b)

(c)

Figure 2.24 Power BJT (Darlington) modules: (a) dual switch, (b) quad switch, (c) six switch.

the lower the current ratings are. Hence the need for the dual-switch module, which can be used as a building block of either of the bridge topologies if the required load current is too high for the available quad-switch or six-switch modules.

A dual power MOSFET module is illustrated in Figure 2.25a and a six-switch module in Figure 2.25b. Figure 2.25c depicts a quad-switch circuit with a front-end diode rectifier to be supplied from. A single-switch module in which, to increase the overall current rating, four power MOSFETs are connected in parallel and simultaneously controlled using a single gate is shown in Figure 2.25d. The resistors connected

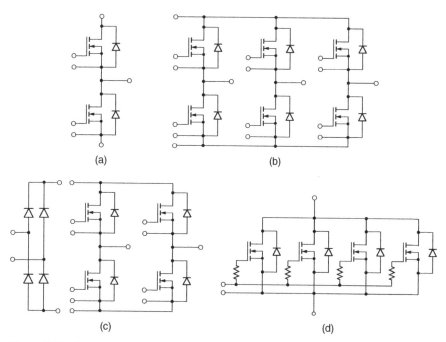

Figure 2.25   Power MOSFET modules: (a) dual switch, (b) six switch, (c) quad switch with diode rectifier, (d) four-transistor switch.

to each internal gate of the four transistors equalize the voltage signals applied to the individual MOSFETs.

Topologies of IGBT modules analogous to those of the BJT in Figure 2.24 are shown in Figure 2.26. Figure 2.27a depicts a part of an IGBT chopper, specifically the "high-end" (connected to plus of the source), while the "low-end" part (connected to minus of the source) is shown in Figure 2.27b. A single leg of a three-level diode-clamped IGBT inverter appears in Figure 2.27c. Finally, an arrangement of 7 IGBTs and 13 diodes constituting the power circuit of a three-phase, ac-to-dc-to-ac converter, a *frequency changer*, is shown in Figure 2.28. Not shown are the required external resistive and capacitive components, too large to incorporate into the module.

The described power modules contain power circuits only. Control of the switches require additional external components. To facilitate the construction of power electronic converters, the so-called intelligent power modules (IPMs) have been introduced by several manufacturers. In IPMs, power components are accompanied by protection circuits and gate drives. This is not an easy task, as the delicate digital devices must co-exist with the high-voltage and high-current semiconductor power switches.

While studying Chapters 4 through 7, the reader is encouraged to return to this section to appreciate the practical usefulness of power modules. The example topologies described represent only a tip of an iceberg of the great variety of modular circuits available on the market of power electronic devices.

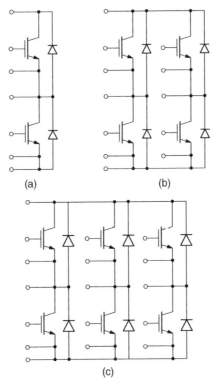

Figure 2.26   IGBT modules: (a) dual switch, (b) quad switch, (c) six switch.

2.7 WIDE BANDGAP DEVICES

As known from the solid-state physics, electrons in atoms occupy certain energy bands, the energy being expressed in electron-volts, eV. *Valence electrons*, that is, those in the *valence band*, are bound to individual atoms, while *conduction electrons* can move freely among the atoms. The range of energy required to free an electron from its bond to an atom is called a *conduction band*. The difference between the highest energy of the valence band and the lowest energy of the conduction band is called a *bandgap*. It is very wide in insulators, narrow in semiconductors, and nonexistent in metals (conductors).

In silicon (Si) and gallium arsenide (GaAs), the commonly used materials for semiconductor power switches, the bandgaps are 1.1 eV and 1.4 eV, respectively. Back in early 1990s research on materials with bandgaps exceeding 3 eV was initiated, and since then the technology of the so-called wide bandgap (WBG) semiconductor devices has significantly progressed. The common WBG materials, the silicon carbide (SiC), zinc oxide (ZnO), and gallium nitride (GaN), have the bandgaps of 3.3–3.4 eV. As of now, the SiC technology is most advanced.

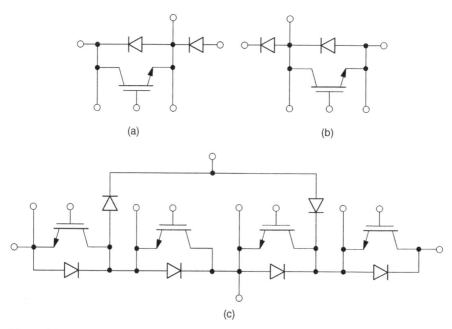

Figure 2.27   IGBT modules: (a) high end of chopper, (b) low end of chopper, (c) leg of a three-level diode-clamped inverter.

Due to the relatively high cost, the WBG devices are still mostly encountered in space and military applications. However, several companies, such as Cree (USA), Fairchild (USA), Infineon (Germany), Powerex (USA/Japan), STMicroelectronics (Switzerland), or Toshiba (Japan), are increasingly involved in the development of the SiC market. The already manufactured SiC-based power devices include diodes,

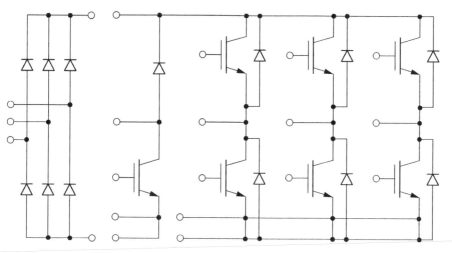

Figure 2.28   IGBT-based modular frequency changer.

power MOSFETs, BJTs, and various power modules. However, several challenges must be overcome to realize the full potentials of WBG devices, to capture a significant share of the market of semiconductor power switches and, possibly, to revolutionize the power electronics.

Indeed, research has shown that in comparison with silicon-based devices, WBG switches can operate with:

(a) more than 10 times higher voltages

(b) more than 10 times higher frequencies

(c) twice as high temperatures

(d) energy losses reduced by up to 90%

It is also worth mentioning that WBG semiconductors, especially GaN, emit visible light, which makes them useful in solid-state lighting, such as very bright LEDs. In the power area, potential applications of WBG devices include variable-speed drives, highly efficient data centers, compact power supplies for consumer electronics, energy systems integration, dc transmission lines, and electric vehicles. It is said that 1970–1990 was a thyristor and power MOSFET era of power electronics, 1990–2010 was an Si IGBT era, and the next period will be SiC era.

SUMMARY

Power electronic converters are based on semiconductor power switches that operate in two states only: the on-state and the off-state. In the on-state, the voltage drop across a switch is low, resulting in low conduction losses. In the off-state, the current through a switch is practically zero, so almost no losses are produced. However, during transitions from one state to another, switching losses are generated, because for a short time both the transient voltage and current are substantial. Each switching is thus associated with energy loss, and the more switchings per second are executed, that is, the higher the switching frequency is, the higher the power loss becomes.

Semiconductor power switches can be classified as uncontrolled, semi-controlled, and fully controlled. Power diodes are uncontrolled switches, which start conducting when forward biased, and cease to conduct when the current changes its polarity. SCRs, triacs, and BCTs are semi-controlled switches that can be triggered into conduction ("fired") when forward biased. Once fired, the devices cannot be extinguished by a control signal. Most common fully controlled switches are GTOs, IGCTs, power BJTs, power MOSFETs, and IGBTs. The GTO and IGCT can be fired in the same way as the SCR, but they can be extinguished by a strong negative gate current pulse. The current in the BJT, MOSFET, and IGBT can be linearly controlled, but to minimize losses they are operated, like all switches, in the on–off regime only. BJTs are current controlled, while power MOSFETs and IGBTs are voltage controlled, thus requiring a negligible amount of gate power.

Except for triacs and BCTs, all semiconductor power switches can conduct current in one direction only. However, not all switches can block voltages of both the forward and reverse polarities. The symmetric blocking capability is typical for SCRs, triacs, and BCTs, some GTOs and IGCTs, and nonpunch-through IGBTs. The other devices may not be subjected to a reverse voltage unless a series-connected diode is employed to block that voltage.

Catalogs and data sheets provide information on semiconductor power switches in the form of restrictive and descriptive parameters, characteristics, and SOAs. Those data, especially the rated voltage and current, help to select the most appropriate devices for a given application. SCRs, GTOs, and IGCTs are the largest and slowest switches, while power MOSFETs are the smallest and fastest. Each type has its advantages and disadvantages, but the IGBTs dominate the field of the most common, low- and medium-power converters. Power semiconductor industry also offers a wide choice of power modules, which, in a single case, combine several switches in a variety of circuit configurations. More advanced solutions, the IPMs, also contain digital components for control and protection of the constituent power switches.

WBG semiconductors, such as silicon carbide, now mostly in advanced development stages, offer significant advantages over the existing silicon-based technology of power switches. The superior operating characteristics include very high voltages, operating temperatures, and frequencies, as well as greatly reduced energy losses.

FURTHER READING

[1] Baliga, B. J., *Advanced High Voltage Power Device Concepts*, Springer, New York, 2012.

[2] Baliga, B. J., *Fundamentals of Power Semiconductor Devices*, Springer, New York, 2008.

[3] Lutz, J., Schlangenotto, H., Scheuermann, U., and De Doncker, R., *Semiconductor Power Devices*, Springer, Berlin, 2011.

[4] Madjour, K., Silicon carbide market update: from discrete devices to modules, *PCIM Europe 2014*, available at: http://apps.richardsonrfpd.com/Mktg/Tech-Hub/pdfs/YOLEPCIM_2014_SiC_Market_ARROW_KMA_Yole-final.pdf

[5] Wide bandgap semiconductors: pursuing the promise, *U.S. DOE Advanced Manufacturing Office*, available at: http://www.manufacturing.gov/docs/wide_bandgap_semiconductors.pdf

[6] Available at: http://new.abb.com/semiconductors

[7] Available at: www.irf.com/

[8] Available at: http://www.ixys.com/ProductPortfolio/PowerDevices.aspx

[9] Available at: www.pwrx.com

3 Supplementary Components and Systems

In this chapter: supplementary components and systems for power electronic converters are reviewed; drivers, protection circuits, snubbers, filters, cooling methods, and control systems are described; and example solutions are presented.

3.1 WHAT ARE SUPPLEMENTARY COMPONENTS AND SYSTEMS?

A practical power electronic converter is a complex systems comprising several subsystems and numerous components. Many of them are not shown in the converter circuit diagrams, which are usually limited to the power circuit and, sometimes, a block diagram of the control system. The supplementary components and systems for modern power electronic converters include:

(1) *Drivers* for individual semiconductor power switches, which provide the switching signals, interfacing the switches with the control system.

(2) *Protection circuits*, which safeguard converter switches and sensitive loads from excessive currents, voltages, and temperatures.

(3) *Snubbers*, which protect switches from transient overvoltages and overcurrents at turn-on and turn-off and reduce the switching losses.

(4) *Filters*, which improve quality of the power drawn from the source or that supplied to the load. As parts of the power circuit, filters are usually shown in circuit diagrams of converters.

(5) *Cooling systems*, which reduce thermal stresses on switches.

(6) *Control systems*, which govern the converter operation, including protection tasks.

Because of its introductory scope, this book is focused on principles of power conversion and control and their realization in power circuits of power electronic converters. However, the reader should be aware of the basic properties of supplementary components and systems and their impacts on operation of converters. Therefore, the

Introduction to Modern Power Electronics, Third Edition. Andrzej M. Trzynadlowski.
© 2016 John Wiley & Sons, Inc. Published 2016 by John Wiley & Sons, Inc.
Companion website: www.wiley.com/go/modernpowerelectronics3e

subsequent sections of this chapter are devoted to a brief coverage of individual types of that equipment. It must be stressed that the great variety of power electronic products offered by scores of manufacturers allows for only a sampling approach to this vast topic.

3.2 DRIVERS

Depending on the type of switches, converter topology, and voltage levels, various driver configurations are employed in power electronic converters. A driver, activated by a logic-level signal from the control system, must be able to provide a sufficiently high voltage or current to the controlling electrode, gate or base, to cause an immediate turn-on. The on-state of the switch must then be safely maintained until turn-off.

The driver must ensure electrical isolation between the low-voltage control system and high-voltage power circuit. It is realized using pulse transformers (PTRs) or optical coupling. The latter is performed by placing a light-emitting diode (LED) in the vicinity of a light-activated semiconductor device. Alternately, instead of transferring light signal through free space, a fiber-optic cable can be employed. Because of the fundamentally different driving signal requirements, different solutions are used for semi-controlled thyristors (SCRs, triacs, BCTs), current-controlled switches (GTOs, IGCTs, power BJTs), and voltage-controlled hybrid devices (power MOSFETs and IGBTs). As a general comment, note that practical commercial drivers are more complicated than the subsequently presented examples, whose purpose is to convey only the basic concepts.

3.2.1 Drivers for SCRs, Triacs, and BCTs

To "fire," that is, to trigger into conduction, an SCR, triac, or BCT, the pulse of gate current, i_G, must have sufficient magnitude and duration, and a possibly short rise time, that is, high di_G/dt. The isolation between the control circuitry and the power circuit is necessary, at least for the switches with ungrounded cathodes. The isolation can be provided by either an optocoupler or a transformer. Both solutions have their advantages and disadvantages. An optocoupler requires a power supply and an amplifier on the thyristor side, which is not needed when a transformer is used. However, extra circuitry must be employed to avoid saturation of the transformer core.

A driver for an SCR, based on a pulse transformer, PTR, and transistor amplifier, TRA, is shown in Figure 3.1. Diode D1 and Zener diode DZ connected across the primary winding provide a freewheeling path for the primary current at turn-off and prevent saturation of the transformer core. Diode D2 in the gate circuit rectifies the secondary current of the transformer.

A simple optically isolated driver for an SCR is shown in Figure 3.2. The optocoupler is composed of a light-emitting diode, LED, and a small light-activated thyristor, LAT. The energy for the gate signal is obtained directly from the power circuit, as it is the voltage across the SCR that produces the gate current when the LAT is activated

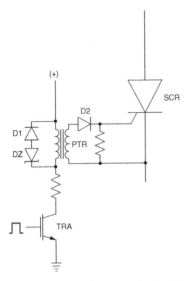

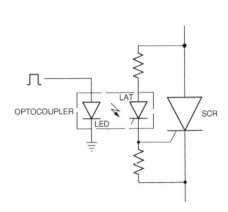

Figure 3.1   Driver for an SCR with transformer isolation.

Figure 3.2   Optically isolated driver for an SCR.

by the LED. The LAT must withstand the same voltage as the driven SCR. This is not a serious problem though, as LATs, also used in high-voltage transmission lines, are among semiconductor devices with the highest voltage ratings.

Figure 3.3 shows a nonisolated driver for a triac. A TRA provides the gate current for the triac. An optically isolated driver using a light-activated triac, LATR, is illustrated in Figure 3.4.

3.2.2   Drivers for GTOs and IGCTs

Although GTOs and IGCTs are turned on similarly to SCRs, drivers for these switches are more complex than those for the semi-controlled thyristors because of the required

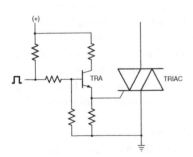

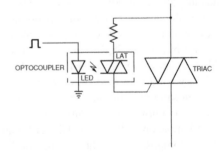

Figure 3.3   Nonisolated driver for a triac.

Figure 3.4   Optically isolated driver for a triac.

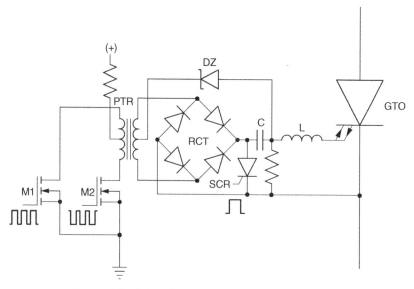

Figure 3.5   Driver for a GTO with transformer isolation.

very high magnitude of the gate current pulse for turn-off. An example gate drive circuit is shown in Figure 3.5. To turn the GTO on, the PTR transmits high-frequency current pulses generated by alternatively switched MOSFETs M1 and M2. The firing current is supplied to the gate through the Zener diode, DZ, and inductor, L, that limits the rate of change, di_G/dt, of the current. Simultaneously, capacitor C is charged via the four-diode rectifier, RCT. Cessation of the impulse train indicates that turn-off is to be performed. The turn-off is initiated by the SCR, which causes a rapid discharge of the capacitor in the gate-cathode circuit.

Drivers with optical isolation are also known. The GTO side of the driver must have its own power supply to provide the necessary gate current, especially for turn-off. Instead of the SCR in the driver in Figure 3.5, a BJT, a power MOSFETs, or combination of these can be used to initiate the turn-on and turn-off gate pulses.

3.2.3   Drivers for BJTs

To generate the base current, the BJT drivers must be of the current-source type. A high-quality driver should have the following characteristics:

(1) High current pulse at turn-on, to minimize the turn-on time.
(2) Adjustable base current in on-state, to minimize losses in the base–emitter junction. The initial boost current should be reduced after turn-on.
(3) Prevention of hard saturation of the transistor. A saturated BJT has a significantly longer turn-off time than a quasi-saturated one.
(4) Reverse base current for turn-off, to further minimize the turn-off time.

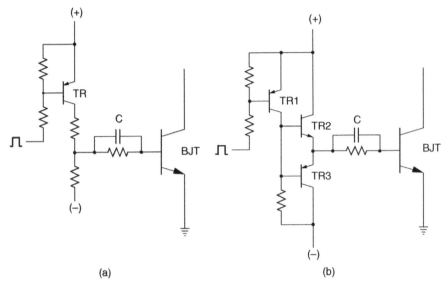

Figure 3.6 Nonisolated drivers for a BJT: (a) single-transistor driver, (b) driver with the class B output stage.

(5) Possibly low impedance between the base and emitter in the on-state and a reverse base–emitter voltage in the off-state. These measures increase the collector–emitter voltage blocking capability of the transistor.

Two simple nonisolated drivers are shown in Figure 3.6. The single-ended driver in Figure 3.6a requires only one transistor, TR, but its performance is inferior to more advanced schemes. Power losses in the driver are reduced in the circuit in Figure 3.6b. Input transistor TR1 drives the so-called *class B output stage* consisting of an npn transistor TR2 and a pnp transistor TR3. Capacitor C in both drivers speeds up switching processes providing a current boost to the base.

To increase the turn-off speed, an antisaturation circuit shown in Figure 3.7 called a *Baker's clamp* can be used. The purpose of the clamp is to shunt the current from the base through diode D0, in dependence on the collector–emitter voltage, to shift the operating point of the transistor from the hard saturation line to the quasi-saturation region (see Figure 2.13). Diodes D1 through D3 produce appropriate bias for the clamping diode D0 (in practice, the number of diodes in the series can be greater or less than three). Diode D4 provides the path for negative base current during turn-off.

If a BJT is to be employed in a high-voltage power circuit, an isolation transformer can be used, as shown in Figure 3.8. However, the range of available duty ratios of the driven BJT is then limited to about 0.1–0.9. Therefore, isolated drivers for BJTs are usually based on optocouplers. A commercial single-chip driver with optical isolation

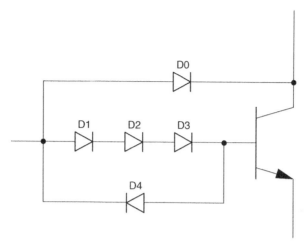

Figure 3.7   Antisaturation Baker's clamp for a BJT.

is shown in Figure 3.9a. Phototransistors are used in the internal optocouplers of the driver, and the BJT is driven by a class B output stage. The waveform of the base current generated by the driver is illustrated in Figure 3.9b. It is not to scale: in reality, the positive and negative peak values of the current are 10–20 times higher than the sustained on-state current.

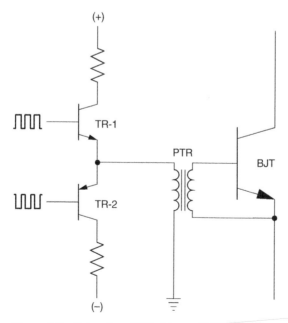

Figure 3.8   Driver for a BJT with transformer isolation.

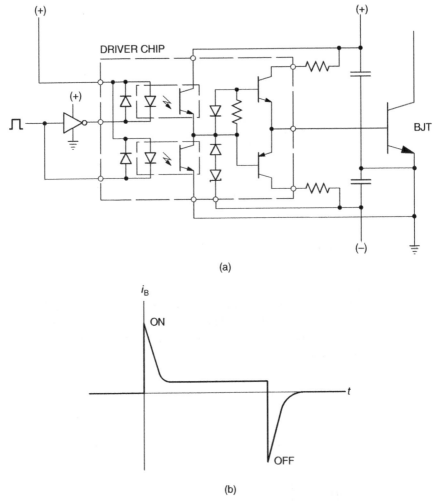

(a)

(b)

Figure 3.9   Driver for a BJT with optical isolation: (a) circuit diagram, (b) waveform of base current.

3.2.4   Drivers for Power MOSFETs and IGBTs

In the steady-state, gates of hybrid semiconductor power switches draw almost no current. As such, they can be directly activated from logic gates. However, if high-frequency switching is desired, an electric charge must quickly be transferred to and from the gate capacitance. This requires high pulses of gate current at the beginning of the turn-on and turn-off signals. Standard logic gates by themselves are incapable of supplying (sourcing) or drawing (sinking) such high currents, sharply limiting the maximum available switching frequency. Therefore, to fully utilize the high-speed potentials of hybrid switches, the very fast power MOSFETs in

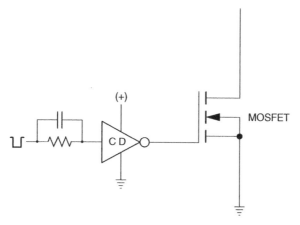

Figure 3.10 Gate drive for a power MOSFET with high-current TTL clock driver.

particular, provisions must be made in the drivers to source or sink transient current pulses.

For uniformity, all the subsequent drivers are shown in application to power MOSFETs although they can be used for IGBTs as well. A simple gate drive circuit with a high-current TTL clock driver, CD, is shown in Figure 3.10. Figure 3.11 illustrates a power MOSFET driven from a PTR. The internal parasitic diode, D, in the auxiliary MOSFET, AM, provides the path for charging current of the main MOSFET's gate capacitance. When the PTR saturates, AM blocks the discharge current from the gate until turn-off, which is initiated by a negative pulse from the transformer that turns AM on. The driver is particularly convenient for switches requiring a floating (ungrounded) gate drive.

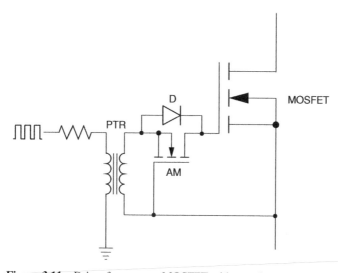

Figure 3.11 Driver for a power MOSFET with transformer isolation.

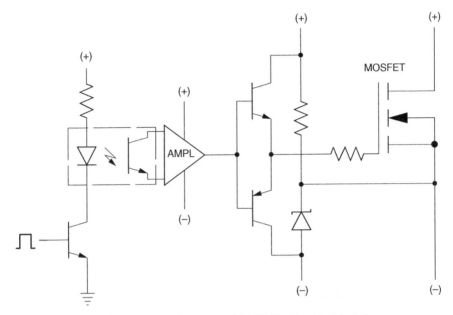

Figure 3.12   Driver for a power MOSFET with optical isolation.

A driver for power MOSFETs and IGBTs with optical isolation is shown in Figure 3.12. In addition to turning the driven switch on and off, the driver provides an overcurrent protection for the switch. The overcurrent condition is detected by sensing the collector–emitter voltage in the on-state. Since the hybrid devices offer an approximately constant resistance in the main circuit, an elevated voltage implies an overcurrent. The switch is then turned off and an indicator circuit is activated.

3.3   OVERCURRENT PROTECTION SCHEMES

Semiconductor power switches can easily suffer a permanent damage if a short circuit occurs somewhere in the converter or the load, or if an overcurrent occurs due to an excessive load, that is, a too low load impedance and/or counter-EMF. The following three basic approaches to overcurrent protection are used in power electronic converters:

(1) Fuses
(2) An SCR "crowbar" arrangement
(3) Turning the switches off when overcurrent is detected

High-power slow-reacting semiconductor power devices, such as diodes, SCRs, or GTOs, are protected by special, fast-melting fuses connected in series with each

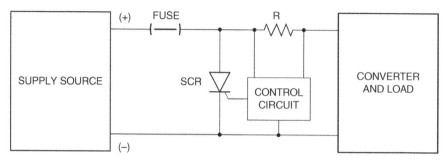

Figure 3.13   SCR "crowbar" for overcurrent protection of power electronic converter.

device. The fuses are made of thin silver bands, not thicker than a tenth of an inch, usually with a number of rows of punched holes. One or more of such bands are packed in sand and enclosed in a cylindrical ceramic body with tinned mounting brackets. The fuse coordination I^2t parameter of a fuse must be less than that of the protected device, but not so low as to cause a breakdown under regular operating conditions. A properly selected fuse should melt within a half cycle of the 60-Hz (or 50-Hz) voltage.

Low-cost power electronic converters can be protected by a single fuse between the supply source and the input to the converter. A more sophisticated solution, illustrated in Figure 3.13, involves an SCR "crowbar," connected across the input terminals of a converter. The input current to the converter is monitored by sensing the voltage drop across the low-resistance resistor R or employing a current sensor. If overcurrent is detected, the SCR is fired shorting the supply source and causing a meltdown of the input fuse. Alternately, a fast circuit breaker can be employed in place of the fuse. In a similar manner, the SCR crowbar can be employed for an overvoltage protection.

Fully controlled semiconductor power switches are best protected by turning them off when overcurrent occurs. The turn-off process is often slowed down to avoid an excessive rate of change, di/dt, of the switch current that could generate a hazardous voltage spike across the device. As exemplified by the system in Figure 3.12, dedicated overcurrent protection circuits are often incorporated in the modern drivers.

A special case of a potentially dangerous short circuit, called a *shoot-through*, is specific for the *bridge topology*, typical for many power converters. In individual legs (branches) of a bridge circuit, two (or more) switches are connected in series, in the so-called *totem pole* arrangement illustrated in Figure 3.14 with two IGBTs. Normally, their states are mutually exclusive, that is, when one switch is on, the other is off, and the current always flows through the load, R. However, if for any reason, for example, due to a driver failure or incorrect timing of switching signals, one switch is turned on before the other has turned off, a short circuit occurs. To reduce probability of an overshoot, the beginning of a turn-on signal for one switch is delayed with respect to the end of a turn-off signal for the other switch by the so-called *dead time*.

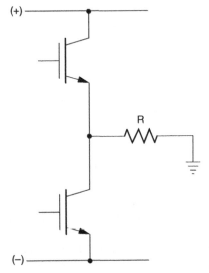

Figure 3.14   Totem pole arrangement of two switches in a leg of the bridge topology.

3.4   SNUBBERS

Of all the possible modes of operation of semiconductor devices, switching between two extreme states is the most trying one, subjecting switches in power electronic converters to various stresses. For example, if no measures were taken, a rapid change of the switched current at turn-off would produce potentially damaging voltages spikes in stray inductances of the power circuit. At turn-on, a simultaneous occurrence of high voltage and current could take the operating point of a switch well beyond the safe operating area (SOA). Therefore, switching-aid circuits called snubbers must often accompany semiconductor power switches. Their purpose is to prevent transient overvoltages and overcurrents, attenuate excessive rates of changes of voltage and current, reduce switching losses, and ensure that the switch does not operate outside its SOA. Snubbers also help to maintain uniform distribution of voltages across the switches that are connected in series to increase the effective voltage rating or currents in the switches that are connected in parallel to increase the effective current rating. Functions and configurations of snubbers depend on the type of switch and converter topology.

Analysis of snubber circuits is tedious at best, and often quite difficult, due to the nonlinear properties of the semiconductor devices involved and relative complexity of certain schemes. Therefore, the subsequent treatment of this topic is mostly qualitative. In practice, besides specialized literature, computer simulations using PSpice or similar software tools are usually employed for snubber development.

To illustrate the need for snubbers, a practical example is considered. A simple BJT-based chopper is shown in Figure 3.15a. High load inductance is assumed, so that the output current is practically constant and equal I_o. Consequently, the load

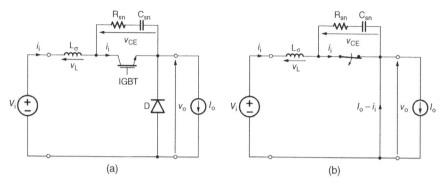

Figure 3.15 BJT-based chopper with RC snubber: (a) circuit diagram, (b) equivalent circuit in the off-state.

can be modeled by a current source. A stray inductance, L_σ, of the power circuit of the chopper is lumped between the source of input voltage, V_i, and the transistor. The snubber circuit composed of resistance R_{sn} and capacitance C_{sn} is connected in parallel with the transistor.

The collector–emitter voltage, v_{CE}, across the BJT is given by

$$v_{CE} = V_i - v_L - v_o, \tag{3.1}$$

where v_L and v_o denote the inductor and output voltages, respectively. With the transistor in the on-state, $v_{CE} \approx 0$. At $t = 0$, the BJT is turned off so that its collector current, i_C, decreases linearly from the initial value of I_o, reaching zero at $t = t_0$. As a result, a transient voltage appears across the stray inductance. The voltage waveform has the shape of a pulse with the duration of t_0 and a peak value, $V_{L,p}$, of

$$V_{L,p} = L_\sigma \frac{di_c}{dt} = -L_\sigma \frac{I_o}{t_0}. \tag{3.2}$$

In the meantime, the freewheeling diode, D, has taken over conduction of the output current, thus $v_o = 0$ and the peak value, $V_{CE,p}$ of the collector–emitter voltage is

$$V_{CE,p} = V_i - V_{L,p} = V_i + L_\sigma \frac{I_o}{t_0}. \tag{3.3}$$

Fast turn-offs are desirable from the point of view of reduction of switching losses, but Eq. (3.3) shows that when t_0 approaches zero, the voltage across the BJT approaches infinity. Even with a finite but short t_0, the voltage can easily be excessive, damaging the transistor.

With the snubber in place, the equivalent circuit of the chopper at turn-off is shown in Figure 3.15b. When switch S representing the BJT opens, a series RLC circuit is created. As known from the theory of such circuits, if $R_{sn} < 2\sqrt{L_\sigma/C_{sn}}$, the current, i_i, in the circuit is given by

$$i_i = I_p e^{-\frac{R_{sn}}{L_\sigma}t} \cos(\omega_d t + \varphi), \tag{3.4}$$

where

$$\omega_d = \sqrt{\frac{1}{L_\sigma C_{sn}} - \left(\frac{R_{sn}}{2L_\sigma}\right)^2}. \tag{3.5}$$

The values of amplitude I_p and angle φ in Eq. (3.4) can be determined from the known initial and final conditions. Since the initial current, $i_i(0)$, equals I_0 and the final, steady-state current through the capacitor is zero, $I_p = I_0$ and $\varphi = 0$, that is,

$$i_i = I_0 e^{-\frac{R_{sn}}{L_\sigma}t} \cos(\omega_d t). \tag{3.6}$$

The collector–emitter voltage across the snubber can be obtained by differentiating i_i and substituting di_i/dt in the equation

$$v_{CE} = V_i - v_L = V_i - L_\sigma \frac{di_i}{dt}, \tag{3.7}$$

which, after some rearrangements, yields

$$v_{CE} = V_i \left[1 - e^{-\frac{R_{sn}}{L_\sigma}t} \cos(\omega_d t)\right]. \tag{3.8}$$

With a properly designed and tuned snubber, the collector–emitter voltage displays only a small overshoot. Example voltage and current waveforms for the both cases considered are shown in Figure 3.16. By combining the $i_C(t)$ and $v_{CE}(t)$ waveforms into the $i_C = f(v_{CE})$ relation, the so-called *switching trajectory* is obtained, which represents a locus of operating points of the transistor. As shown in Figure 3.17, using the SOA of the BJT as a background, it can easily be checked whether the snubber protects the transistor from unsafe operating conditions. A large portion of the switching trajectory in Figure 3.17a for the snubberless transistor lies outside the SOA. However, that in Figure 3.17b, when the snubber is installed, it lies within the SOA with significant safety margins.

Snubbers are not inherently necessary, as it is often possible to select power switches of such high ratings that even large transient voltages and current would not be dangerous. However, there is always a price to pay for oversized semiconductor devices: not only in the literal sense, but also in terms of the increased weight, bulk, and losses. On the other hand, snubbers also contribute to the cost, weight, and size of a power electronic converter, and they are not free of losses. Therefore, a search for

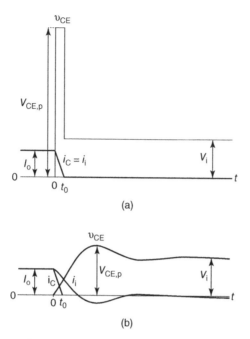

Figure 3.16   Voltage and current waveforms in the chopper of Figure 3.15: (a) without snubber, (b) with snubber.

snubbers that are optimal for a given application constitutes a true test of diligence and expertise of a designer.

Snubbers for individual power switches are disposed of in *resonant converters*, where a single resonant circuit is used to provide safe and low loss switching conditions for all switches of the converter. Resonant power electronic converters belong mostly to the family of low-power dc-to-dc converters, covered in Section 8.4. Certain concepts of resonant operation have been extended on high-power converters. The so-called *resonant dc link inverter* is described in Section 7.5.

3.4.1   Snubbers for Power Diodes, SCRs, and Triacs

The rate of rise of the reverse recovery current in power diodes is high, so that an overvoltage at turn-off is very likely, even with a small amount of stray inductance. Therefore, simple RC snubbers, such as that in Figure 3.15, are connected in parallel with the diode. This type of switching aid circuits are often termed *turn-off snubbers*, since they alleviate the voltage stresses at turn-off.

RC snubbers are also employed in SCR- and triac-based converters, mostly for the purpose of preventing false triggering (firing) from excessive *dv/dt*. In certain applications, firing an SCR at a wrong time may prove catastrophic. SCRs and triacs should also be protected from extreme *di/dt* values. This is accomplished by placing an inductor in series with the device. The required inductance of such a *turn-on*

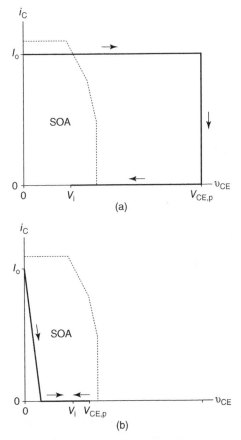

Figure 3.17   Switching trajectories of the BJT of Figure 3.15: (a) without snubber, (b) with snubber.

snubber is low, so that just the stray inductance of wiring of the power circuit is often sufficient. Snubbers for the power diode and SCR are illustrated in Figure 3.18.

3.4.2 Snubbers for GTOs and IGCTs

Simple turn-on and turn-off snubbers for the GTO are shown in Figure 3.19. The turn-on snubber protects the GTO from overcurrents when it takes over conduction of the current from another slow device, that is, a high-power freewheeling diode. Also, the *di/dt* is attenuated. The diode–resistor circuit allows fast dissipation of the energy stored in the inductor when the GTO turns off.

The RDC (resistor–diode–capacitor) snubber reduces the anode–cathode voltage at turn-off, limiting the switching loss. Similarly to that for the SCR, the snubber also prevents the GTO from re-firing due to supercritical values of *dv/dt*. At turn-off, the input current, i_i, is diverted into the snubber capacitor through the diode, while

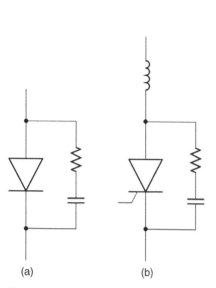

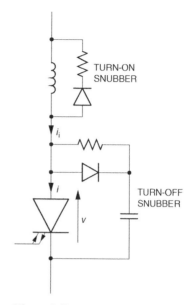

Figure 3.18 Snubbers for: (a) power diode, (b) SCR.

Figure 3.19 GTO with turn-on and turn-off snubbers.

the anode current, i, in the GTO decreases. The anode is clamped to the capacitor, whose initial voltage is zero and rising. Thus, a simultaneous occurrence of high anode voltage, v, and current, i, at turn-off is avoided. The capacitor gets quickly discharged in the resistor–GTO circuit at the following turn-on.

3.4.3   Snubbers for Transistors

In the figures illustrating this section, snubbers for transistors (BJTs, MOSFETs, and IGBTs) are shown with the IGBT, but the same solutions can be applied to the other devices. Snubbers similar to those for the GTO (see Figure 3.19) are often employed. The inductive turn-on snubber ensures that the collector–emitter voltage, v_{CE}, drops to the saturation level prior to the collector current reaching its full on-state value.

Figure 3.20 shows another, related solution in the form of a combined, turn-on and turn-off snubber. At turn-on, the series inductor slows down the rate of increase of the collector current, i_C. Simultaneously, the capacitor discharges through the resistor, inductor, and transistor. At turn-off, the input current, i_i, is diverted from the transistor to the diode–capacitor bypass. When the capacitor is fully charged, the remaining electromagnetic energy of the inductor is dissipated in the resistor.

It must be pointed out that when selecting a snubber, the whole converter topology should be taken into account, since other components of the power circuit can interfere with proper operation of snubbers. Common snubbers for popular bridge converters based on switches with antiparallel freewheeling diodes are shown in Figure 3.21 for a single leg of the bridge.

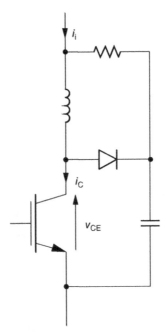

Figure 3.20   Combined on-and-off snubber for a transistor.

In recent years, there has been a tendency to design snubberless converters. The accumulated practical experience how to minimize the stray inductance and the robust SOAs of modern devices allow the turn-off snubbers to be disposed of. Yet, to protect the internal freewheeling diodes and reduce the electromagnetic interference (EMI), simple RC snubbers in parallel with the switches are still recommended. Also, in bridge converters, a single inductor can be placed in series with the input terminals of the bridge. If a converter is directly supplied from a transformer, the secondary leakage inductance can play the role of the turn-on snubber, limiting the rate of change of currents in the converter.

3.4.4   Energy Recovery from Snubbers

Snubbers presented in the previous sections modify the switching trajectory and reduce switching losses. However, the energy temporarily stored in the inductive and capacitive components is dissipated in resistors and irretrievably lost. In high-frequency and high-power converters, the amount of energy lost in snubbers is substantial, straining cooling systems and reducing efficiency of the converters. Therefore, measures have been developed to recover energy from snubbers and direct it to the load or return to the supply source.

The energy recovery systems can be passive or active, the latter involving an auxiliary power converter. An example scheme of passive recovery of energy from a capacitive snubber is shown in Figure 3.22. Transistor T operates in the PWM mode

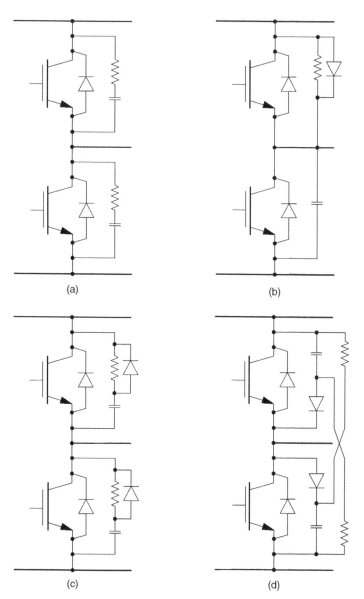

Figure 3.21   Snubbers for transistors in bridge converters: (a) RC, (b) RCD, (c) charge and discharge RCD, (d) discharge-suppressing RCD.

to control the output voltage, v_o, across a load, which is assumed to have an inductive component. Hence there is a need for a freewheeling diode, D_{fw}, that provides a path for the load current when the transistor turns off. The simple capacitive turn-off snubber is composed of diode D_{sn} and capacitor C_{sn}. The energy recovery circuit is based on diodes D1 and D2, inductor L, and capacitor C. At turn-off, the snubber

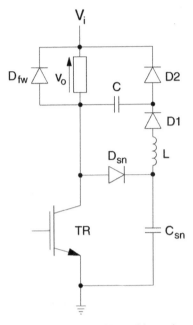

Figure 3.22 Turn-off capacitive snubber with passive energy recovery.

capacitor is charged to the input (supply) voltage, V_i. When the transistor turns on, the charge stored in the capacitor is transferred to capacitor C by means of electric resonance in the C_{sn}–L–D1–C–T circuit. At the subsequent turn-off, capacitor C_{sn} charges again, while capacitor C discharges through diode D2 into the load. Since no resistors are employed, most of the energy from the snubber capacitor is recovered and consumed by the load.

In high-power, GTO-based converters, auxiliary step-up choppers, described in Section 6.3, are used to transfer energy from snubbers to the power supply. A step-up chopper is a PWM dc-to-dc converter whose pulsed output voltage has an adjustable magnitude, which is higher than that of the input voltage. An example active energy recovery system is shown in Figure 3.23. When the GTO turns on, the charge of the snubber capacitor, C_{sn}, is resonated into a large storage capacitor C via the GTO, inductor L, and diode D. The storage capacitor acts as a voltage source for the step-up chopper, which boosts the voltage and transfers the energy to the supply line.

3.5 FILTERS

Filters are indispensable circuit components of most practical power electronic converters. Generally, depending on their placement, filters can be classified as input, output, and intermediate ones. Input filters, also called *line* or *front-end* filters, screen the supply source from the harmonic currents drawn by converters and improve the

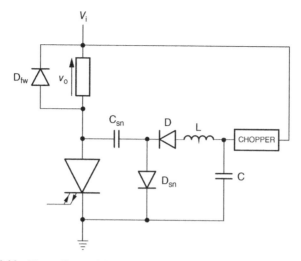

Figure 3.23   Turn-off capacitive snubber for a GTO with active energy recovery.

input power factor of the converters. Output, or load, filters are employed to improve quality of the power supplied to the load. The intermediate filters, usually called *links*, interface two converters forming a cascade, such as that of a rectifier and inverter in the ac-to-dc-to-ac power conversion scheme.

A few simple filter topologies involving inductive (L) and capacitive (C) components only serve the whole area of power electronics. Those are covered in next chapters along the converters in which they are applied. Usually, filter inductors carry the whole input or output current. Therefore, their resistance must be keep at a possibly low level, implying a low number of turns, N_t, of a thick-wire winding. Since the coefficient of inductance is proportional to N_t^2 and inversely proportional to the reluctance of the inductor's magnetic circuit, high-permeability cores with large cross-sectional areas are employed. Still, the filter inductances tend to be low, typically of the order of several to tens of millihenries. Inductor cores are made of thin laminations insulated from each other with varnish or shellac to minimize the eddy current losses.

To make up for the usually low inductances, filter capacitors must have large capacitances. Aluminum-foil electrolytic capacitors are therefore used, with the available coefficients of capacitance up to 500 mF. However, the high capacitances are accompanied by low voltage ratings. For capacitors in power electronic applications, it is desirable to have low stray inductance and low equivalent series resistance. Data sheets of capacitors specify maximum allowable values of the working voltage, nonrepetitive surge voltage, voltage and current ripple, and temperature. The highest capacitance-per-volume ratios are obtained in *polarized* electrolytic capacitors which may not operate under a reversed voltage. Nonpolarized capacitors have capacitances about half as high as those of polarized capacitors of the same physical size. In three-phase systems, capacitors are connected in delta to maximize the line-to-line effective capacitance to one-and-a-half of that of the single capacitor.

A filter at the dc side of a power converter is employed to reduce the voltage and the current ripple. Depending on specific requirements, the filter consists of a capacitor connected in parallel with the converter terminals, inductor connected in series, or both these components. If a dc current I_C is drawn from a capacitor C during a time interval Δt, the voltage across the capacitor drops by

$$\Delta V_C = \frac{I_C}{C} \Delta t. \tag{3.9}$$

It means that the higher the filter capacitance, the more stable the voltage across it. Analogously, if a dc voltage V_L is applied to an inductor L over an interval Δt, the current in the inductor increases by

$$\Delta I_L = \frac{V_L}{L} \Delta t, \tag{3.10}$$

which implies a stabilizing impact of the inductance on the current.

Filters in ac circuits are used for blocking or shunting the ripple currents but pass the fundamental currents. If the harmonic frequencies are much higher than the fundamental frequency, as in PWM converters, filter inductors can be connected in series with converter terminals, carrying the whole input or output current. Since the impedance, Z_L, of an inductor is given by

$$Z_L = 2\pi f L, \tag{3.11}$$

then the inductor can serve as a *low-pass filter* for the current. In contrast, a capacitor, whose impedance, Z_C, is

$$Z_C = \frac{1}{2\pi f C}, \tag{3.12}$$

constitutes a *high-pass filter*. Consequently, as in dc filters, capacitors in ac filters are placed across converter terminals, shunting the high-frequency currents. In order to avoid a dangerous resonance overvoltages, care should be taken for the resonance frequency of an LC filter to be significantly higher than the supply frequency (60 Hz or 50 Hz).

The series–parallel placement of the inductive and capacitive filter components is not practical if the undesired ac current components have low frequencies, as those in input currents of phase-controlled converters. In that case, since the dominant ripple frequency is not much higher than the fundamental frequency, the fundamental current would be suppressed by the filter, too. Therefore, as subsequently described in Sections 4.1.2 and 4.2.1, *resonant filters* are placed between ac lines supplying diode- and SCR-based rectifiers. Each filter is tuned to a specific harmonic frequency, usually the 5th, 7th, and 12th multiples of fundamental frequency, requiring a total of at least nine inductors and nine capacitors.

Special filters are needed to protect the power system from the conducted EMI. If high-frequency currents, particularly those of radio-level frequencies, were allowed to spread in the system, they would severely pollute the electromagnetic environment and disturb operation of sensitive communication systems. The *EMI filters*, also called *radio-frequency filters*, are particularly indispensable in power converters characterized by high-switching frequencies. High-order EMI filters, involving several resistors, inductors, and capacitors, are common.

Power electronic converters are also sources of the *radiated EMI*, that is, electromagnetic waves generated due to high *di/dt* rates resulting from the switching mode of operation. In practice, two factors help to minimize the environmental impact of this phenomenon. These are the snubbers, which reduce the rate of change of switched currents and voltages, and the metal cabinets in which power electronic converters are usually enclosed and which act as effective electromagnetic shields.

3.6   COOLING

Losses in power electronic converters produce heat that must be transferred away and dissipated to the surroundings. The major losses occur in semiconductor power switches, while their small size limits their thermal capacity. High temperatures of the semiconductor structures cause degradation of electrical characteristics of switches, such as the maximum blocked voltage or the turn-off time. Serious overheating can lead to destruction of a semiconductor device in a short time. To maintain a safe temperature, a power switch must be equipped with a heat sink (radiator) and be subjected to at least the *natural convection* cooling. The heat generated in the switch is transferred via the heat sink to the ambient air, which then tends to move upward and away from the switch.

Forced air cooling, very common in practice, is more effective than the natural cooling. The cooling air is propelled by a fan, typically placed at the bottom of the cabinet that houses the power electronic converter. Slotted openings at the top part of the cabinet allow the heated air to escape to the surroundings. Energy consumption of the low-power fan does not tangibly affect overall efficiency of the converter.

If power density of a converter, that is, the rated power-to-weight ratio, is very high, *liquid cooling* may be needed. Water, automotive coolant, or oil can be used as the cooling medium. The fluid is forced through extended hollow copper or aluminum bars to which semiconductor power switches are bolted. Thanks to the high specific heat of water, water cooling is very effective although it poses a danger of corrosion. On the other hand, oil, whose specific heat is less than half of that of water, has much better insulating and protecting properties.

Thermal equivalent circuits facilitate design and analysis of cooling systems. The concept of those circuits is based on the formal similarity between heat transfer and electrical phenomena. Thermal quantities and their electric equivalents are listed in Table 3.1. As an example, a power diode with a heat sink and the corresponding

Table 3.1 Comparison of Thermal and Electrical Quantities

Thermal Quantity	Electrical Quantity
Amount of heat (energy), Q (J)	Electric charge, Q (C)
Heat current (power), P (W)	Electric current, I (A)
Temperature, Θ (°K)	Electric voltage, V (V)
Thermal resistance, R_Θ (°K/W)	Electric resistance, R (Ω)
Thermal capacity, C_Θ (J/°K)	Electric capacitance, C (F)
Thermal time constant, $\tau_\Theta = R_\Theta C_\Theta$ (s)	Electrical time constant, $\tau = RC$ (s)

thermal equivalent circuit are illustrated in Figure 3.24. The variables indicated on the circuit diagram are:

P_l — power loss in the diode, W

Θ_J, Θ_C, Θ_S, and Θ_A — absolute temperatures of the pn junction, case, sink, and ambient air, respectively, K

$R_{\Theta JC}$, $R_{\Theta CS}$, and $R_{\Theta SA}$ — junction to case, case to sink, and sink to ambient air thermal resistances, respectively, K/W

$C_{\Theta J}$, $C_{\Theta J}$, and $C_{\Theta J}$ — junction, case, and sink thermal capacities, respectively, J/K.

Values of thermal resistances are listed in data sheets of semiconductor power switches and heat sinks, and the thermal capacity of a component can be determined

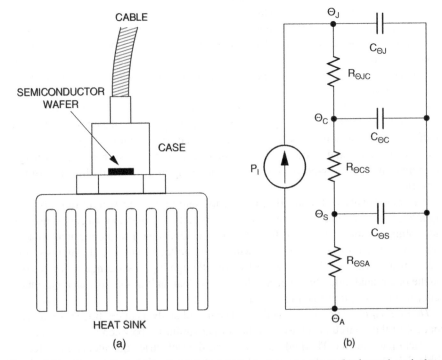

Figure 3.24 Power diode with a heat sink: (a) physical arrangement, (b) thermal equivalent circuit.

from its specific heat and mass. Thermal capacities are used for the calculation of transient temperatures, and knowledge of thermal resistances is sufficient for computation of steady-state temperatures. The goal of cooling system design is to ensure that the maximum temperatures allowed for the converter switches are not exceeded.

3.7   CONTROL

To function efficiently and safely, a power electronic converter must be properly controlled, that is, the process of power conversion must be accompanied by concurrent information processing. Various control technologies have been employed over the several last decades, starting with the analog electronic circuits with discrete components and progressing to the contemporary integrated microelectronic digital systems. It is worth mentioning that simple analog control is still employed in many low-power dc–dc converters, especially those with very high switching frequencies.

In practical applications, a power electronic converter usually constitutes a part of a larger engineering system, such as an adjustable speed drive or active power filter. The control system of the converter is then subordinated to a master controller, for instance, that of speed of a motor fed from the converter. The main task of the converter control system is to generate signals for semiconductor power switches that result in the desired fundamental output voltage or current. Other "housekeeping chores" are also performed, for example, control of electromechanical circuit breakers connecting the converter with the supply system. Often, especially in expensive, high-power converters, the control system also monitors operating conditions. When a failure is detected, the system turns off the converter and displays the diagnosis. Generally, the system draws information from the human operator, sensors, and master controllers and converts it into switching signals for the converter switches and external circuit breakers. The control system also supplies information about operation of the converter back to the operator, via displays, indicators, or recorders.

Microcontrollers, digital signal processors (DSPs), and *field-programmable logic arrays* (FPGAs) are now firmly established as the tools of choice for control of modern power electronic converters. The awesome and continuously increasing computational power of those devices allows a single processor to control a whole power conversion system, which may include several converters. Prices of digital processors are falling, so that they are economical even in low-cost applications.

Microprocessors employed as microcontrollers come from many manufacturers and vary widely. As the name indicates, microcontrollers are designed for control applications. A microcontroller has a central processing unit (CPU) that executes programs stored in a read-only memory (ROM), a random-access memory (RAM) for data storage, and input and output devices for communication with the outside world. Typically, a microcontroller has a smaller instruction set than a comparable microcomputer, but it may have certain additional functional blocks. The number of data and instruction bits (8, 12, 16, 24, 32, or 64), as well as the memory size, depends on the complexity of required tasks. Microcontrollers are low-power devices, often ruggedized in some way to withstand harsh operating conditions such as high temperatures or vibration.

In power electronic systems, the so-called *DSP controllers* are often employed. Similar to microcontrollers, in addition to the basic components, they have specialized functional blocks, such as A/D and D/A converters, embedded timers, or generators of PWM switching signals. Modern microcontrollers and DSPs can be quite complex due to the tendency to increase their speed and functionality by placing more memory and peripheral circuitry on a single chip. The differences between DSPs and microcontrollers are more functional than structural, as both have the data processing units, memories, and input/output circuitry. However, CPUs of DSPs are capable of extremely fast execution of a small set of simple instructions. The most advanced DSPs utilize the floating point arithmetic and can be programmed with higher level languages instead of the cumbersome assembly language. Hardware multipliers and shift registers dramatically reduce the number of steps required to perform such algebraic operations as multiplication, division, or square root calculation.

FPGAs are programmable digital logic chips, which contain thousands of small logic blocks with flip-flops (memory elements) and programmable interconnects. After defining and compiling the logic function to be implemented, it can be downloaded in the form of a binary file into the FPGA. If another logic function is needed, the FPGA can easily be re-programmed. Thus, an FPGA can be considered a virtual breadboard, which requires no component changes and re-soldering, and which greatly facilitates fast prototyping. In general, FPGAs are somewhat inferior to application-specific integrated circuits (ASICs), but they are less expensive and more convenient to use.

As an example of a digital control system, a block diagram of an adjustable-speed ac drive governed by a digital control system based on a DSP controller is illustrated in Figure 3.25. The adjustable-frequency and -magnitude currents for the three-phase ac motor are produced by an inverter, whose three phases are independently controlled by current regulators, CR_A through CR_C. Each current regulator receives a reference signal from the controller, compares it with the signal obtained from current sensors, and generates appropriate switching signals for the inverter. The control system reacts to the speed control signal from a speed sensor. Most signals are digital, so if the inexpensive analog current sensors are employed, analog-to-digital (A/D) converters must be used.

As an alternative, the currents could be measured by digital sensors, with the current control incorporated in the operating algorithm of the DSP controller. In that case, the input signals to the controller would include those from the current sensors. If the drive system in question is a part of a larger process, the reference speed signal can come from another, hierarchically superior controller.

Concluding this brief overview of converter control systems, the growing tendency to integrate semiconductor power switches and control circuits in a common package has led to the so-called integrated assemblies (IA). They represent an extension of the intelligent power modules, already mentioned in Section 2.6. An IA, for example, can include a three-phase high-power inverter bridge, the dc-link capacitor bank, optically

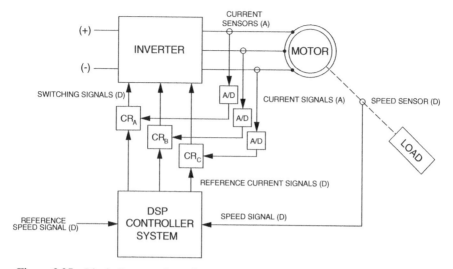

Figure 3.25   Block diagram of an adjustable-speed ac drive: (A) = analog, (D) = digital.

isolated drivers with their power supplies, fault monitoring and protection circuits, current sensors, and simple user interface for an external controller.

SUMMARY

A complete power electronic converter includes a number of supplementary components and systems, most of them not shown in the typical power circuit diagrams. On command from the control system, drivers generate switching voltages and currents for gates/bases of converter switches. Overcurrent protection schemes, based on fast-melting fuses or dedicated circuitry, safeguard the switches from damage caused by short circuits in the converter or load. Snubbers relieve transient voltage and current stresses on switches at turn-on and turn-off and reduce switching losses.

Power filters, based on inductors and capacitors, are placed at the input and output terminals of converters to improve the quality of power drawn from the power system or supplied to the load. Intermediate filters provide a match between two converters in a cascade arrangement. EMI filters are placed on the supply side of fast-switching converters to reduce the conducted electromagnetic noise. Cooling systems remove heat from switches to protect them from excessive temperatures.

Overall operation of a power electronic converter is governed by a control system. Today's control systems are based on microcontrollers, DSP controllers, and FPGAs, all of which are capable of implementing sophisticated control algorithms. Modern power electronic converters are complex engineering systems, which combine efficient energy conversion with advanced information processing. Team effort of engineers of various specialties is required for successful design of those high-quality apparatus.

FURTHER READING

[1] Rashid, M. H. (editor), *Power Electronics Handbook*, 3rd ed., Butterworth-Heinemann, Burlington, MA, 2011.
[2] Sozanski, K., *Digital Signal Processing in Power Electronics Control Circuits*, Springer, London, 2013.

4 AC-to-DC Converters

In this chapter: three-phase ac-to-dc converters are introduced, beginning with the uncontrolled, diode-based rectifiers and progressing to phase-controlled rectifiers, dual converters, and pulse width modulation (PWM) rectifiers, circuit topologies, control principles, and operating characteristics are presented; device selection for ac-to-dc converters is explained; and typical applications of rectifiers are described.

4.1 DIODE RECTIFIERS

Single-phase diode rectifiers have already been expounded in Chapter 1. Widely employed in low-power electronic circuits, they are not feasible for ac-to-dc conversion at the medium- and high-power levels. The output voltage of those rectifiers is of poor quality due to the low dc component and high ripple factor. Therefore, the realm of medium- and high-power electronics is dominated by three-phase, six-pulse rectifiers. For completeness, a three-pulse diode rectifier, although impractical, will briefly be discussed first.

4.1.1 Three-Pulse Diode Rectifier

The circuit diagram of a three-pulse (three-phase, half-wave) diode rectifier is shown in Figure 4.1. Supplied from a three-phase, four-wire ac power line, the rectifier consists of three power diodes, DA through DC. The load is connected between the common-cathode node of the diode set and the neutral, N, of the supply line.

It can be shown that at a given instant only the diode that is supplied with the highest line-to-neutral voltage is conducting the output current, i_o. A situation when the highest voltage is that of phase B is illustrated in Figure 4.2. As $v_{BN} > v_{AN}$ and $v_{BN} > v_{CN}$, then it is diode DB that is conducting, while diodes DA and DC are reverse biased by line-to-line voltages v_{BA} and v_{BC}, respectively.

Each of the three line-to-neutral voltages is higher than the other two for one-third of the cycle of input voltage. Consequently, in the continuous conduction mode, each diode conducts the current within a 120°-wide angle interval. Voltage and current waveforms of the three-pulse rectifier with a resistive load (R-load) are shown in Figure 4.3. The waveform pattern of the output voltage is repeated every 120° ($2\pi/3$ rad).

Introduction to Modern Power Electronics, Third Edition. Andrzej M. Trzynadlowski.
© 2016 John Wiley & Sons, Inc. Published 2016 by John Wiley & Sons, Inc.
Companion website: www.wiley.com/go/modernpowerelectronics3e

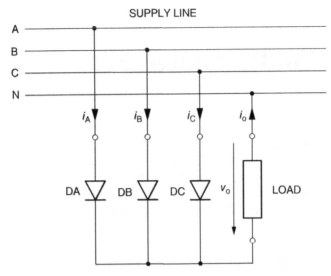

Figure 4.1   Three-pulse diode rectifier.

As in all considerations concerning electronic converters, idealized, lossless semiconductor devices are assumed, so that the voltage drops across conducting switches are neglected. Since for $0 \leq \omega t \leq 2\pi/3$,

$$v_{\mathrm{o}} = v_{\mathrm{AN}} = V_{\mathrm{LN,p}} \sin\left(\omega t + \frac{\pi}{6}\right), \tag{4.1}$$

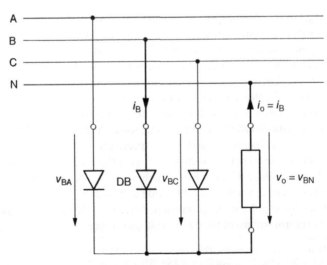

Figure 4.2   Example current path and voltage distribution in a three-pulse diode rectifier.

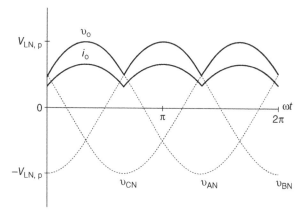

Figure 4.3   Waveforms of output voltage and current in a three-pulse diode rectifier (R-load).

where $V_{LN,p}$ is the peak value of the supply line-to-neutral voltage, then the dc component, $V_{o,dc(C)}$, of the output voltage is

$$V_{o,dc}(C) = \frac{1}{\frac{2}{3}\pi} \int_0^{\frac{2}{3}\pi} V_{LN,p} \sin\left(\omega t + \frac{\pi}{6}\right) d\omega t = \frac{3}{2\pi} V_{LN,p} \left[\cos\left(\omega t + \frac{\pi}{6}\right)\right]_{\frac{2}{3}\pi}^{0}$$

$$= \frac{3\sqrt{3}}{2\pi} V_{LN,p} \approx 0.827 V_{LN,p}. \tag{4.2}$$

The "(C)" subscript indicates the continuous conduction mode.

The output current with a large dc component flows through the neutral wire. However, a dc component equal to one-third of the dc-output currents appears also in currents drawn from wires A, B, and C. This property disqualifies the three-pulse rectifier from practical applications. In a power system supplying the rectifier, the dc current would cause saturation of transformer cores, resulting in distortion of voltage waveforms in the system. Another potential trouble stems from the fact that all components of the power system are designed for ac sinusoidal currents of a specific frequency (60 Hz in the USA). The waveform of the phase-A line current, i_A, is shown in Figure 4.4. Clearly, it is completely different from a sinewave. Such currents drawn by the rectifier would disturb the operation of protection systems.

4.1.2   Six-Pulse Diode Rectifier

The six-pulse (three-phase, full-wave) diode rectifier, depicted in Figure 4.5 with an RLE load, is the most commonly used ac-to-dc power converter producing a fixed dc voltage. The power circuit of the rectifier consists of six power diodes in a three-phase bridge configuration. Diodes DA, DB, and DC form the common-cathode group, and diodes DA′, DB′, and DC′ constitute the common-anode group. At any instant, only one pair of diodes conducts the current, one in the common-cathode group and one in

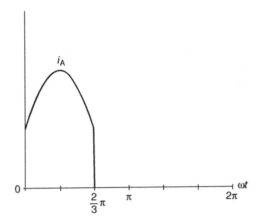

Figure 4.4 Waveform of input current in a three-pulse diode rectifier (R-load).

the common-anode group, the two diodes belonging to different phases of the bridge. Consequently, six combinations of conducting diodes are possible, each pair being active for one-sixth of the cycle of the supply voltage, that is, for a 60°-wide angle interval.

The conducting diode pair is that supplied with the highest line-to-line voltage. This is illustrated in Figure 4.6 for voltage v_{AC} being higher than the remaining five line-to-line voltages, v_{AB}, v_{BC}, v_{BA}, v_{CA}, and v_{CB}. Note that six line-to-line voltages are distinguished here as, for instance, v_{AB} is considered separately from v_{BA}. Diodes DA and DB′ form a path for the output current. The other four diodes are subjected to voltages v_{AC} (diode DC), v_{CB} (diode DC′), and v_{AB} (diodes DA′ and DB). The phasor diagram of the ac voltages is shown in Figure 4.7 at the instant when voltage

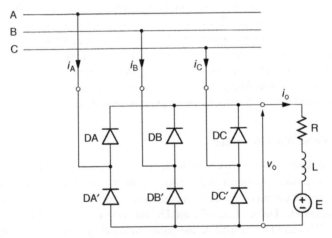

Figure 4.5 Six-pulse diode rectifier.

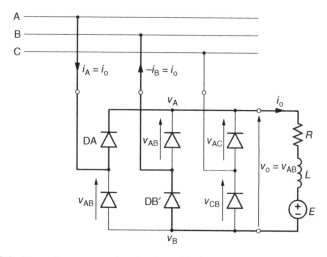

Figure 4.6 Example current path and voltage distribution in a six-pulse diode rectifier.

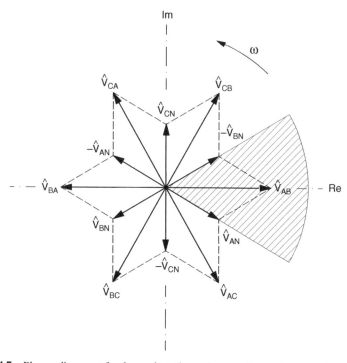

Figure 4.7 Phasor diagram of voltages in a three-phase ac line and the maximum-voltage area.

v_{AB} (the real part of phasor $\hat{V}_{AB}$) is higher than the other line-to-line voltages, that is, when phasor $\hat{V}_{AB}$ is located in the shaded 60°-wide sector of the complex plane.

It can be seen that voltages v_{AC}, v_{CB}, and v_{AB} are positive (the respective phasors have positive real parts), imposing the reverse bias on the nonconducting diodes. The phasor diagram also allows determination of the sequence of conducting diode pairs, which is DA and DB′, DA and DC′, DB and DC′, DB and DA′, DC and DA′, DC and DB′, and so on. Thus, each diode conducts the current for one-third of the cycle of supply voltage. The process of a diode taking over conduction of current from another diode is called *natural commutation*.

Output Voltage and Current. Under most operating conditions, the output current is continuous, as illustrated in Figure 4.8. The output voltage within the 0 to $\pi/3$ interval equals to the line-to-line voltage v_{AB} is given by

$$v_{AB} = V_{LL,p} \sin\left(\omega t + \frac{\pi}{3}\right), \tag{4.3}$$

where $V_{LL,p}$ denotes the peak value of the supply line-to-line voltage. The output voltage waveform repeats itself every $\pi/3$ radians, so the average output voltage, $V_{o,dc(C)}$, can be found as

$$V_{o,dc(C)} = \frac{1}{\frac{\pi}{3}} \int_0^{\frac{\pi}{3}} V_{LL,p} \sin\left(\omega t + \frac{\pi}{3}\right) d\omega t = \frac{3}{\pi} V_{LL,p} \left[\cos\left(\omega t + \frac{\pi}{3}\right)\right]_{\frac{\pi}{3}}^{0}$$

$$= \frac{3}{\pi} V_{LL,p} \approx 0.955 V_{LL,p}. \tag{4.4}$$

The output voltage is not only of higher quality than that of the three-pulse rectifier, but it also has a much bigger dc component. Since those are the line-to-line voltages that are rectified here, a dc-output voltage is twice as high as that of a three-pulse rectifier supplied from the line-to-neutral voltages.

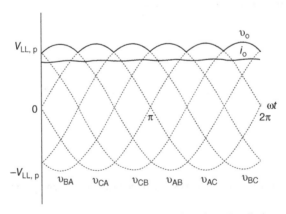

Figure 4.8 Waveforms of output voltage and current in a six-pulse diode rectifier in the continuous conduction mode (RLE load).

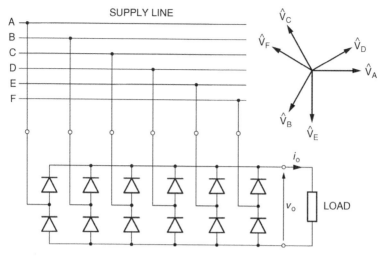

Figure 4.9   Twelve-pulse diode rectifier.

A general formula for a p-pulse uncontrolled (diode) rectifier in the continuous conduction mode is

$$V_{o,dc(unc)} = \frac{p}{\pi} V_{i,p} \sin\left(\frac{\pi}{p}\right), \quad p = 2, 3, \ldots \tag{4.5}$$

The integer value of p is not limited to 2, 3, or 6. Indeed, both the half-wave and full-wave rectifier topologies described can be expanded by increasing the number of phases of the ac supply source and the corresponding number of diode pairs. As an example, a full-wave, twelve-pulse rectifier is shown in Figure 4.9. The six-phase supply is obtained from a three-phase line using a transformer with six secondary windings. Clearly, if p increases, the dc-output voltage approaches the peak line-to-line input voltage and the voltage ripple factor approaches zero. Such "super rectifiers" can mostly be found in electrochemical plants, but otherwise the field of ac-to-dc power conversion is dominated by six-pulse bridge rectifiers because of the common availability of the three-phase electrical power.

Equation of the output current waveform, $i_{o(C)}(\omega t)$, in the continuous conduction mode for the first cycle of this current, that is, for the 0 to $\pi/3$ interval of ωt, can be derived using the method explained in Chapter 1. For generality, an RLE load is considered. The circuit equation is

$$V_{AB} = L\frac{di_o}{dt} + Ri_o + E, \tag{4.6}$$

where L, R, and E denote the load inductance, resistance, and EMF, respectively. The forced component, $i_{o(F)}(\omega t)$, generated by voltage v_{AB} given by Eq. (4.3) is

$$i_{o(F)}(\omega t) = \frac{V_{LL,p}}{Z} \sin\left(\omega t + \frac{\pi}{3} - \varphi\right) - \frac{E}{R}, \tag{4.7}$$

where Z is the load impedance, equals to

$$Z = \sqrt{R^2 + (\omega L)^2}, \tag{4.8}$$

and φ denotes the load angle, given by

$$\varphi = \tan^{-1}\left(\frac{\omega L}{R}\right) = \cos^{-1}\left(\frac{R}{Z}\right). \tag{4.9}$$

Since

$$\frac{E}{R} = \frac{V_{LL,p}}{Z}\frac{E}{V_{LL,p}}\frac{Z}{R} = \frac{V_{LL,p}}{Z}\frac{\varepsilon}{\cos(\varphi)}, \tag{4.10}$$

where $\varepsilon \equiv E/V_{LL,p}$ is a *load EMF coefficient*, Eq. (4.7) can be rewritten as

$$i_{o(F)}(\omega t) = \frac{V_{LL,p}}{Z}\left[\sin\left(\omega t + \frac{\pi}{3} - \varphi\right) - \frac{\varepsilon}{\cos(\varphi)}\right]. \tag{4.11}$$

The natural component, $i_{o(N)}(\omega t)$, specific for a resistive-inductive load, is

$$i_{o(N)}(\omega t) = A_{(C)}e^{-\frac{R}{L}t} = A_{(C)}e^{-\frac{R}{\omega L}\omega t} = A_{(C)}e^{-\frac{\omega t}{\tan(\varphi)}}, \tag{4.12}$$

where $A_{(C)}$ is a constant to be determined from the initial and final conditions. Consequently,

$$\begin{aligned} i_{o(C)}(\omega t) &= i_{o(F)}(\omega t) + i_{o(N)}(\omega t) \\ &= \frac{V_{LL,p}}{Z}\left[\sin\left(\omega t + \frac{\pi}{3} - \varphi\right) - \frac{\varepsilon}{\cos(\varphi)}\right] + A_{(C)}e^{-\frac{\omega t}{\tan(\varphi)}}. \end{aligned} \tag{4.13}$$

To find $A_{(C)}$, advantage can be taken from the fact that the initial value of the current, at $\omega t = 0$, equals the final value, at $\pi/3$. Thus,

$$i_{o(C)}(0) = \frac{V_{LL,p}}{Z}\left[\sin\left(\frac{\pi}{3} - \varphi\right) - \frac{\varepsilon}{\cos(\varphi)}\right] + A_{(C)} \tag{4.14}$$

equals

$$i_{o(C)}\left(\frac{\pi}{3}\right) = \frac{V_{LL,p}}{Z}\left[\sin\left(\frac{2\pi}{3} - \varphi\right) - \frac{\varepsilon}{\cos(\varphi)}\right] + A_{(C)}e^{-\frac{\pi}{3\tan(\varphi)}}. \tag{4.15}$$

Comparing the right-hand sides of Eqs. (4.14) and (4.15) yields

$$A_{(C)} = \frac{V_{LL,p}}{Z} \frac{\sin(\varphi)}{1 - e^{-\frac{\pi}{3\tan(\varphi)}}},$$ (4.16)

which, when substituted in Eq. (4.13), gives

$$i_{o(C)}(\omega t) = \frac{V_{LL,p}}{Z} \left[\sin\left(\omega t + \frac{\pi}{3} - \varphi\right) - \frac{\varepsilon}{\cos(\varphi)} + \frac{\sin(\varphi)}{1 - e^{-\frac{\pi}{3\tan(\varphi)}}} e^{-\frac{\omega t}{\tan(\varphi)}} \right].$$ (4.17)

The discontinuous conduction occurs when the load EMF, E, exceeds the lowest instantaneous value of output voltage, v_o. Diodes of the rectifier become reverse biased and the load current cannot pass through them. Consequently, until the input voltage increases above E and the currents starts flowing, $v_o = E$. The lowest output voltage occurs at $\omega t = 0$ (see Figure 4.8) and, according to Eq. (4.3), equals $(\sqrt{3}/2)V_{LL,p}$. Therefore, the discontinuous conduction mode of operation of the rectifier may only happen when $\varepsilon > \sqrt{3}/2$.

This conclusion can be confirmed analytically. At the borderline between the continuous and discontinuous conduction modes, the minimal value of the output current, occurring at the ends of the considered zero to $\pi/3$ interval of ωt reaches zero. Consequently, if the current is to be continuous, it must be greater than zero at $\omega t = 0$ which, according to Eq. (4.17), is tantamount to the condition

$$\sin\left(\frac{\pi}{3} - \varphi\right) - \frac{\varepsilon}{\cos(\varphi)} + \frac{\sin(\varphi)}{1 - e^{-\frac{\pi}{3\tan(\varphi)}}} > 0,$$ (4.18)

that is,

$$\varepsilon < \left[\sin\left(\frac{\pi}{3} - \varphi\right) + \frac{\sin(\varphi)}{1 - e^{-\frac{\pi}{3\tan(\varphi)}}} \right] \cos(\varphi).$$ (4.19)

The last relation is illustrated in Figure 4.10. Indeed, the minimum value of the load EMF coefficient, ε, for the discontinuous conduction is seen to be $\sqrt{3}/2 \approx 0.866$. Above $\varepsilon = 0.955$, no conduction is possible because of the permanent reverse bias of the diodes by the load EMF.

Derivation of the expression for the discontinuous output current waveform, $i_{o(D)}(\omega t)$, proceeds similarly to that for the continuous current. The current starts flowing when the waveform of input voltage crosses over the level of load EMF, that is, at

$$\omega t = \sin^{-1}(\varepsilon) - \frac{\pi}{3} = \alpha_c.$$ (4.20)

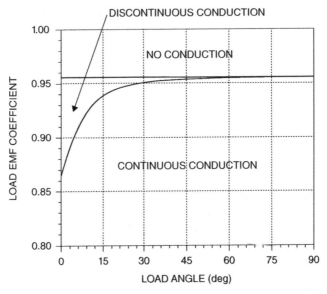

Figure 4.10   Conduction mode areas of the six-pulse diode rectifier.

Angle α_c will subsequently be referred to as a *crossover angle*. The general equation for the current waveform, similar to Eq. (4.13), is

$$i_{o(D)}(\omega t) = \frac{V_{LL,p}}{Z}\left[\sin\left(\omega t + \frac{\pi}{3} - \varphi\right) - \frac{\varepsilon}{\cos(\varphi)}\right] + A_{(D)}e^{-\frac{\omega t}{\tan(\varphi)}}. \qquad (4.21)$$

Constant $A_{(D)}$ can be found from the initial condition

$$i_{o(D)}(\alpha_c) = 0, \qquad (4.22)$$

which yields

$$A_{(D)} = -\frac{V_{LL,p}}{Z}\left[\sin\left(\omega t + \frac{\pi}{3} - \varphi\right) - \frac{\varepsilon}{\cos(\varphi)}\right]e^{-\frac{\omega t}{\tan(\varphi)}}. \qquad (4.23)$$

This, when substituted in Eq. (4.21), gives

$$i_{o(D)}(\omega t) = \frac{V_{LL,p}}{Z}\left\{\sin\left(\omega t + \frac{\pi}{3} - \varphi\right) - \frac{\varepsilon}{\cos(\varphi)}\right.$$
$$\left. - \left[\sin\left(\alpha_c + \frac{\pi}{3} - \varphi\right) - \frac{\varepsilon}{\cos(\varphi)}\right]e^{-\frac{\omega t - \alpha_c}{\tan(\varphi)}}\right\}. \qquad (4.24)$$

Output voltage and current waveforms in an example case of discontinuous conduction are shown in Figure 4.11. A current pulse begins at the crossover angle, α_c,

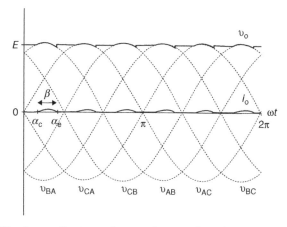

Figure 4.11 Waveforms of output voltage and current in a six-pulse diode rectifier in the discontinuous conduction mode (RLE load).

and ends at the *extinction angle*, α_e. The length, β, of the pulse in the angle domain, called a *conduction angle*, is given by

$$\beta = \alpha_e - \alpha_c. \tag{4.25}$$

The extinction angle, which depends on the load EMF coefficient, ε, and load angle, φ, can best be calculated using a "brute force" approach, that is, computing the output current according to Eq. (4.24) for sequential values of ωt, starting at α_c and proceeding until the current crosses zero in the negative direction. Clearly, this is only valid when the conditions for discontinuous conduction are satisfied (see diagram in Figure 4.10).

All the equations derived for the RLE load can easily be adapted for simpler loads by substituting $\varphi = 0$ for an RE load, $\varepsilon = 0$ for an RL load, and $\varphi = 0$ and $\varepsilon = 0$ for an R load. An RE load, for example, may represent a battery, RL load an electromagnet, and R load an electrochemical process. The RLE load is usually employed to model a dc motor.

The dc component, $V_{o,dc(D)}$, of output voltage in the discontinuous conduction mode of operation of the rectifier can be found by averaging the output voltage waveform, $v_o(\omega t)$. As

$$v_o(\omega t) = \begin{cases} v_{AB}(\omega t) & \text{for} \quad \alpha_c < \omega t < \alpha_e \\ E & \text{otherwise} \end{cases}, \tag{4.26}$$

then

$$V_{o,dc(D)} = \frac{1}{\frac{\pi}{3}} \left[\int_0^{\alpha_c} E d\omega t + \int_{\alpha_c}^{\alpha_e} V_{LL,p} \sin\left(\omega t + \frac{\pi}{3}\right) d\omega t + \int_{\alpha_e}^{\frac{\pi}{3}} E d\omega t \right]$$

$$= \frac{3}{\pi} V_{LL,p} \left[2 \sin\left(\alpha_c + \frac{\beta}{2} + \frac{\pi}{3}\right) \sin\left(\frac{\beta}{2}\right) + \varepsilon \left(\frac{\pi}{3} - \beta\right) \right]. \tag{4.27}$$

As seen in Figure 4.11, the complex expression for $V_{o,dc(D)}$ notwithstanding, the dc-output voltage of the rectifier is approximately equal to the peak value of the supply line-to-line voltage. In both conduction modes, the average output current, $I_{o,dc}$, is given by the simple expression

$$I_{o,dc} = \frac{V_{o,dc} - E}{R}, \tag{4.28}$$

because no dc voltage can appear across the load inductance.

The superiority of the continuous conduction mode over the discontinuous conduction mode is evident. The average output voltage does not depend on the load, and the ripple factors of output voltage and current are low. Also, a continuous current drawn by the rectifier from the ac supply system is of higher quality with respect to both the harmonic content and the input power factor than a discontinuous one.

Input Current and Power Factor. Limiting the subsequent considerations to the continuous conduction mode, and assuming ideal output current, $i_o = I_{o,dc}$, the phase-A line current, i_A, is given by

$$i_A = \begin{cases} I_{o,dc} & \text{for} \quad 0 < \omega t < \frac{2}{3}\pi \\ -I_{o,dc} & \text{for} \quad \pi < \omega t < \frac{5}{3}\pi \\ 0 & \text{otherwise} \end{cases} \tag{4.29}$$

as depicted in Figure 4.12, which also shows the fundamental line current, $i_{A,1}$. The rms value, I_A, of i_A is

$$I_A = \sqrt{\frac{1}{2\pi}\left[\int_0^{\frac{2}{3}\pi} I_{o,dc}^2 \, d\omega t + \int_\pi^{\frac{5}{3}\pi} I_{o,dc}^2 \, d\omega t + \right]} = \sqrt{\frac{2}{3}} I_{o,dc} = 0.82 I_{o,dc}, \tag{4.30}$$

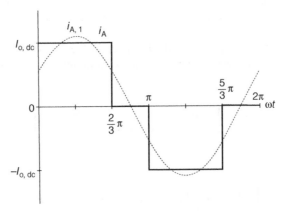

Figure 4.12 Waveform of input current in a six-pulse diode rectifier (assuming ideal dc-output current).

while the rms fundamental current, $I_{A,1}$, can be calculated as

$$I_{A,1} = \frac{1}{\sqrt{2}} \sqrt{I_{A,1c}^2 + I_{A,1s}^2}, \tag{4.31}$$

where

$$I_{A,1c} = \frac{1}{\pi} \left[\int_0^{\frac{2}{3}\pi} I_{o,dc} \cos(\omega t) \, d\omega t + \int_\pi^{\frac{5}{3}\pi} -I_{o,dc} \cos(\omega t) \, d\omega t \right] = \frac{3}{\pi} I_{o,dc} \tag{4.32}$$

and

$$I_{A,1s} = \frac{1}{\pi} \left[\int_0^{\frac{2}{3}\pi} I_{o,dc} \sin(\omega t) \, d\omega t + \int_\pi^{\frac{5}{3}\pi} -I_{o,dc} \sin(\omega t) \, d\omega t \right] = \frac{\sqrt{3}}{\pi} I_{o,dc}. \tag{4.33}$$

Thus,

$$I_{A,1} = \frac{1}{\sqrt{2}} \sqrt{\left(\frac{3}{\pi} I_{o,dc}\right)^2 + \left(\frac{\sqrt{3}}{\pi} I_{o,dc}\right)^2} = \frac{\sqrt{6}}{\pi} I_{o,dc} \approx 0.78 \, I_{o,dc}. \tag{4.34}$$

The harmonic content, $I_{A,h}$, of the line current is

$$I_{A,h} = \sqrt{I_A^2 - I_{A,1}^2} = \sqrt{\left(\sqrt{\frac{2}{3}} I_{o,dc}\right)^2 - \left(\frac{\sqrt{6}}{\pi} I_{o,dc}\right)^2}$$

$$= \sqrt{\frac{2}{3} - \frac{6}{\pi^2}} I_{o,dc} \approx 0.24 \, I_{o,dc}, \tag{4.35}$$

and the total harmonic distortion, THD, is

$$\text{THD} = \frac{I_{A,h}}{I_{A,1}} = \frac{\sqrt{\frac{2}{3} - \frac{6}{\pi^2}} I_{o,dc}}{\frac{\sqrt{6}}{\pi} I_{o,dc}} = \sqrt{\frac{\pi^2}{9} - 1} \approx 0.31, \tag{4.36}$$

which is quite high, even though the rectifier operates in the continuous conduction mode. Recall the less-than-5% desired value of THD mentioned in Chapter 1.

Both the conversion efficiency, η_c, and the input power factor, PF, of a six-pulse diode rectifier in the continuous conduction mode are high. Here, because of the assumed ideal dc-output current, the conversion efficiency is 100%. The power factor, determined as

$$
\text{PF} = \frac{P_i}{S_i} = \frac{P_o}{S_i} = \frac{V_{o,dc}I_{o,dc}}{\sqrt{3}V_{LL}I_L} = \frac{\frac{3}{\pi}V_{LL,p}I_{o,dc}}{\sqrt{3}\frac{V_{LL,p}}{\sqrt{2}}\sqrt{\frac{2}{3}}I_{o,dc}} = \frac{3}{\pi} \approx 0.955 \qquad (4.37)
$$

is also close to unity. Comparing Figures 4.11 and 4.12, it can be seen that the fundamental input current, $i_{A,1}$, lags line-to-line voltage v_{AB} by 30°, that is, according to Figure 4.7, it is in phase with line-to-neutral voltage v_{AN}. If current i_A were sinusoidal, this zero-phase shift would result in a unity input power factor.

All the figures of merit deteriorate dramatically in the discontinuous conduction mode due to the significant ac component in the output current. For better understanding of power relationships in the rectifier, the rms value, $I_{A,1}$, of fundamental line current will now be determined from the power balance instead from the Fourier-series expressions leading to Eq. (4.34). Assuming a balanced set of the input voltages, all three phases of the ac supply source contribute the same amount of input real power, equal to a third of the output power. Hence,

$$
V_{AN}I_{A,1} = \frac{1}{3}V_{o,dc}I_{o,dc}, \qquad (4.38)
$$

where V_{AN} denotes the rms value of line-to-neutral voltage v_{AN}. As $V_{AN} = V_{LL,p}/\sqrt{6}$, and, according to Eq. (4.4), $V_{o,dc} = 3V_{LL,p}/\pi$, then

$$
I_{A,1} = \frac{V_{o,dc}I_{o,dc}}{3V_{AN}} = \frac{\frac{3}{\pi}V_{LL,p}I_{o,dc}}{3\frac{V_{LL,p}}{\sqrt{6}}} = \frac{\sqrt{6}}{\pi}I_{o,dc}, \qquad (4.39)
$$

which confirms the result of Eq. (4.34).

The nonsinusoidal, quasi-square wave line currents drawn from the power system (grid) are rich in low-order harmonics, as illustrated in Figure 4.13, which shows the spectrum of current in Figure 4.12. Harmonic amplitudes are expressed in the per-unit format, the dc-output current, $I_{o,dc}$, taken as the base current. The harmonic currents have many adverse effects on the supply system. In particular, they produce extra losses in electric machines, transformers, and capacitors, interfere with communication systems, and may cause metering errors, malfunctions of control systems, and excitation of resonances with capacitor banks in the system. The increasingly stringent standards of power quality in power systems enforce use of input filters (line filters) to reduce the harmonic currents.

It must be pointed out that currents drawn by three-phase rectifiers are free from triple harmonics, that is, those with harmonic numbers being multiples of three. This property can, for example, be discerned in the spectrum in Figure 4.13. The sum of line currents i_A, i_B, and i_C is zero at all instants of time because of the lack of the

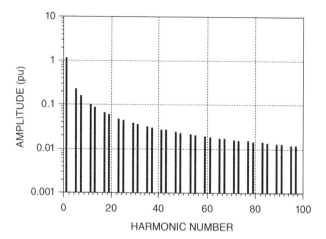

Figure 4.13   Harmonic spectrum of input current in a six-pulse diode rectifier (assuming ideal dc-output current).

fourth (neutral) wire. Since the fundamentals of these currents differ from each other by the 120° phase shift, then the corresponding phase shift for any triple harmonic would be a multiple of 360°. It means that if these harmonic currents were present, they would all be in phase and their sum would not be zero, which would contradict the Kirchhoff Current Law. The currents in question are also void of even harmonics because of the half-wave symmetry of the waveforms. Therefore, the most prominent harmonics present are the 5th, 7th, 11th, and 13th.

A filter for preventing currents of frequency $k\omega$ from propagating in the system consists of three inductors and capacitors connected between the three wires of the supply line. The line-to-line inductance and capacitance are denoted by L_f and C_f, respectively. Assuming lossless filter components, if

$$L_f C_f = \frac{1}{(k\omega)^2}, \tag{4.40}$$

then the impedance of the filter for the kth harmonic current is zero and the current is shunted from the power system. A typical practical input filter, also called a *harmonic trap*, is shown in Figure 4.14. It is composed of two resonant LC filters designated Filter 1 and Filter 2, for the fifth and seventh harmonics, respectively, and a damped resonant Filter 3 that presents a low impedance over a wide frequency band, with the minimum around 12ω. The reactors and capacitors of each filter are connected in a way that maximizes their utilization. Specifically, in each filter the circuit between each two wires of the three-phase line constitutes a series connection of two inductors and one capacitor shunted by two capacitors in series. Consequently, each inductor needs to have the inductance of only half of the required resonant inductance, L_f, and each capacitor must have the capacitance of two-thirds of the resonant capacitance C_f.

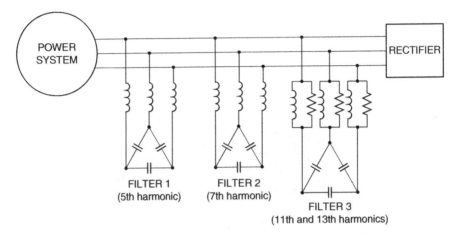

Figure 4.14   Input filter (harmonic trap) for a three-phase rectifier.

4.2 PHASE-CONTROLLED RECTIFIERS

Phase-controlled rectifiers, which allow adjustment of the average output voltage, $V_{o,dc}$, have the same topologies as diode rectifiers, with the diodes replaced with SCRs. The dc component of the output voltage is adjustable, with the maximum available value equal to that of the corresponding diode rectifier and given by Eq. (4.5).

Most practical controlled rectifiers are of the three-phase, six-pulse type. As semicontrolled switches, the SCRs are impractical for PWM, but they are perfectly suited for phase control. If a forward-biased SCR is fired, it starts conducting a current. Depending on the conduction mode, the SCR ceases to conduct when either the current drops to zero or it is taken over by another SCR. The more the firing instant is delayed with respect to the instant when the SCR became forward biased, the lower average output voltage is obtained.

With the load EMF, E, of negative polarity, a negative dc-output voltage can be produced. The output current cannot be negative because of the unidirectionality of semiconductor switches. Thus, such a situation represents a negative power flow, from the load EMF to the ac supply source. The voltage reversal property, characteristic for controlled rectifiers only, can be augmented by a current reversal feature using two rectifiers in the antiparallel configuration. Such dual converters, capable of producing positive and negative dc-output voltage and current, are particularly useful in control of dc motors.

4.2.1 Phase-Controlled Six-Pulse Rectifier

The power circuit of a phase-controlled six-pulse rectifier based on SCRs is shown in Figure 4.15. SCRs are often called thyristors, hence the SCRs of the rectifier in question are designated TA through TC′. The switching sequence of the SCR pairs is the same as that in the diode rectifier described in Section 4.1.2. However, the turn-on

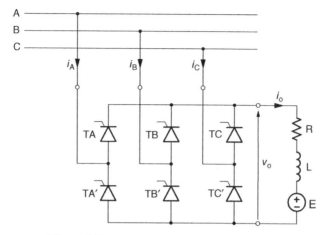

Figure 4.15   Phase-controlled six-pulse rectifier.

of an SCR can be delayed with respect to the natural commutation instant, provided that the SCR is forward biased. As already explained in Section 1.4, that delay in the angle domain is called a *firing angle* and denoted by α_f.

In the presented considerations, these are SCRs TA and TB' that are turned first, that is, when $0 \leq \omega t < \pi/3$. Therefore, the firing angle for these SCRs is measured from $\omega t = 0$, with the SCRs in question being fired at α_f. Other SCR pairs are fired with appropriate delays that are multiples of 60°. Hence, for instance, SCRs TA and TC' are fired next, at $\omega t = \alpha_f + \pi/3$.

Output Voltage and Current. The full sequence of firing pulses is shown in Figure 4.16 for $\alpha_f = 45°$. The corresponding output voltage and current waveforms of

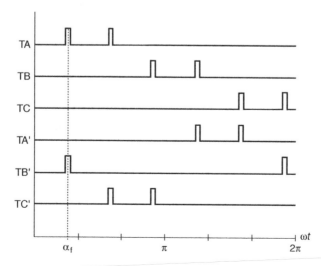

Figure 4.16   Firing pulses in a phase-controlled six-pulse rectifier.

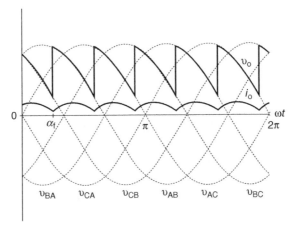

Figure 4.17   Waveforms of output voltage and current in a phase-controlled six-pulse rectifier in the continuous conduction mode ($\alpha_f = 45°$, RLE load).

the rectifier in the continuous conduction mode are illustrated in Figure 4.17. It can be seen that delaying the firing of SCRs has resulted in the reduction of the average output voltage as compared with that of a diode rectifier. The dc-output voltage is given by

$$V_{o,dc(C)} = \frac{1}{\frac{\pi}{3}} \int_{\alpha_f}^{\alpha_f + \frac{\pi}{3}} V_{LL,p} \sin\left(\omega t + \frac{\pi}{3}\right) d\omega t = \frac{3}{\pi} V_{LL,p} \cos(\alpha_f). \qquad (4.41)$$

Equation (4.41) constitutes a special case of a general formula for the dc-output voltage, $V_{o,dc(cntr)}$, of all multi-pulse ($p > 1$) controlled rectifiers in the continuous conduction mode, which is

$$V_{o,dc(cntr)} = V_{o,dc(unc)} \cos(\alpha_f), \qquad (4.42)$$

where $V_{o,dc(unc)}$ is the dc-output voltage of the corresponding diode rectifier, given by Eq. (4.5).

The voltage control characteristic expressed by Eq. (4.41) is shown in Figure 4.18. As mentioned before, the dc-output voltage can be negative, that is, the power can flow from the load EMF to the ac supply source. For a rectifier to operate in this so-called *inverter mode*, two conditions must be met: (1) a negative load EMF and (2) the firing angle greater than 90°. Then, as illustrated in Figure 4.19, it is the load EMF that delivers the power, with the load current having the same polarity as the EMF. Example waveforms of the output voltage and current in the inverter mode, with $\alpha_f = 105°$, are shown in Figure 4.20.

It must be stressed that because of the load EMF affecting the bias of the SCRs, not all values of the firing angle are feasible. Considering, for instance, SCRs TA and TB′, which apply voltage v_{AB} to the output terminals of the rectifier, it is obvious

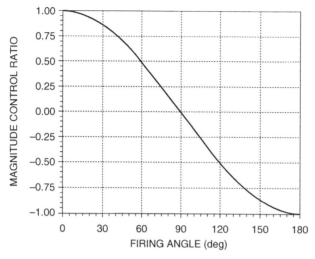

Figure 4.18 Control characteristic of a phase-controlled six-pulse rectifier in the continuous conduction mode.

that they cannot be fired when the load EMF, E, is equal to or greater than v_{AB}, that is, when $\alpha_f \leq \alpha_c$ or $\alpha_f \geq \pi/3 - \alpha_c$. Therefore, based on Eq. (4.20), feasible values of the firing angle are those that satisfy the condition

$$\sin^{-1}(\varepsilon) - \frac{1}{3}\pi < \alpha_f < \frac{2}{3}\pi - \sin^{-1}(\varepsilon), \tag{4.43}$$

which is illustrated in Figure 4.21.

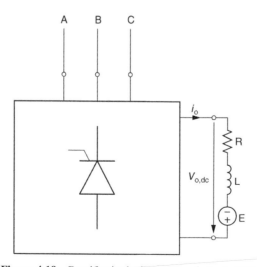

Figure 4.19 Rectifier in the inverter mode ($\alpha_f > 90°$).

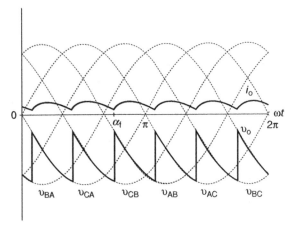

Figure 4.20   Waveforms of output voltage and current in a phase-controlled six-pulse rectifier in the continuous conduction mode ($\alpha_f = 105°$).

An equation for the continuous output current, $i_{o(C)}(\omega t)$, within the 0-to-$\pi/3$ interval can be derived similarly to that for the diode rectifier, the only difference being that $i_{o(C)}\left(\alpha_f\right) = i_{o(C)}\left(\alpha_f + \pi/3\right)$, instead of $i_{o(C)}(0) = i_{o(C)}(\pi/3)$. The current waveform is given by

$$
i_{o(C)}(\omega t) = \frac{V_{LL,p}}{Z}\left[\sin\left(\omega t + \frac{\pi}{3} - \varphi\right) - \frac{\varepsilon}{\cos(\varphi)} + \frac{\sin(\varphi - \alpha_f)}{1 - e^{-\frac{\pi}{3\tan(\varphi)}}}e^{-\frac{\omega t - \alpha_f}{\tan(\varphi)}}\right], \qquad (4.44)
$$

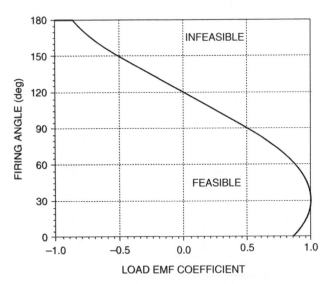

Figure 4.21   Area of feasible firing angles.

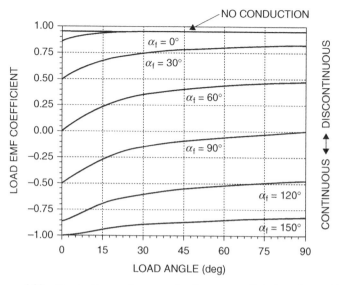

Figure 4.22 Conduction mode areas of a phase-controlled six-pulse rectifier.

and the condition for the continuous conduction is $i_{o(C)}(\alpha_f) > 0$. By virtue of Eq. (4.44), this condition can be expressed as

$$\varepsilon < \left[\sin\left(\alpha_f + \frac{\pi}{3} - \varphi\right) + \frac{\sin(\varphi - \alpha_f)}{1 - e^{-\frac{\pi}{3\tan(\varphi)}}} \right] \cos(\varphi). \qquad (4.45)$$

The relation above is illustrated in Figure 4.22. Clearly, the diagram in Figure 4.10 for the diode rectifier represents the $\alpha_f = 0$ case. Analogously, if a zero is substituted for α_f in Eqs. (4.44) and (4.45), the corresponding formulas (4.17) and (4.19) for the uncontrolled rectifier are obtained.

In the discontinuous conduction mode, a pulse of the output current begins at $\omega t = \alpha_f$, similarly to the start of current flow at $\omega t = \alpha_c$ in the diode rectifier. Therefore, expressions for the current waveform, $i_{o(D)}(\omega t)$, and average output voltage, $V_{o,dc(D)}$, can be found directly from Eqs. (4.24) and (4.27) by substituting α_f for α_c. This yields

$$i_{o(D)}(\omega t) = \frac{V_{LL,p}}{Z} \left\{ \sin\left(\omega t + \frac{\pi}{3} - \varphi\right) - \frac{\varepsilon}{\cos(\varphi)} \right. $$
$$\left. - \left[\sin\left(\alpha_f + \frac{\pi}{3} - \varphi\right) - \frac{\varepsilon}{\cos(\varphi)} \right] e^{-\frac{\omega t - \alpha_f}{\tan(\varphi)}} \right\} \qquad (4.46)$$

and

$$V_{o,dc(D)} = \frac{3}{\pi} V_{LL,p} \left[2\sin\left(\alpha_f + \frac{\beta}{2} + \frac{\pi}{3}\right) \sin\left(\frac{\beta}{2}\right) + \varepsilon\left(\frac{\pi}{3} - \beta\right) \right], \qquad (4.47)$$

where the conduction angle, β, is

$$\beta = \alpha_c - \alpha_f. \tag{4.48}$$

Note that in the continuous conduction mode, $\beta = \pi/3$, which when substituted in Eq. (4.47) yields Eq. (4.41). The discontinuous conduction mode with a positive and negative output voltage is illustrated in Figure 4.23.

Input Current and Power Factor. The waveform of the input, line current, i_A, and its fundamental, $i_{A,1}$, in the continuous conduction mode of the rectifier are shown in Figure 4.24. As before, an ideal dc-output current is assumed. The waveforms

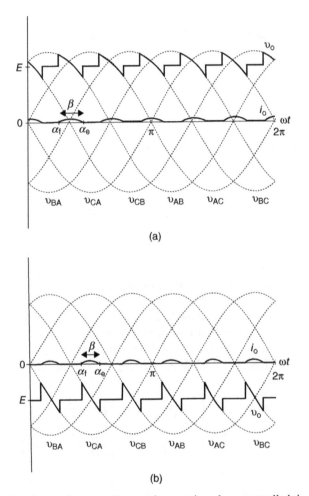

(a)

(b)

Figure 4.23 Waveforms of output voltage and current in a phase-controlled six-pulse rectifier in the discontinuous conduction mode: (a) rectifier operation ($\alpha_f = 45°$), (b) inverter operation ($\alpha_f = 135°$).

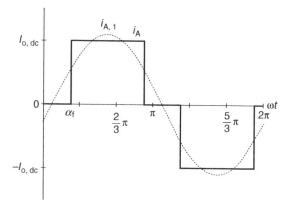

Figure 4.24 Waveform of input current in a phase-controlled six-pulse rectifier (ideal dc-output current).

are similar to those in a diode rectifier (see Figure 4.12), but with one important difference, which is a phase shift equals to the firing angle. This shift results in a reduced input power factor. Indeed, replacing $V_{o,dc}$ in Eq. (4.37) with $V_{o,dc} \cos(\alpha_f)$, the power factor of the controlled rectifier is found to be

$$ \text{PF} = \frac{V_{o,dc}}{V_{LL,p}} = \frac{3}{\pi} \cos(\alpha_f) \approx 0.95 \cos(\alpha_f). \tag{4.49} $$

To improve the power factor and shunt the harmonic currents from the power system, input filters such as those in Figure 4.14 are recommended.

Impact of Source Inductance. For simplicity, an ideal ac source was assumed in the considerations presented. In reality, the power system feeding a rectifier introduces certain amount of resistance and inductance on the supply side. These are mainly resistances and inductances of the system transformers and power lines. In fact, from the utilities point of view, the source inductance is desirable as it reduces the high-frequency current harmonics and increases the short-circuit impedance.

For the analysis of a six-pulse phase-controlled rectifier supplied through an inductance, it is convenient to employ the network shown in Figure 4.25. The actual full-wave bridge topology of the rectifier has been replaced here with an equivalent six-pulse half-wave configuration, with the actual line-to-line voltages appearing as phase voltages. The total source resistance and inductance in each path of the actual supply current are represented by lumped parameters R_s and L_s, respectively.

The input current produces a voltage drop across the source resistance, which reduces the output voltage of the rectifier. In practice, this effect is insignificant and comparable with the usually negligible impact of voltage drops across the conducting power switches. Therefore, in the subsequent considerations, a zero source resistance will be assumed. However, the source inductance, which prevents rapid changes of the supply currents when one SCR ceases to conduct and another SCR takes over

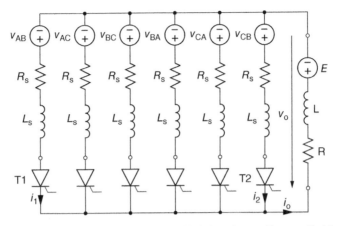

Figure 4.25   Equivalent circuit of a phase-controlled six-pulse rectifier supplied from a practical dc voltage source.

the conduction of the output current, strongly affects the operation of a rectifier. This process is called *line-supported commutation*, which is initiated by firing an SCR and results in a turnoff of another SCR. This type of commutation happens in the continuous conduction mode only, while in the discontinuous conduction mode, SCRs cease to conduct, without any external action, when their currents drop to zero.

As illustrated in Figure 4.24, when no source inductance exists, conduction of the input current by an SCR begins and ends rapidly. However, rapid current changes are impossible in inductors, and the commutation in rectifiers fed from practical ac sources is not instantaneous. As a result, during the commutation, both the incoming and outgoing SCRs share the current. Figure 4.25 pertains to a situation, when SCR T1 is gradually taking over the output current, i_o, from SCR T2. Since T1 is connected to the source of the line-to-line voltage v_{AB} and T2 to the source of v_{CB}, this case represents the commutation between SCRs TA and TC in the actual bridge rectifier. It can be seen in Figures 4.17 and 4.20 that this happens at $\omega t = \alpha_f$, when TA is fired causing extinction of TC.

Assuming an ideal dc-output current, $i_o = I_{o,dc}$, the circuit in Figure 4.25 is described by equations

$$v_o = v_{AB} - L_s \frac{di_1}{dt} \tag{4.50}$$

$$v_o = v_{CB} - L_s \frac{di_2}{dt} \tag{4.51}$$

and

$$i_o = i_1 + i_2 = I_{o,dc} \tag{4.52}$$

where i_1 and i_2 denote currents conducted by SCRs T1 and T2, respectively, voltage v_{AB} is given by Eq. (4.2), and v_{CB} by

$$v_{CB} = V_{LL,p} \sin \left(\omega t + \frac{2}{3}\pi \right).$$ (4.53)

Adding Eqs. (4.50) and (4.51) side by side gives

$$2v_o = v_{AB} + v_{CB} - L_s \left(\frac{di_1}{dt} + \frac{di_2}{dt} \right) = v_{AB} + v_{CB},$$ (4.54)

because

$$\frac{di_1}{dt} + \frac{di_2}{dt} = \frac{d}{dt}(i_1 + i_2) = \frac{d}{dt}I_{o,dc} = 0.$$ (4.55)

Consequently,

$$v_o = \frac{v_{AB} + v_{CB}}{2} = \frac{1}{2}\left[V_{LL,p} \sin\left(\omega t + \frac{1}{3}\pi\right) + V_{LL,p} \sin\left(\omega t + \frac{2}{3}\pi\right) \right]$$
$$= \frac{\sqrt{3}}{2}V_{LL,p} \cos(\omega t).$$ (4.56)

To derive equations for currents $i_1(\omega t)$ and $i_2(\omega t)$ during the commutation interval, Eqs. (4.50) and (4.51) are subtracted side by side, yielding

$$\frac{di_1}{dt} = -\frac{di_2}{dt} = \frac{1}{2L_s}(v_{AB} - v_{CB})$$
$$= \frac{1}{2L_s}\left[V_{LL,p} \sin\left(\omega t + \frac{1}{3}\pi\right) - V_{LL,p} \sin\left(\omega t + \frac{2}{3}\pi\right) \right]$$
$$= \frac{V_{LL,p}}{2L_s} \sin(\omega t).$$ (4.57)

Now, $i_1(\omega t)$ can be calculated as

$$i_1(\omega t) = \int \frac{V_{LL,p}}{2L_s} \sin(\omega t)dt + A_1 = \frac{V_{LL,p}}{2X_s} \cos(\omega t) + A_1,$$ (4.58)

where X_s is the source reactance, equals to ωL_s, and A_1 is an integration constant. Since $i_1(\alpha_f) = 0$, then

$$A_1 = \frac{V_{LL,p}}{2X_s} \cos(\alpha_f)$$ (4.59)

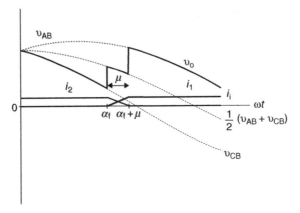

Figure 4.26 Waveforms of voltage and current in a phase-controlled six-pulse rectifier during commutation.

and

$$i_1(\omega t) = \frac{V_{LL,p}}{2X_s}[\cos(\alpha_f) - \cos(\omega t)], \tag{4.60}$$

while

$$i_2(\omega t) = I_{o,dc} - i_1(\omega t) = I_{o,dc} - \frac{V_{LL,p}}{2X_s}[\cos(\alpha_f) - \cos(\omega t)]. \tag{4.61}$$

The commutation process is illustrated in Figure 4.26. Beginning at the firing angle, α_f, current i_1 gradually increases and current i_2 decreases, reaching $I_{o,dc}$ and zero, respectively, at $\omega t = \alpha_f + \mu$. Angle μ, called an *overlap angle* or *commutation angle*, represents the length of the commutation interval in the angle domain. Its value can be determined using Eq. (4.60) and taking into account that $i_1(\alpha_f + \mu) = I_{o,dc}$. Thus,

$$\frac{V_{LL,p}}{2X_s}[\cos(\alpha_f) - \cos(\alpha_f + \mu)] = I_{o,dc} \tag{4.62}$$

and

$$\mu = \left| \cos^{-1}\left[\cos(\alpha_f) - 2\frac{X_s I_{o,dc}}{V_{LL,p}}\right] - \alpha_f \right|. \tag{4.63}$$

Example waveforms of the output voltage and actual, nonideal current of a six-pulse phase-controlled rectifier with a nonzero load inductance are shown in

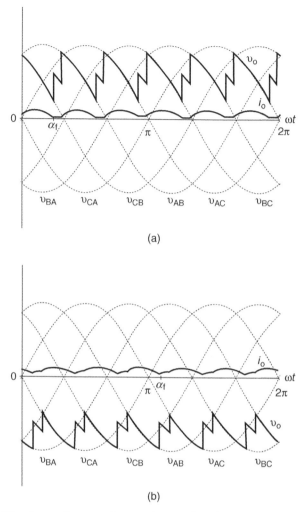

Figure 4.27   Waveforms of output voltage and current in a phase-controlled six-pulse rectifier supplied from a source with inductance: (a) rectifier mode ($\alpha_f = 45°$), (b) inverter mode ($\alpha_f = 135°$).

Figure 4.27 for $V_{o,dc} > 0$ (rectifier mode) and for $V_{o,dc} < 0$ (inverter mode). The operating conditions of the rectifier are the same as in Figures 4.17 and 4.20. It can be seen that the noninstantaneous commutation has significantly affected both the output voltage and the current.

The output voltage, which during commutation is the arithmetical mean of the line-to-line voltages involved, has its dc component reduced. The difference, $\Delta V_{o,dc}$, between the dc-output voltage with $X_s = 0$ and with $X_s > 0$ can be calculated by spreading the area between the respective waveforms over the $\pi/3$ interval of ωt.

Using partial results of Eqs. (4.57) and (4.62), an expression for $\Delta V_{o,dc}$ is obtained as

$$\Delta V_{o,dc} = \frac{3}{\pi} \int_{\alpha_f}^{\alpha_f + \mu} v_{AB} - \frac{v_{AB} + v_{CB}}{2} d\omega t = \frac{3}{\pi} \int_{\alpha_f}^{\alpha_f + \mu} \frac{v_{AB} - v_{CB}}{2} d\omega t$$

$$= \frac{3}{2\pi} \int_{\alpha_f}^{\alpha_f + \mu} V_{LL,p} \sin(\omega t) \, d\omega t$$

$$= \frac{3}{2\pi} V_{LL,p} [\cos(\alpha_f) - \cos(\alpha_f + \mu)] = \frac{3}{\pi} X_s I_{o,dc}. \qquad (4.64)$$

The output current has also been reduced and its lowest instantaneous value has become closer to zero. This implies certain reduction of the areas of continuous conduction mode in Figure 4.20.

The output voltage of a practical rectifier is also affected, albeit to a lesser degree, by voltage drops across the source resistance, conducting power switches, and wiring of the rectifier including the cable connecting it to the load. Thus, the apparent internal resistance, R_r, of a bridge rectifier is given by

$$R_r = \frac{3}{\pi} X_s + R_s + 2R_{ON} + R_w, \qquad (4.65)$$

Where R_{ON} is the equivalent on-state resistance of power switches employed in the rectifier (in bridge converters, two switches in series conduct the output current) and R_w is the wiring resistance. Based on Eqs. (4.41) and (4.64), the dc-output voltage of a practical rectifier in the continuous conduction mode can be expressed as

$$V_{o,dc} = \frac{3}{\pi} V_{LL,p} \cos(\alpha_f) - R_r I_{o,dc}, \qquad (4.66)$$

which, unsurprisingly, indicates that the rectifier constitutes a practical dc voltage source whose terminal voltage decreases with the increase in the drawn current.

Another adverse effect of the source inductance is the so-called *line voltage notching*. Consider for instance the commutation interval in Figure 4.26. The sinusoidal waveforms of v_{AB} and v_{CB} shown there are those of the source EMFs. However, during commutation, the corresponding line-to-line voltages, v_{ab} and v_{cb}, at the very input to the rectifier substantially differ from the source EMFs. They both acquire the waveform of the output voltage, that is,

$$v_{ab} = v_{bc} = v_o = \frac{1}{2}(v_{AB} + v_{CB}), \qquad (4.67)$$

because SCR TB′ connects supply line B to the negative output terminal of the rectifier, while the simultaneously conducting SCRs TA and TC connect lines A and C to the positive terminal. Consequently, the v_{ab} waveform, as shown in Figure 4.28, becomes seriously distorted, the notches appearing at all the commutation intervals involving SCRs TA, TA′, TB, and TB′. Similar distortions affect, of course, the other

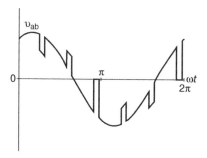

Figure 4.28 Notched waveform of input voltage in a phase-controlled six-pulse rectifier supplied from a source with inductance.

input voltages as well. Although not a problem for the rectifier itself, the notched voltages may disturb the operation of other equipment supplied in parallel with the rectifier. The source inductance affects the operation of all multipulse rectifiers, both controlled and uncontrolled.

4.2.2 Dual Converters

It has been shown in the preceding section that controlled rectifiers can generate a negative dc-output voltage if the load includes an EMF of such polarity and magnitude that the electrical power can be transferred to the ac supply source. However, under no circumstances a negative output current can be produced, as it would have to flow from the cathode to the anode in the SCRs. It is said that a controlled rectifier can only operate in two quadrants of an operation plane, which is shown in Figure 4.29. The values, $I_{o,dc}$ and $V_{o,dc}$, of the dc-output current and voltage of a rectifier represent coordinates of an *operating point* in the plane, and all the allowable operating points

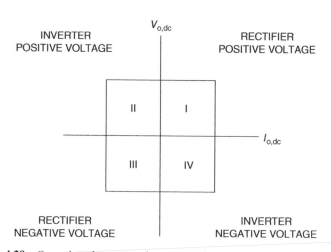

Figure 4.29 Operation plane, operating area, and operating quadrants of a rectifier.

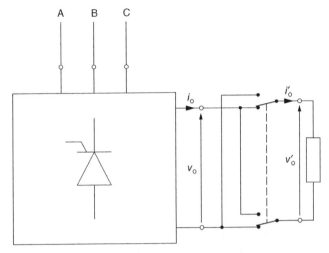

Figure 4.30   Controlled rectifier with an electromechanical cross-switch.

range out an *operating area*. Clearly, a controlled rectifier can only operate in the first and fourth quadrants, while diode rectifiers can operate in the first quadrant only.

Extension on four quadrants of operation can be accomplished by installing an electromechanical cross-switch at the output of a controlled rectifier, as shown in Figure 4.30. With the cross-connection between the rectifier and the load terminals, the load current, i'_o, equals $(-i_o)$ and the load voltage, v'_o, equals $(-v_o)$. In this way, the load can operate in the second or third quadrant, depending on the polarity of the dc-output voltage, $V_{o,dc}$. However, this solution is practical only when the switching required is infrequent, such as in dc motor driven vehicles, because of the limited life span and low operating frequency of electromechanical switches. Therefore, in most practical applications, the so-called *dual converters* are used.

A dual converter, shown in Figure 4.31, is an antiparallel combination of two controlled rectifiers. For generality, different ac supply sources for each rectifier are

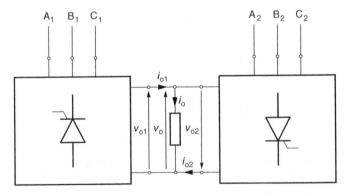

Figure 4.31   Antiparallel connection of two controlled rectifiers.

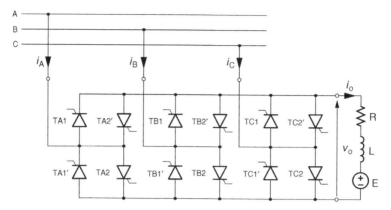

Figure 4.32   Six-pulse circulating current-free dual converter.

indicated. There are two basic types of dual converters: the *circulating current-free* and *circulating current-conducting* ones. The meaning of the "circulating current" will be explained later.

Circulating Current-Free Dual Converter. The power circuit of a circulating current-free dual converter is shown in Figure 4.32. The single SCRs of the regular rectifier are replaced here with antiparallel pairs of SCRs. If a positive output current is needed, SCRs TA1 through TC1' are in operation, while SCRs TA2 through TC2' are all off (not fired). Vice-versa, to produce a negative output current, SCRs TA2 through TC2' are fired in an appropriate sequence while SCRs TA1 through TC1' remain unused.

The circulating current-free dual converter is simple and compact, but it has two serious disadvantages. Clearly, only one of the antiparallel connected SCR can conduct the current, as the other SCR is reverse biased by the voltage drop across the conducting one. However, if both SCRs are off, one of them is always forward biased, and improper firing may cause a short circuit. For example, when SCRs TB1 and TC1' are conducting, the voltage, v_{BC}, across SCRs TC1 and TC2' produces forward bias of the latter SCR. If fired, TC2' would short lines B and C. This can easily be prevented by appropriate control of the firing signals, but when a change of polarity of the output current is required, the incoming rectifier must wait until the current in the outgoing one dies out and the conducting SCRs turn off. This mandatory delay slows down the response of the converter to current control commands, which in certain application is unacceptable.

The other disadvantage, common for all phase-controlled rectifiers, consists in the transition to the discontinuous conduction mode at large firing angles (see Figure 4.22). The continuous conduction area of operation can be expanded by adding inductance to the load. Such inductance would also reduce the current ripple, but it is a costly solution as the inductor carrying the whole load current would have to be large and expensive. Another solution consists in keeping currents conducted by the SCRs at a sufficiently high level, no matter how large the firing angle is. This idea is ingeniously employed in the circulating current-conducting dual converters.

Circulating Current-Conducting Dual Converter. In a circulating current-conducting dual converter, both the constituent rectifiers operate simultaneously, one in the rectifier mode, with the firing angle, α_{f1}, less than 90°, and the other in the inverter mode, with the firing angle, α_{f2}, equal to $180° - \alpha_{f1}$. According to Eq. (4.41), if both rectifiers operate in the continuous conduction mode, their output voltages have the same dc component, since $\cos(\alpha_{f1}) = -\cos(\alpha_{f2})$. However, the instantaneous values of these voltages differ from each other, and the difference, v_o, produces a current circulating between the rectifiers. If the rectifiers were directly connected as in Figure 4.31, the circulating current, i_{cr}, limited only by the resistance of wires and conducting SCRs, would be excessive. Therefore, inductors are placed between the rectifiers and the load to attenuate the ac component of the circulating current.

One version of a circulating current-conducting dual converter is shown in Figure 4.33. To avoid short circuits between the ac supply lines, the converter is fed from two secondary windings of a three-phase transformer, with the input voltages for both rectifiers equal with respect to the amplitude and phase. Two inductors, L_1 and L_2, separate the rectifiers from the load. Current paths at $\omega t = \pi/3$ are indicated by thick lines. Waveforms of the load current and output voltages of both rectifiers are shown in Figure 4.34 for $\alpha_{f1} = 45°$ and $\alpha_{f2} = 135°$. The vertical dashed line indicates the considered instant of operation.

Waveforms of the differential output voltage, $\Delta v_o = v_{o1} + v_{o2}$, and circulating current, i_{cr}, are illustrated in Figure 4.35. The average differential voltage is zero, and the waveform of the circulating current indicates a borderline continuous conduction mode. In practice, the sum of the firing angles of the constituent rectifiers is made slightly less than 180°, which results in a non-zero dc component of Δv_o.

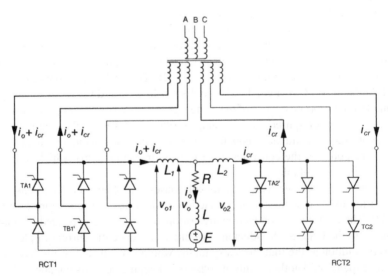

Figure 4.33   Six-pulse circulating current-conducting dual converter supplied from two separate ac sources.

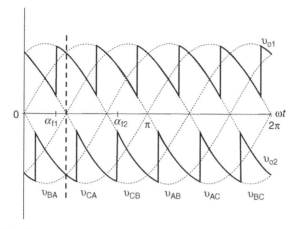

Figure 4.34   Waveforms of output voltages of constituent rectifiers in a circulating current-conducting dual converter ($\alpha_{f1} = 45°$, $\alpha_{f2} = 135°$).

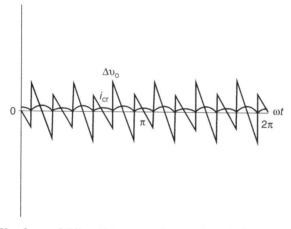

Figure 4.35   Waveforms of differential output voltage and circulating current in a circulating current-conducting dual converter: $\alpha_{f1} + \alpha_{f2} = 180°$.

In practical dual converters, a closed-loop control circuit maintains the average circulating current at the desired value, typically at 10–15% of the rated load current. The control circuit adjusts the firing angle (here α_{f2}) of the rectifier that conducts the circulating current only, while control of the load current is accomplished by adjusting the firing angle (here α_{f1}) of the other rectifier. Waveforms of the differential output voltage and the circulating current when $\alpha_{f1} = 45°$ and $\alpha_{f2} = 134°$ are shown in Figure 4.36.

Circulating current-conducting dual converters have excellent dynamic characteristics as the transitions between operating quadrants can be made almost instantaneously. The absence of the discontinuous conduction mode allows wide-range

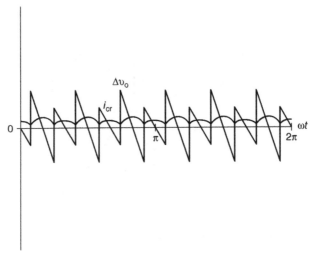

Figure 4.36   Waveforms of differential output voltage and circulating current in a circulating current-conducting dual converter: $\alpha_{f1} + \alpha_{f2} = 179°$.

control of the output voltage and current. Another version of such converter is shown in Figure 4.37. Here, both rectifiers are supplied from the same ac line, but the four separating inductors, L_1 through L_4, prevent an interphase short circuit. Two circulating currents, i_{cr1} and i_{cr2}, flow in this converter.

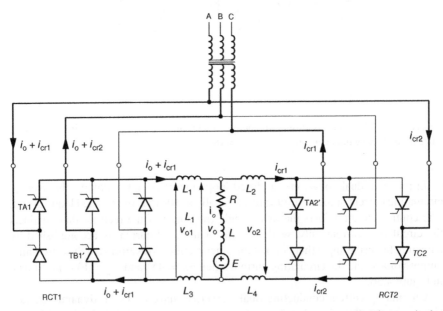

Figure 4.37   Six-pulse circulating current-conducting dual converter supplied from a single ac source.

PWM dual converters are feasible but impractical. Instead, high-quality four-quadrant dc power is obtained from a four-quadrant dc-to-dc converter (chopper) covered in Chapter 6. The input dc source for the chopper can be obtained from a controlled rectifier.

4.3 PWM RECTIFIERS

The uncontrolled and phase-controlled rectifiers are classified as *line-commutated converters*, since the turn-off conditions for their switches are determined by the voltages of the ac supply line. As explained in Section 4.2.1, the nonsinusoidal supply currents and dependence of the input power factor on the firing angle are major disadvantages of those converters. The availability of fully controlled power switches, such as IGBTs or power MOSFETs, allows to replace the phase control of output voltage by PWM. In contrast to line-commutated power electronic converters, those based on fully controlled switches are called *force-commutated*. It will be shown that the adverse effects on the ac supply system can greatly be mitigated employing force-commutated PWM rectifiers with small LC input filters.

4.3.1 Impact of Input Filter

The enhancing impact of an LC input filter on the quality of the supply current can best be explained considering a single-phase PWM rectifier shown in Figure 4.38. The indicated distribution of currents at the input to the rectifier represents an ideal situation, not fully attainable in practice. The input current, i_i, consists of a

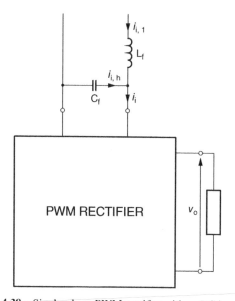

Figure 4.38   Single-phase PWM rectifier with an LC input filter.

fundamental current, $i_{i,1}$, and a harmonic current, $i_{i,h}$. Radian frequencies of the harmonic components of $i_{i,h}$ are multiples of the fundamental frequency, ω, that is, they are many times higher than this frequency. As a result, higher harmonics of the input current are much stronger attenuated by the filter inductance, L_f, than the fundamental; the inductive reactance of the filter for the kth current harmonic being k times *higher* than that for the fundamental.

In contrast to L_f, the filter capacitance, C_f, offers the current harmonics an easier passage in comparison with that for the fundamental, since the capacitive reactance for the kth harmonic is k times *lower* than that for the fundamental. Therefore, the rectifier draws most of the harmonic current from the capacitor. If all significant harmonics of the input current had frequencies approaching infinity or, more realistically, if the low-order harmonics were negligible, the actual distribution of currents would be close to the ideal case illustrated in Figure 4.38.

The filter described is impractical for phase-controlled rectifiers, in which the largest harmonics of the input current are the low-frequency ones. In contrast, frequencies of harmonic currents in PWM rectifiers are high enough to be effectively attenuated using small inductances and capacitances of the input filter. In practice, the same source inductance that causes unwanted commutation effects in phase-controlled rectifiers is often sufficient as the inductive part of the input filter for PWM rectifiers. Then, only the input capacitors need to be installed as discrete elements of the filter to provide the capacitive part.

4.3.2 Principles of PWM

The general idea of PWM has already been explained in Chapter 1. In most considerations thereof, the duty ratios of switches of the generic converter were assumed constant, while the variable duty ratios of switches in the generic PWM inverter (see Figure 1.23) were only briefly mentioned. PWM rectifiers are the first practical PWM power electronic converters described in this book, and, as in the mentioned inverter, duty ratios of their switches vary throughout the cycle of input voltage. To cover the issue of PWM in depth, certain important concepts will first be introduced and explained.

Switching variables, whose values determine the state of a converter, can be defined as the minimum set of binary variables that allows representation of each allowed state. Considering as an example the generic converter in Figure 1.2, three switching variables, $x_{12}, x_{34},$ and x_5, suffice to describe all three allowed states. Specifically, the switching variables can be defined as

$$x_{12} = \begin{cases} 0 & \text{if } S1 \text{ and } S2 = \text{OFF} \\ 1 & \text{if } S1 \text{ and } S2 = \text{ON} \end{cases} \tag{4.68}$$

$$x_{34} = \begin{cases} 0 & \text{if } S3 \text{ and } S4 = \text{OFF} \\ 1 & \text{if } S3 \text{ and } S4 = \text{ON} \end{cases} \tag{4.69}$$

$$x_5 = \begin{cases} 0 & \text{if } S5 = \text{OFF} \\ 1 & \text{if } S5 = \text{ON} \end{cases} . \tag{4.70}$$

Then, $x_{12}x_{34}x_5 = 001$, 100, and 010 imply states 0, 1, and 2, respectively. Three binary variables can, in general, describe $2^3 = 8$ states, but in this case the remaining five states are forbidden as resulting in either a short circuit or an open circuit ("floating" output). Note also that the number of switching variables is less than that of switches in the generic converter, which could actually produce as many as $2^5 = 32$ possible states. Such minimal representation of converter states is even more economical in many practical power electronic systems, such as in the popular voltage-source inverters presented in Chapter 7. However, certain control algorithms may require that each switch of a converter must be assigned a switching variable, equals to 0 if the switch is OFF (open) and 1 if the switch is ON (closed).

Clearly, waveforms of switching variables are trains of rectangular pulses with the unity magnitude, and the term PWM means modulation of widths (durations) of individual pulses. Operation of ac-input and ac-output power electronic converters is cyclic, the length of a cycle being determined by the respective input or output frequency. Typically, in PWM converters, the cycle of the input or output ac voltage consists of sub-cycles, called *switching intervals* or *switching cycles*. The reciprocal, f_{sw}, of a switching period (length of a switching cycle), T_{sw}, is called a *switching frequency*. Switching frequencies in PWM power electronic converters are usually in the range of 4–20 kHz. The average switching frequency is N times higher than the input or output frequency of the converter. It means that a cycle of the input or output frequency contains N switching cycles, where generally N does not have to be an integer.

Although there is only one method of phase control, the number of PWM techniques is countless, and new algorithms, usually modifications of the known ones, are still being published. In most cases, a single pulse of each switching variable occupies one switching interval. Since the duty ratio, d_x, of switching variable x can be adjusted from zero to unity, it is possible that no pulse is present in the switching interval or that a pulse of x fills the whole interval.

In modern PWM techniques, pulses of switching variables, which are closely associated with actual switching signals of converter switches, are seldom generated directly and independently from each other. Instead, the most common PWM algorithms produce a timed sequence of *converter states*, each state defined by the corresponding values of switching variables.

As an example, consider a simple hypothetical PWM converter described by two switching variables, x_1 and x_2. All possible states are allowed, and each state is designated by the decimal equivalent of the binary number $(x_1x_2)_2$. Thus, the converter in question can assume any of the following states: state 0, when $x_1x_2 = 00$; state 1, when $x_1x_2 = 01$; state 2, when $x_1x_2 = 10$; or state 3, when $x_1x_2 = 11$. Let us say that the PWM algorithm dictates the following sequence and timing of states within certain switching interval: 0 (50 μs)–2 (40 μs)–3 (20 μs)–2 (40 μs)–0 (50 μs). State 1 is not used. Clearly, the switching period is 200 μs, which corresponds to the switching frequency of 5 kHz. The switching pattern in question is shown in Figure 4.39. It can be seen that the pulse width of x_1 is 100 μs and that of x_2 is 20 μs, which means that the duty ratio of x_1 is 0.5 and that of x_2 is 0.1. Both pulses are centered about the middle of the switching interval. All that information can be derived from the timed state sequence, without analyzing the individual switching variables.

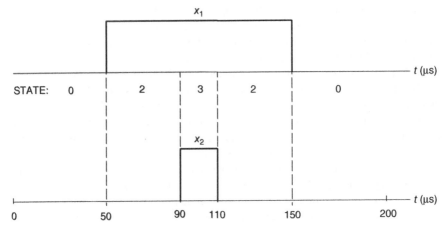

Figure 4.39 Example switching pattern of a hypothetical four-state PWM converter.

Other important concepts in control of PWM converters are *voltage and current space vectors*. Proposed in the 1920s for dynamic analysis of electric machines, they have been successfully adapted for use in power electronic converters in the 1980s. *Space vector PWM* (SVPWM) techniques are now most popular. Space vectors superficially resemble phasors, commonly used in the analysis of ac circuits. However, they differ from phasors in several aspects, and these two types of electric quantities should not be confused.

To explain the origin of voltage and current space vectors, a cross-section of stator of a simple three-phase electric ac machine is shown in Figure 4.40. Each of the three-phase windings is composed of two conductors perpendicular to the page plane

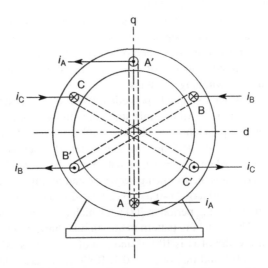

Figure 4.40 Stator of a three-phase electric ac machine.

and connected in the back of the stator. The front ends of the conductors of individual phases are connected in wye or delta (this connection is not shown). These six conductors, marked A, B, C, A′, B′, and C′, form three coils displaced by 120° from each other. Stator currents, i_A, i_B, and i_C are considered positive when they enter the stator winding at front ends of the corresponding conductors.

Currents in the coils generate magnetomotive forces (MMFs), $\vec{F}_A$, $\vec{F}_B$, and $\vec{F}_C$, which add up to the stator MMF, $\vec{F}_s$. A case of balanced currents (same magnitude, 120° mutual displacement) is illustrated in Figure 4.41. As the phasor diagram indicates, currents i_A and i_B are positive with their instantaneous values equal to half of the peak value, and current i_C is at its negative peak. The MMFs, each perpendicular to the coil that produces it, are true vectors as they have specific magnitude, orientation, and polarity in the physical space of the stator. It can easily be shown that as time progresses and the currents follow the positive phase sequence, $\vec{F}_s$ rotates counter-clockwise with the angular velocity, ω, equals to the radian frequency of the stator currents. Thus, stationary vectors $\vec{F}_A$, $\vec{F}_B$, and $\vec{F}_C$ yield a revolving vector $\vec{F}_s$.

Each of the considered space vectors can be described by its horizontal (direct-axis) and vertical (quadrature-axis) components. If the direct, d, axis of stator is assumed as real and the quadrature, q, axis as imaginary, an MMF vector can be represented by a complex number. For example, as illustrated in Figure 4.42,

$$\vec{F}_s = F_{ds} + jF_{qs} = F_s e^{j\Theta_s}, \tag{4.71}$$

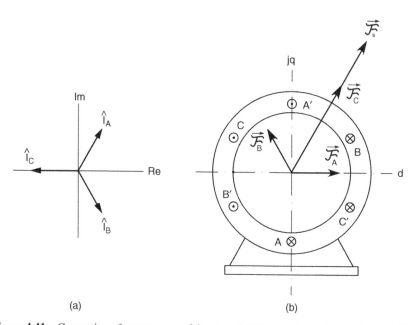

(a)   (b)

Figure 4.41   Generation of space vector of the stator MMFs in a three-phase stator: (a) phasor diagram of stator currents, (b) vectors of MMFs.

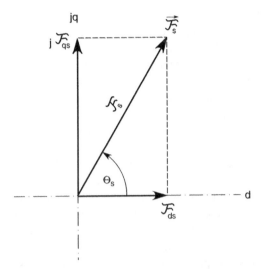

Figure 4.42 Space vector of stator MMFs and its components.

where $\mathcal{F}_{ds}$ and $\mathcal{F}_{qs}$ denote the direct and quadrature components of the vector of stator MMF, respectively, and $\mathcal{F}_s$ and Θ_s are the magnitude and phase angle of that vector.

The space vector $\vec{\mathcal{F}}_s$ is given by

$$\vec{\mathcal{F}}_s = \mathcal{F}_{as} + \mathcal{F}_{bs}e^{j120°} + \mathcal{F}_{bs}e^{j240°}, \tag{4.72}$$

where $\mathcal{F}_{as}$, $\mathcal{F}_{bs}$, and $\mathcal{F}_{cs}$ are the values of the individual phase MMFs. The MMFs in Figure 4.41 and in the per-unit system (assuming the largest possible value as the reference) are $\mathcal{F}_{as} = \mathcal{F}_{bs} = 0.5$ p.u. and $\mathcal{F}_{cs} = -1$ p.u. Substitution of these values in Eq. (4.72) yields $\vec{\mathcal{F}}_s = 0.75 + j1.299 = 1.5e^{j60°}$ p.u., that is, $\mathcal{F}_{ds} = 0.75$ p.u., $\mathcal{F}_{qs} = 1.299$ p.u., $\mathcal{F}_s = 1.5$ p.u, and $\Theta_s = 60°$. These results clearly agree with those seen in Figures 4.41 and 4.42.

An MMF is a product of number of turns in a coil and current in the coil. Therefore, dividing an MMF space vector by the turn number (which carries no physical units) gives a current space vector, $\vec{i}$. Equation (4.72) can be used to express components of that vector, i_d and i_q, in terms the phase currents, i_A, i_B, and i_C:

$$\vec{i} = \begin{bmatrix} i_d \\ i_q \end{bmatrix} = \begin{bmatrix} 1 & -\frac{1}{2} & -\frac{1}{2} \\ 0 & \frac{\sqrt{3}}{2} & -\frac{\sqrt{3}}{2} \end{bmatrix} \begin{bmatrix} i_A \\ i_B \\ i_C \end{bmatrix}. \tag{4.73}$$

This abc → dq conversion, called Park transformation, is only valid for three-wire three-phase systems, in which the phase currents add up to zero, that is, only two currents are independent. Consequently, the three currents, i_A, i_B, and i_C, can be converted into two currents, i_d and i_q, without any loss of information.

The concept of current space vectors is somewhat abstract, as it applies to any three-phase currents, not necessarily those in the stator of an ac machine. The extension of that concept on the space vector of line-to-neutral voltages given by

$$\vec{v} = \begin{bmatrix} v_d \\ v_q \end{bmatrix} = \begin{bmatrix} 1 & -\frac{1}{2} & -\frac{1}{2} \\ 0 & \frac{\sqrt{3}}{2} & -\frac{\sqrt{3}}{2} \end{bmatrix} \begin{bmatrix} v_{AN} \\ v_{BN} \\ v_{CN} \end{bmatrix} \tag{4.74}$$

is even more abstract, as it is difficult to think about voltage "pointing out" in a specific direction. However, the current and voltage space vectors are very convenient tools for the analysis and control of three-phase electric machines and three-phase power electronic converters. A space vector of line-to-line voltages can be defined similarly to that of the line-to-neutral ones.

The presented considerations have illustrated differences between space vectors and phasors. A phasor, used to simplify arithmetic operations on *sinusoidal* ac quantities, is a complex number describing a single such quantity, usually current or voltage. Thus, the current phasor shares the magnitude and phase (but not the frequency) with the corresponding current, and its real part represents the actual value of the current at a given instant of time. In contrast, a current space vector captures the information about all three currents in a three-phase three-wire system. Importantly, these currents do not have to be sinusoidal, as the current vector, an instantaneous quantity, is determined by their instantaneous values.

As illustrated in Figure 4.42, the definition of a space vector expressed by Eq. (4.72) yields a true vector, whose magnitude is 1.5 times larger than that of the corresponding phasor. Therefore, an alternative definition of space vectors includes multiplication by 2/3 of the right-hand side of the relevant equation. In such case, the example MMF space vector $\vec{F}_s$ given by a modified Eq. (4.72) would have the magnitude, F_s, of 1 p.u. as does the phasor of that MMF. The difference between those two definitions of space vectors is insignificant as long as a given definition is consistently used in all considerations.

Depending on the type of three-phase PWM converter, given currents or voltages are controllable, which means that their space vector can be made to follow a reference vector $\vec{i}^*$. As the time progresses, the reference vector moves in the complex plane dq. In particular, in the steady state, $\vec{i}^*$ rotates with a constant angular velocity equal to the radian frequency of the represented currents. As already mentioned, the control is realized by imposing a specific timed sequence of states of the converter.

For illustration, consider an unspecified converter, whose state X produces current vector $\vec{I}_X$, state Y produces vector $\vec{I}_Y$, and state Z produces vector $\vec{I}_Z$, whose magnitude is zero. These three vectors are to be employed to generate the reference vector $\vec{i}^*$ located between $\vec{I}_X$ and $\vec{I}_Y$, as shown in Figure 4.43.

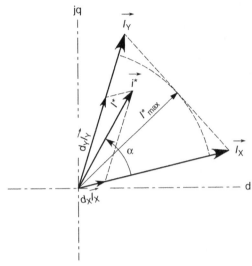

Figure 4.43   Synthesis of a rotating space vector $\vec{i}^*$ from stationary vectors $\vec{I}_X$ and $\vec{I}_Y$.

As time progresses, the position of reference vector changes. Also its magnitude can vary. However, switching intervals are practically so short that within each of them the reference vector can be assumed stationary and unchanging. As seen in Figure 4.43, the reference vector can be represented as

$$\vec{i}^* = d_X \vec{I}_X + d_Y \vec{I}_Y, \tag{4.75}$$

where d_X and d_Y are duty ratios of state X and state Y, respectively, that is, relative durations of these states with respect to the length, T_{sw}, of the switching interval. The maximum realizable magnitude, I^*_{max}, of the reference vector equals the radius of the arc inscribed in the triangle formed by the "framing" vectors $\vec{I}_X$ and $\vec{I}_Y$. In such case the sum of d_X and d_Y equals 1. Otherwise, to fill up the switching interval, a zero vector, I_Z, is enforced with such a duty ratio d_Z that

$$d_X + d_Y + d_Z = 1. \tag{4.76}$$

In this way, a revolving vector $\vec{i}^*$ is generated as a *time average* of fractions of stationary vectors, $\vec{I}_X$, $\vec{I}_Y$, and $\vec{I}_Z$, that is, by maintaining state X for a total of $d_X T_{sw}$ seconds, state Y for a total of $d_Y T_{sw}$ seconds, and the zero state Z for a total of $d_Z T_{sw}$ seconds. In practice, the switching cycle is usually divided into more than three sub-cycles. Most often six sub-cycles are used, each one holding a half of the total time allocated for a given state.

Formulas for the calculation of d_X, d_Y, and d_Z are derived from complex number equations describing the vector diagram in Figure 4.43. In particular, if current vectors

$\vec{I}_X$ and $\vec{I}_Y$ have the same magnitude and differ in phase by 60°, which is a common situation in practical three-phase converters, then

$$d_X = m \sin(60° - \alpha), \tag{4.77}$$

$$d_Y = m \sin(\alpha), \tag{4.78}$$

where m denotes the so-called *modulation index* defined as

$$m \equiv \frac{I^*}{I^*_{\max}}. \tag{4.79}$$

Symbol I^* denotes the magnitude of the reference vector $\vec{i}^*$, and $I^*_{\max}$ is the maximum available value of this magnitude. Duty ratio of the zero state, according to Eq. (4.76), is given by

$$d_Z = 1 - d_X - d_Y. \tag{4.80}$$

In certain PWM converters (but not all), the modulation index has the same meaning as the previously introduced magnitude control ratio, M; (see Eq. 1.40). The term "modulation index" has been inherited from terminology describing modulation in communication systems.

The sequence of states is so determined as to obtain the best quality or efficiency of operation of the converter. The former condition is satisfied when individual switches of the converter are switched in a possibly regular manner, that is, when time intervals between consecutive switchings are possibly uniform. The highest efficiency is achieved when the number of switchings per cycle of the ac voltage is minimal.

Certain control schemes for PWM converters utilize the concept of *rotating reference frame* for space vectors of currents and voltages. Note that as these vectors rotate in the stationary dq set of coordinates, their d and q components are sinusoidal functions of time, that is, ac variables. They are inconvenient to use in the control schemes, which usually employ dc quantities. If a given revolving vector is described by D and Q coordinates of a frame rotating with an angular velocity ω equals or close to that of the vector, these coordinates become dc variables. Traditionally, the superscript "e" is used to denote space vectors defined in a rotating reference frame, which in the theory of electric machinery is often referred to as the *excitation frame*.

Consider a voltage space vector $\vec{v}$ expressed in the stationary dq reference frame as

$$\vec{v} = v_d + jv_q. \tag{4.81}$$

The same vector in the rotating DQ reference frame is given by

$$\vec{v}^e = \vec{v} e^{-j\omega t} = v_D + jv_Q, \tag{4.82}$$

and the relation between dq and DQ components is described by a matrix equation:

$$\begin{bmatrix} v_D \\ v_Q \end{bmatrix} = \begin{bmatrix} \cos(\omega t) & \sin(\omega t) \\ -\sin(\omega t) & \cos(\omega t) \end{bmatrix} \begin{bmatrix} v_d \\ v_q \end{bmatrix}. \tag{4.83}$$

Equations (4.82) and (4.83) can be re-written as

$$\vec{v} = \vec{v}^e e^{j\omega t} \tag{4.84}$$

and

$$\begin{bmatrix} v_d \\ v_q \end{bmatrix} = \begin{bmatrix} \cos(\omega t) & -\sin(\omega t) \\ \sin(\omega t) & \cos(\omega t) \end{bmatrix} \begin{bmatrix} v_D \\ v_Q \end{bmatrix}. \tag{4.85}$$

The described concept of rotating reference frame is illustrated in Figure 4.44.

4.3.3 Current-Type PWM Rectifier

Circuit diagram of the current-type PWM rectifier is shown in Figure 4.45. In the technical literature, it is sometimes referred to as a current-source rectifier, which can be misleading because the input capacitors imply that the converter is supplied from a voltage source (see Figure 1.6). The converter is of the *buck* type, which in this case means that the output dc voltage, V_o, can only be stepped down from a specific maximum value, $V_{o,max}$. As indicated by Eq. (4.4), the maximum value of V_o in a three-phase diode rectifier, or an SCR rectifier with the firing angle of zero, equals

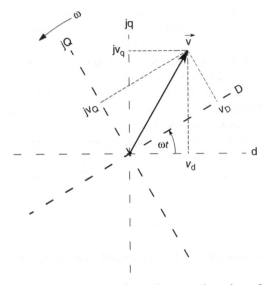

Figure 4.44 Voltage space vector in the stationary and rotating reference frames.

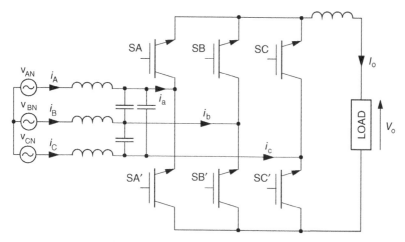

Figure 4.45   Current-type PWM rectifier.

to about 95% of the peak value of the supply line-to-line ac voltage. As illustrated in Figure 1.20a, even at $V_o = V_{o,max}$, PWM rectifiers produce certain notches in the output voltage waveform. As a result, the maximum available dc-output voltage reaches some 92% of the peak line-to-line input voltage.

The reference to current in the name of the rectifier comes from the fact that, similarly to a current source, the output current, i_o, cannot be reversed. Thus, as in the phase-controlled rectifiers, a negative output voltage is required for a negative, load to source, power flow. The inductance connected in series with the load smoothes the output current. It does not have to be a distinct device if the internal load inductance is deemed sufficiently large.

Two major advantages of the PWM rectifier are (1) feasibility of sinusoidal input currents with the unity input factor and (2) continuous output current, thanks to the narrow pulses and notches of the waveform of output voltage. With a sufficiently high switching frequency, the input currents and the output currents are only slightly rippled. The magnitude of output voltage is controlled by the value of modulation index, equals here to the magnitude control ratio, $M = m = V_o/V_{o,max}$. For the unity input power factor, the space vector of input currents is made to follow the reference vector $\vec{i}^*$ revolving in synchronism with the space vector, $\vec{v}_i$, of input line-to-neutral voltages. Note that it is only the phase angle, β^*, of $\vec{i}^*$ that is of interest, as the magnitude, I^*, depends on the load-dependent output current. In the subsequent considerations, this current is assumed constant and equal to its average value, I_o, thanks to the ripple-attenuating load inductance.

Two and only two switches of the rectifier are allowed to conduct at any time, one in the upper row and the other in the lower row. If, for example, Switches SA and SB were ON, the potential of the upper bus of the rectifier would be undetermined, as the bus would be simultaneously connected to the supply lines A and B. Also, the split of the output current, I_o, between those switches would be undefined. Thus, switching variables a, b, c, a', b', and c' of Switches SA through SC$'$ must satisfy the condition

$$a + b + c = a' + b' + c' = 1. \tag{4.86}$$

The above condition limits the number of allowable states of the rectifier to 9, namely:

State 1: $a = b' = 1$ (conducting switches: SA and SB$'$)
State 2: $a = c' = 1$ (conducting switches: SA and SC$'$)
State 3: $b = c' = 1$ (conducting switches: SB and SC$'$)
State 4: $b = a' = 1$ (conducting switches: SB and SA$'$)
State 5: $c = a' = 1$ (conducting switches: SC and SA$'$)
State 6: $c = b' = 1$ (conducting switches: SC and SB$'$)
State 7: $a = a' = 1$ (conducting switches: SA and SA$'$)
State 8: $b = b' = 1$ (conducting switches: SB and SB$'$)
State 9: $c = c' = 1$ (conducting switches: SC and SC$'$)

In state 1, currents i_A, i_B, and i_C are equal to I_0, $-I_0$, and 0, respectively. Thus, according to Eq. (4.73), the space vector of input currents in this state is

$$\vec{I_1} = \frac{3}{2}I_0 - j\frac{\sqrt{3}}{2}I_0. \tag{4.87}$$

Current vectors associated with the remaining states can be determined similarly. The active (nonzero) vectors for states 1 through 6 of the rectifier are shown in Figure 4.46. States 7, 8, and 9 produce zero vectors of input currents:

$$\vec{I_7} = \vec{I_8} = \vec{I_9} = 0. \tag{4.88}$$

It can be seen that the active vectors have the magnitude, I, of $\sqrt{3}I_0$ and that the circle limiting the magnitude, I^*, of the reference current vector, $\vec{i}^*$, has a radius of $1.5I_0$. The vectors divide the dq plane into six sectors (sextants) designated I through VI. The reference current vector is given by

$$\vec{i}^* = I^* e^{j\beta} \tag{4.89}$$

while the in-sector angle of it is denoted by α. If the rectifier is so controlled that $\vec{i}^*$ is in phase with $\vec{v}_i$, it operates in the first quadrant with the unity power factor, while if $\vec{i}^*$ lags $\vec{v}$ by 180°, the unity power factor is maintained too, but the rectifier operates in the second quadrant, that is, in the inverter mode. The voltage vector can be easily determined using voltage sensors at the rectifier's input.

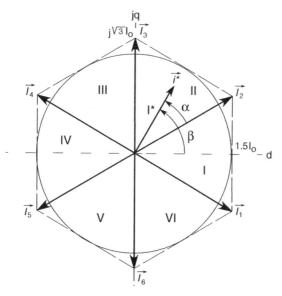

Figure 4.46 Reference current vector in the vector space of input currents of the current-type PWM rectifier.

Based on formulas (4.77) through (4.79), durations of states X and Y framing a sector in which the reference current vector is currently located are given by

$$T_X = mT_{sw} \sin\left(60° - \alpha\right) \tag{4.90}$$

$$T_Y = mT_{sw} \sin(\alpha) \tag{4.91}$$

and the duration of a zero-vector state Z by

$$T_Z = T_{sw} - T_X - T_Y. \tag{4.92}$$

To minimize the number of commutations (switchings), the following state sequences are used in individual sectors of the dq plane:

Sector I   States $1 - 2 - 7 - 2 - 1 - 7 \ldots$
Sector II   States $2 - 3 - 9 - 3 - 2 - 9 \ldots$
Sector III   States $3 - 4 - 8 - 4 - 3 - 8 \ldots$
Sector IV   States $4 - 5 - 7 - 5 - 4 - 7 \ldots$
Sector V   States $5 - 6 - 9 - 6 - 5 - 9 \ldots$
Sector VI   States $6 - 1 - 8 - 1 - 6 - 8 \ldots$

Thus, within a switching cycle each state appears twice, and each appearance lasts half of the allotted time. An example switching cycle is illustrated in Figure 4.47. It represents a situation when $m = 0.65$ and $\beta = 70°$, which implies the location of the

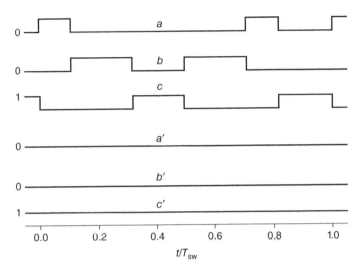

Figure 4.47   Example waveforms of switching variables in one switching cycle of the current-type PWM rectifier.

reference current vector in sector II. Thus, $\alpha = 40°, X = 2, Y = 3, Z = 9$, and, according to Eqs. (4.90) through (4.92), $T_2 = 0.22T_{sw}$, $T_3 = 0.42T_{sw}$, and $T_9 = 0.36T_{sw}$.

Analysis of waveforms of the switching variables in Figure 4.47 indicates that three switches, here SA, SB, and SC, turn on and off twice per switching cycle, while the remaining three switches, SA′, SB′, and SC′, do not change their state. Generally, in even sectors the commutating switches are those in the upper row of the rectifier bridge, while in odd sectors switches that are in the lower row perform commutations. On average, there is one turn-on and one turn-off per switch per switching cycle. The described control scheme is illustrated in Figure 4.48. If the inverter mode of operation is required, it can be signalized by a negative value of the desired modulation index, m, with $|m|$ substituted for m in Eqs. (4.90)–(4.92) for computation of intervals T_X, T_Y, and T_Z.

To illustrate the impact of modulation index, example waveforms of the output voltage, v_o, and current, i_o, in a current-type PWM rectifier operating in the rectifier mode are shown in Figure 4.49, and waveforms of the input current, i_a, and its fundamental, $i_{a,1}$, in Figure 4.50. The corresponding line current, i_A, (not shown) is similar to $i_{a,1}$, but with certain amount of ripple depending on the size of the filter capacitors. Inverter operation of the rectifier is illustrated in Figure 4.51. Example harmonic spectra of the input current are shown in Figure 4.52 to demonstrate the impact of switching frequency, f_{sw}, on current harmonics.

Sensors of input voltages provide information about the angle of the reference current vector. Two sensors are sufficient, as in the three-wire system the phase voltages add up to zero, so only two of them are independent variables. In practical rectifiers, additional sensors can be employed if required by the control algorithm. A sensor of the output dc voltage is very common, this voltage being the main controlled quantity.

RECTIFIER

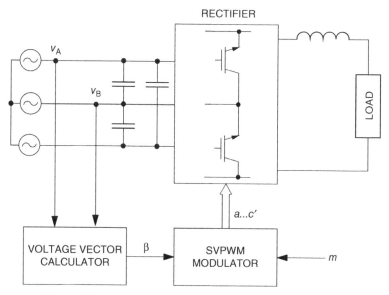

Figure 4.48   Control scheme of the current-type PWM rectifier.

On the other hand, sensors increase the cost and decrease reliability of the converter. Various efforts have thus been made to estimate, rather than measure, certain variables in the so-called *sensorless control schemes*.

4.3.4   Voltage-Type PWM Rectifier

In the voltage-type PWM rectifier shown in Figure 4.53, the output current is reversible but the output voltage is always positive. The rectifier is supplied through input inductors, and the capacitor across the load smoothes the output current. Note that the fully controlled switches of the rectifier are in the upside-down position with respect to that in the current-type PWM rectifier in Figure 4.45, and that each switch is equipped with a freewheeling diode. It is a boost converter as the output dc voltage can only be stepped up from a specific minimum value, $V_{o,min}$.

Voltage-type PWM rectifiers are characterized by "true" dc-output voltage waveform, whose ripple is maintained at a low level by a properly sized capacitor. The boost property allows high values of the output voltage without a step-up transformer at the input. Two basic methods of control of the converter will be subsequently presented. Importantly, except for very low values of the modulation index, m, the magnitude control ratio, M, in the voltage-type PWM rectifier is a *reciprocal* of that index. As a rule of thumb, when $M = m = 1$, then the dc-output voltage of the rectifier equals the peak value of the line-to-line input voltage.

Voltage-Oriented Control with Space Vector PWM. The phase-A branch (or leg) of the rectifier is shown in Figure 4.54. Note that thanks to the existence of two switches, SA and SA′, the branch can theoretically assume a total of four states.

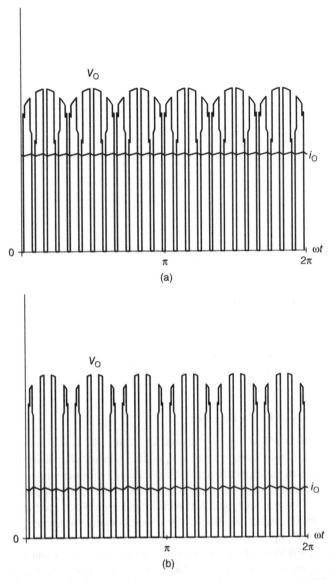

Figure 4.49 Waveforms of output voltage and current in a current-type PWM rectifier: (a) $m = 0.75$, (b) $m = 0.35$ ($f_{sw}/f_o = 24$, RLE load).

However, if both switches were on, the output capacitor would be dangerously short-circuited. On the other hand, if both switches were off, the voltage of point A would depend on the polarity of the input current i_A. It would flow through either DA or DA', making the terminal voltage equal to zero or V_o, respectively. Thus, if the relation between the terminal voltage and the output voltage is to be determined by the

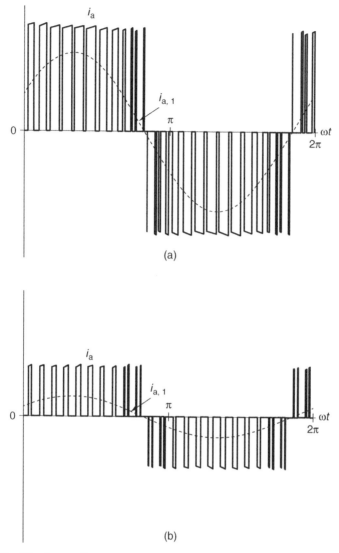

Figure 4.50 Waveforms of input current and its fundamental in a current-type PWM rectifier: (a) $m = 0.75$, (b) $m = 0.35$ ($f_{sw}/f_o = 24$, RLE load).

state of switches, only two states of the rectifier branch can be allowed: SA = ON and SB = OFF, or, vice-versa, SA = OFF and SB = ON. Consequently, a single switching variable, a, defined as

$$a = \begin{cases} 0 & \text{if} \quad \text{SA} = \text{OFF} \quad \text{and} \quad \text{SA}' = \text{ON} \\ 1 & \text{if} \quad \text{SA} = \text{ON} \quad \text{and} \quad \text{SA}' = \text{OFF} \end{cases} \tag{4.93}$$

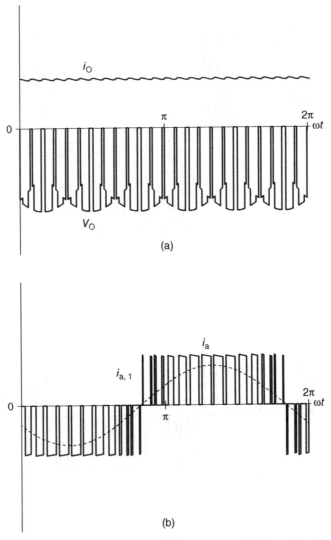

Figure 4.51 Inverter mode of operation of the current-type PWM rectifier: (a) waveforms of output voltage and current, (b) waveforms of input current and its fundamental ($m = 0.75$, $f_{sw}/f_o = 24$, RLE load).

is sufficient to describe the state of the branch. Similarly defined switching variables b and c apply to the other two branches of the rectifier.

In the subsequent considerations, the low-case subscripts indicate voltages at the A′B′C′ terminal set. When $a = 0$ and $i_A > 0$, then $v_a = 0$ as switch SA′ connects terminal A′ to ground. When $a = 0$ and $i_A < 0$, then the current must flow through diode DA′ as SA is off. Again, terminal A′ is connected to ground and $v_a = 0$.

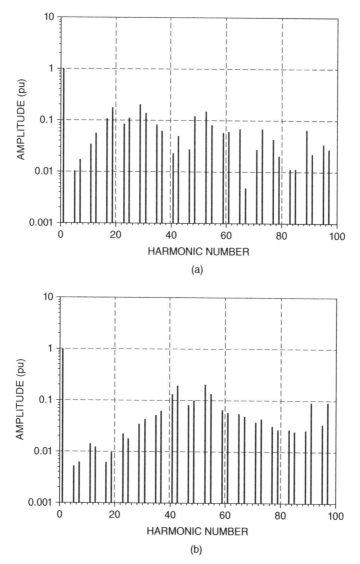

Figure 4.52   Harmonic spectra of input current in a current-type PWM rectifier: (a) $f_{sw}/f_o = 24$, (b) $f_{sw}/f_o = 48$ ($m = 1$, ideal dc-output current).

Vice-versa, when $a = 1$, then independently of the polarity of i_A, $v_a = V_o$. Extension of these observations on the other two branches gives

$$\begin{bmatrix} v_a \\ v_b \\ v_c \end{bmatrix} = V_o \begin{bmatrix} a \\ b \\ c \end{bmatrix}. \tag{4.94}$$

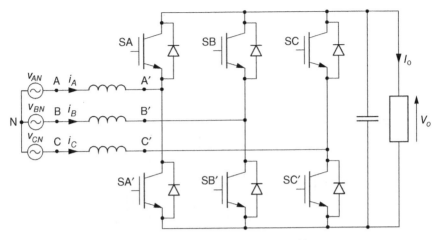

Figure 4.53 Voltage-type PWM rectifier.

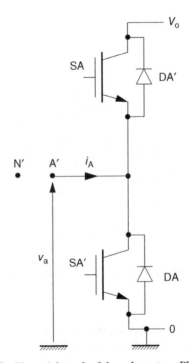

Figure 4.54 Phase-A branch of the voltage-type PWM rectifier.

As $v_{ab} = v_a - v_b$, $v_{bc} = v_b - v_c$, and $v_{ca} = v_c - v_a$, then $v_{an} = (v_{ab} + v_{ac})/3$, $v_{bn} = (v_{ba} + v_{bc})/3$, and $v_{cn} = (v_{ca} + v_{cb})/3$. Here, v_{ab}, v_{bc}, and v_{ca} denote the line-to-line input voltages to the rectifier and v_{an}, v_{bn}, and v_{cn} are the line-to-neutral input voltages. Thus,

$$\begin{bmatrix} v_{ab} \\ v_{bc} \\ v_{ca} \end{bmatrix} = V_0 \begin{bmatrix} 1 & -1 & 0 \\ 0 & 1 & -1 \\ -1 & 0 & 1 \end{bmatrix} \begin{bmatrix} a \\ b \\ c \end{bmatrix} \tag{4.95}$$

and

$$\begin{bmatrix} v_{an} \\ v_{bn} \\ v_{cn} \end{bmatrix} = \frac{V_0}{3} \begin{bmatrix} 2 & -1 & -1 \\ -1 & 2 & -1 \\ -1 & -1 & 2 \end{bmatrix} \begin{bmatrix} a \\ b \\ c \end{bmatrix}. \tag{4.96}$$

The three switching variables imply eight states of the rectifier. Naming a state with a decimal number abc_2, the states can be listed as 0 ($abc = 000$) through 7 ($abc = 111$). States 0 and 7 will subsequently be referred as zero states, because they make all voltages equal to zero. All three input terminals of the rectifier are then simultaneously connected to either the top or the bottom dc bus. Based on the ABC → dq transformation (Eq. 4.74), voltage space vectors corresponding to the active states, 1 through 6, can be determined. The vectors of line-to-line voltages (marked by the "prime" sign) are shown in Figure 4.55a and those of line-to-neutral voltages in

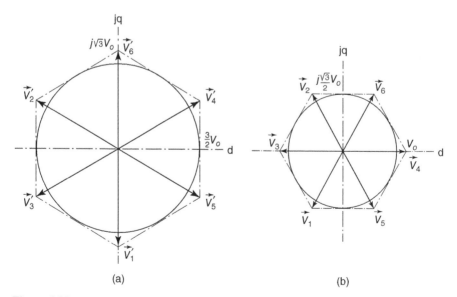

Figure 4.55 Input-voltage space vectors of the voltage-type PWM rectifier: (a) line-to-line voltages, (b) line-to-neutral voltages.

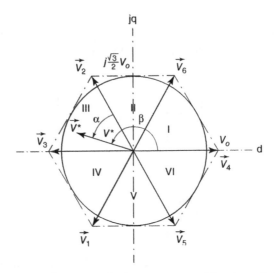

Figure 4.56 Reference voltage vector in the vector space of line-to-neutral input voltages of the voltage-type PWM rectifier.

Figure 4.55b. Notice the similarity of the diagram of the line-to-line voltage vectors with that of the line current vectors in Figure 4.46.

The SVPWM principle with respect to the control of voltages at the input to the rectifier is illustrated in Figure 4.56, which again shows the diagram of space vectors of the line-to-neutral input voltages. The reference voltage vector, $\vec{v}^* = V^* \angle \beta$ (here in sector III), is synthesized from vectors $\vec{V}_X$ (here $\vec{V}_2$), $\vec{V}_Y$ (here $\vec{V}_3$), and $\vec{V}_Z$ ($\vec{V}_0$ or $\vec{V}_7$) using formulas (4.91)–(4.93).

The so-called *voltage-oriented control*, an ingenious concept that employs the rotating reference frame DQ, is employed in the rectifier. Figure 4.57 shows space

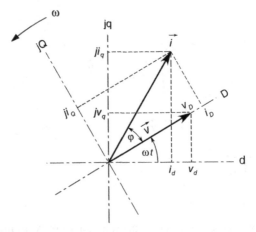

Figure 4.57 Principle of voltage-oriented control in the voltage-type PWM rectifier.

vectors $\vec{v}$ and $\vec{i}$ of the input voltage and current, respectively, in both the stationary and revolving reference frames. It can be seen that to maintain a unity power factor, those two vectors must be aligned. Then, the angle φ between $\vec{v}$ and $\vec{i}$ equals zero and the input power factor, $PF = \cos(\varphi)$, equals one. Thus, with the D-axis aligned with vector $\vec{v}$, that is, v_Q equal zero, the Q-component, i_Q, of vector $\vec{i}$ should be forced to zero, too.

A block diagram of the control system of the rectifier is shown in Figure 4.58. Input voltages and currents are sensed and converted into space vectors. Components i_d and i_q of the current vector $\vec{i}$ are then transformed to components i_D and i_Q in the rotating reference frame. The reference value, i_Q^*, of the Q-component of current vector is set to zero, while that of the D-component, i_D^*, is obtained at the output of a proportional-plus-integral (PI) controller of the output voltage, V_o, of the rectifier. The reference value, V_o^*, of that voltage is compared with the feedback signal of V_o at the input to the controller. Similar PI controllers produce signals v_D^* and v_Q^*, which are

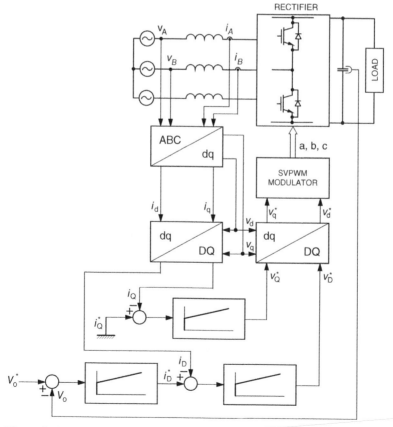

Figure 4.58   Voltage-oriented control system of the voltage-type PWM rectifier.

then transformed into corresponding components v_d^* and v_q^* of the reference voltage vector to be realized by the rectifier using the SVPWM. Components v_d and v_q of the input voltage vector $\vec{v}$ allow determination of the running angle, ωt, for the dq → DQ and DQ → dq transformations using Eqs. (4.83) and (4.85).

Direct Power Control. The unity power factor condition is tantamount to zero reactive power at the input to the rectifier. As known from the theory of three-phase ac circuits, the phasor, $\bar{S}$, of complex power can be calculated as

$$\bar{S} = 3\bar{V}_{AN}\bar{I}_A^* = P + jQ, \tag{4.97}$$

where $\bar{V}_{AN}$ denotes an rms phasor of the line-to-neutral voltage v_{AN}, $\bar{I}_A^*$ is the rms conjugate phasor of the line current i_A, and P and Q denote the average values of the real and reactive power, respectively. An analogous equation for a vector, $\vec{s}$, of complex power is

$$\vec{s} = \frac{2}{3}\vec{v}\vec{i}^* = p + jq, \tag{4.98}$$

where p and q are the instantaneous values of the real and reactive power (do not confuse this q with that denoting the quadrature axis). The coefficient 2/3 in Eq. (4.98) is 4.5 times smaller than its counterpart of 3 in Eq. (4.97). It is so because, as seen in Figure 4.41, the magnitude of a space vector is 1.5 greater than those of its three-phase components, and this magnitude represents a peak value. Each of the rms phasors in Eq. (4.97) is thus $1.5\sqrt{2}$ smaller than the corresponding vector, and their product is $(1.5\sqrt{2})^2 = 4.5$ smaller than that of the vectors in Eq. (4.98).

Substituting in Eq. (4.98) $v_d + jv_q$ for $\vec{v}$ and $i_d - ji_q$ for $\vec{i}^*$ yields

$$p = \frac{2}{3}(v_d i_d + v_q i_q) \tag{4.99}$$

and

$$q = \frac{2}{3}(v_q i_d - v_d i_q). \tag{4.100}$$

Applying the ABC → dq transformation given by Eqs. (4.73) and (4.74), the instantaneous real and reactive powers can now be expressed as

$$p = v_{AN}i_A + v_{BN}i_B + v_{CN}i_C \tag{4.101}$$

and

$$q = \frac{1}{\sqrt{3}}(v_{BC}i_A + v_{CA}i_B + v_{AB}i_C). \tag{4.102}$$

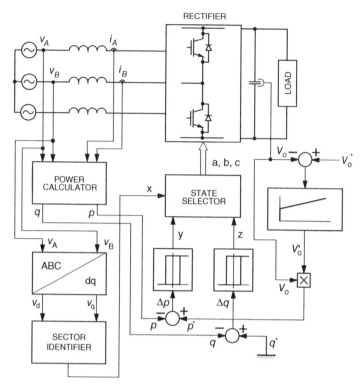

Figure 4.59   DPC system of the voltage-type PWM rectifier.

Thus, sensing the input voltages and currents allows calculation of the real and reactive powers drawn by the rectifier from the supply line.

A block diagram of the direct power control (DPC) of the voltage-type PWM rectifier is shown in Figure 4.59. The next state of the rectifier is determined by the values of three control variables, x, y, and z. Variable x, an integer in the 1–12 range, indicates the 30°-wide sector of the dq plane in which the space vector, $\vec{v}$, of the input voltage is currently located. If the phase of that vector is denoted by β, then

$$x = \text{int}\left(\frac{\beta}{30°}\right) + 1 \tag{4.103}$$

For instance, if $\beta = 107°$, then $x = 4$, which means that $\vec{v}$ is in sector 4. Angle β is determined from the v_d and v_q components of the voltage vector. Logic variables y and z are obtained at the outputs of bang-bang controllers of the real and reactive powers. The width of hysteresis loop of the controllers constitutes a tolerance band for the control errors, Δp and Δq, of these powers. Thus, it is so adjusted as to result in the desired average switching frequency.

The real power, p, is compared with the reference value, p^*, of this power, which is obtained from the control circuit of the output voltage, V_o, of the rectifier. The real

Table 4.1   State Selection in the Voltage-Type PWM Rectifier with DPC

x:		*1*	*2*	*3*	*4*	*5*	*6*	*7*	*8*	*9*	*10*	*11*	*12*
$y = 0$	$z = 0$	6	4	4	5	5	1	1	3	3	2	2	6
	$z = 1$	2	6	6	4	4	5	5	1	1	3	3	2
$y = 1$	$z = 0$	0	4	7	5	0	1	7	3	0	2	7	6
	$z = 1$	0	0	7	7	0	0	7	7	0	0	7	7

power input to the rectifier is closely related to the output power, which is proportional to a square of the output voltage. Consequently, the reference signal p^* is taken as a product of V_o and the output signal, V_o', of a PI controller of that voltage. The reference value, q^*, of the instantaneous reactive power is set to zero and compared with the actual reactive power, q. Based on the values of x, y, and z, the rectifier state that best counteracts the control errors is selected according to Table 4.1.

Voltage and Current Waveforms. As already mentioned, the voltage-type PWM rectifier is a "true" rectifier, in the sense that the input currents and voltages are sinusoidal while both the output voltage and current are true dc waveforms, minor ripple notwithstanding. Example waveforms of the input voltages and currents satisfying the unity power factor condition are shown in Figure 4.60, while Figure 4.61 depicts the corresponding output voltage and current waveforms.

As an aside, it is worth mentioning that both the rotating reference frame and DPC employed in the voltage-type PWM rectifiers are concepts that have been first used in control of three-phase ac motors. The rotating reference frame is a fundamental tool of the so-called field orientation technique, which allows shaping the supply currents in such a way that the magnetic flux and mechanical torque developed in the motor can

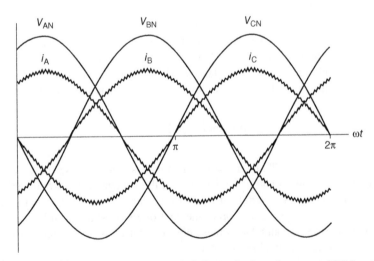

Figure 4.60   Waveforms of input voltage and current in the voltage-type PWM rectifier at unity power factor.

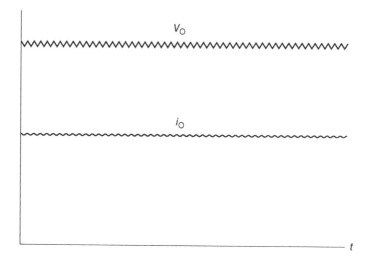

Figure 4.61   Waveforms of output voltage and current in the voltage-type PWM rectifier.

be controlled independently from each other. In the so-called direct torque control, similar to the DPC described in Section 4.3.4, sequential states of the inverter feeding the motor are selected on a basis of the flux and torque control errors.

4.3.5   Vienna Rectifier

Vienna rectifier, the name derived from the city of its origin, was developed in the early 1990s for application in telecommunication power supplies. Thanks to its attributes the Vienna rectifier has since found a solid place among modern PWM ac-to-dc converters. It is characterized by high power density, efficiency, and reliability, while only three semiconductor power switches are employed. The switches are subjected to only half of the output voltage, so that the voltage rating of the rectifier can be significantly higher from that of the switches. The Vienna rectifier is a boost converter with a high switching frequency (tens of kHz). However, the voltage gain, defined as the ratio of output voltage to the peak line-to-line input voltage, typically does not exceed 2.

Version I of the Vienna rectifier is shown in Figure 4.62. Each of the three IGBTs is surrounded by four diodes resulting in three bidirectional switches, SA, SB, and SC. The other six diodes, D_{1A} through D_{2C}, prevent short circuits endangering the output capacitors. For instance, diode D_{1A} blocks the path of a short-circuit current, which would rapidly discharge capacitor C_1 when switch SA were closed.

The two output capacitors, C_1 and C_2, make the rectifier a so-called *three-level converter*. Here, it means that each of the rectifier inputs, A′, B′, and C′, can attain three levels of voltage. Considering, for example, phase A, depending on the polarity of the input current i_A and state of switch SA, inductor L_A can be connected to the upper dc bus, the midpoint M, or the lower dc bus. The idea of multilevel converters

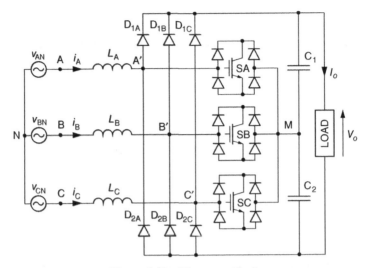

Figure 4.62   Vienna rectifier I.

is expanded in Chapter 7. One of the advantages of such topology is that the voltage across any semiconductor device does not exceed the voltage of a single capacitor, C_1 or C_2. Thus, assuming that the output voltage, V_o, is equally split between the two capacitors, the voltage rating of the rectifier can be twice as high as that of the semiconductor devices.

Ternary switching variables, a, b, and c, can be used for the description of operation of switches SA, SB, and SC. Specifically,

$$a = \begin{cases} 2 & \text{if SA is OFF and } i_A > 0 \\ 1 & \text{if SA is ON} \\ 0 & \text{if SA is OFF and } i_A < 0 \end{cases}. \tag{4.104}$$

Then, assuming that the midpoint M of the rectifier is grounded and the output capacitors equally share the output voltage, the voltage of input terminal A′ is given by

$$V_{A'} = (a - 1) V_o. \tag{4.105}$$

Operation of the other two switches can be described analogously.

Since all three input currents, i_A, i_B, and i_C, cannot have simultaneously the same polarity, the number of states is lower than the theoretical number of combinations of three ternary variables, that is, $3 \times 3 \times 3 = 27$. If, for example, i_A and i_B are positive and i_C is negative, variables a and b can only assume values 1 and 2 each, while variable c values 0 and 1. Designating a state as abc, it can be seen that states 000 and 222 are not possible, while state 111 is a zero state. Thus, the number of possible

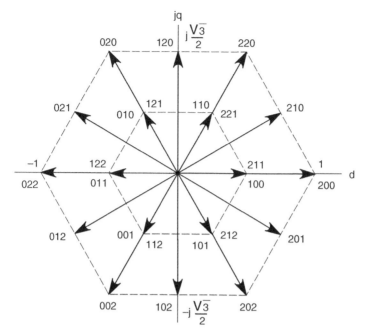

Figure 4.63   Voltage space vectors of Vienna rectifier.

states is 25. Each of this states can be assigned a space vector of input voltages. Voltage vectors of the Vienna rectifier are shown in Figure 4.63 in the per-unit format with V_O as the reference. There are six large vectors, six medium vectors, and six small vectors. Each of the latter corresponds to two states. For instance, vector 0.5 $\angle 60°$ is associated with either state 110 or state 221. Numerous space-vector PWM techniques have been proposed over the years by various researchers.

As for all PWM rectifiers, the two control objectives are: (1) the desired level of output voltage and (2) sinusoidal input currents in phase with the corresponding grid voltages. Originators of the Vienna rectifier proposed a control method that employs simple hysteresis (relay) controllers of the input currents. In addition to the mentioned two objectives, the method also provides balancing of the capacitor voltages, so that midpoint M can be used as an output terminal for reduced voltage loads. Block diagram of the control system is shown in Figure 4.64. The slanting marks denote flows of single signals, while the unmarked lines carry triple signals for individual phases, symbol k denoting A, B, and C. Symbols $\hat{V}_{LN}$ and $\hat{I}_L$ denote peak values of the line-to-neutral grid voltage and line (input) current, respectively, and the asterisks indicate reference values.

The hysteresis controller turns the corresponding switch on if the current error, Δi_k, is greater than the width h of the hysteresis loop. Vice-versa, the switch is turned off if Δi_k is less than h. Binary switching variables x'_k are assigned to those two situations, that is, $x'_k = 1$ when $\Delta i_k > h$ and $x'_k = 0$ otherwise. Switching variables x_k,

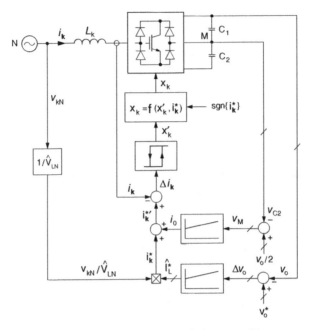

Figure 4.64 Control system of Vienna rectifier.

also binary, control the rectifier switches, with $x_k = 1$ turning switch Sk on and $x_k = 0$ turning it off. They are generated using the following rule:

$$x_k = \begin{cases} x'_k & if \quad i^*_k \geq 0 \\ \bar{x}'_k & if \quad i^*_k < 0 \end{cases} \tag{4.106}$$

As a result, the input current waveforms are rippled sinusoids contained within the h-wide tolerance bands, resembling those of the voltage-type rectifier in the preceding section (see Figures 4.60 and 4.61). Version II of Vienna rectifier, which is functionally equivalent to version I, is shown in Figure 4.65.

4.4 DEVICE SELECTION FOR RECTIFIERS

The voltage rating, V_{rat}, of semiconductor power switches in rectifiers (and, for that matter, in all power electronic converters) should exceed the highest instantaneous voltage possible to appear between any two points of the power circuit. In rectifiers, it is the peak value, $V_{i,p}$, of input voltage. Since semiconductor devices are vulnerable to overvoltages, even those of very short duration, substantial safety margins should

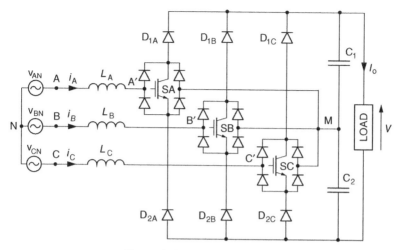

Figure 4.65   Vienna rectifier II.

be assumed in the design. The safety margins depend on the voltage level, the higher ones used in low-voltage converters. Thus, denoting a voltage safety margin by s_V, the condition to be satisfied is

$$V_{rat} \geq (1 + s_V)V_{i,p}. \tag{4.107}$$

In six-pulse rectifiers, $V_{i,p} = V_{LL,p}$.

The current rating, I_{rat}, which is the maximum allowable average current in a power switch, must be greater than the actual maximum average current, $I_{ave(max)}$. In six-pulse rectifiers, each power switch conducts the output current within one-third of the cycle of input voltage. Therefore, $I_{ave(max)} = I_{o,dc(rat)}/3$, where $I_{o,dc(rat)}$ denotes the rated dc-output current, usually determined as the ratio of the rated power of a rectifier to the rated output voltage. Consequently, denoting the current safety margin by s_I, the condition to meet is

$$I_{rat} \geq \frac{1}{3}(1 + s_I)I_{o,dc(rat)}. \tag{4.108}$$

Typically, lower safety margins are used for currents than for voltages. Temporary current overloads are less destructive than overvoltages, thanks to the thermal inertia of the devices always equipped with heat sinks. In dual converters, an additional allowance should be made for the circulating current.

Fully controlled semiconductor power switches for PWM rectifiers are switched many times within a single cycle of input voltage. Therefore, dynamic characteristics of these devices must be taken into account. Specifically, for crisp voltage and current pulses to be produced, the shortest on-time of the switches should be significantly longer than their turn-on time. An analogous rule applies to the off- and turn-off times.

4.5 COMMON APPLICATIONS OF RECTIFIERS

Diode rectifiers are used to provide a fixed voltage to dc-powered equipment, such as electromagnets or electrochemical plants. They are also employed as supply sources for dc-input power electronic converters, that is, inverters and choppers. Phase-controlled rectifiers are primarily used for dc motor control, in high-voltage dc transmission lines, as battery chargers, exciters for synchronous ac machines, and in certain technological processes requiring dc current control, such as electric arc welding. Also, if an inverter or a chopper is to provide a bidirectional flow of energy, it must be supplied from a controlled rectifier which, when operated in the inverter mode, allows transmission of power back to the ac supply system.

The PWM rectifiers have only recently entered the mainstream power electronics market. The current-type rectifiers are a worthy alternative to phase-controlled rectifiers, whose input currents are far from ideal (see Figure 4.24). The voltage-type rectifiers cannot be employed for control of dc motors, because the output voltage can only be adjusted up from the minimum level, roughly equal to the peak value of the supply line-to-line voltage. Therefore, they are mostly used in frequency changers for control of ac motors. The fixed, three-phase, supply ac voltage is first rectified under the unity power factor condition and then, in an inverter, converted into an adjustable three-phase voltage applied to the stator of the controlled motor.

Note that besides controlled rectifiers, other power electronic converters, such as inverters, choppers, and cycloconverters, can operate electrical machinery, too. The converters and machines share the common purpose, which is conversion of one form of energy into another. Therefore, operation of a controlled rectifier supplying a dc motor will be described to illustrate the basic relations between operating modes of these two subsystems.

A dc machine, here assumed to be of the separately excited type, can be represented as in Figure 4.66. The armature circuit is composed of the armature resistance, R_a, armature inductance, L_a, and armature EMF, E_a. The latter quantity is proportional to the speed, n, of the machine (in r/min), and the developed torque, T, is proportional to the armature current, i_a. For reference, the speed and torque are assumed positive when clockwise. Using T and n as operating variables of the machine, an operation plane can be set up as shown in Figure 4.67. The four quadrants of the plane correspond to the four possible modes of operation, from motoring clockwise (first quadrant) to generating counterclockwise (fourth quadrant). The idea of operating quadrants extends on all revolving machines, not necessary electrical ones.

A dc machine and the converter that feeds it form a dc drive system, usually equipped with automatic control of the speed and torque. In the steady state of the drive, the dc component, $V_{a,dc}$, of the armature voltage has the same polarity as the armature EMF, E_a, the two quantities differing only by the small voltage drop across

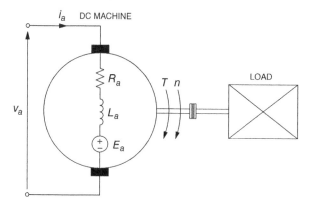

Figure 4.66   Electromechanical representation of a dc machine.

the armature resistance. As the armature voltage and current are supplied by the converter, the operating quadrants of the machine correspond to their counterparts in the converter plane of operation in Figure 4.29. In particular, the motoring mode of the machine is associated with the rectifier mode of the converter, with the power flow from the ac source, through the converter and machine, to the load. Conversely, the generating mode of the machine results from the inverter mode of the converter, with the power flow reversed.

As an example, an electric locomotive supplied from a three-phase overhead line and driven by dc motors is considered. Today, dc motors in traction have largely been superseded by ac motors, but an electric dc locomotive was one of the first high-power applications of SCRs. Several motors operate the wheels of a practical locomotive. For simplicity, only one of these motors, assumed to be fed from a rectifier with an electromechanical switch, is considered.

Figure 4.68 shows a rectifier–switch–motor cascade when the locomotive hauls a train with the motor rotating clockwise, that is, with positive speed and torque. The switch provides direct connection between the rectifier and the motor. Thus, the average armature voltage, V_a, and instantaneous armature current, i_a, of the motor are

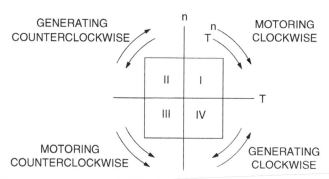

Figure 4.67   Operation plane, operating area, and operating quadrants of a rotating machine.

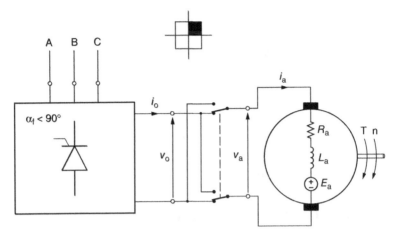

Figure 4.68   A dc motor supplied from a rectifier with cross-switch: first-quadrant operation.

equal to the dc-output, $V_{o,dc}$, and output current, i_o, of the rectifier, respectively. Both the converter and the machine operate in the first quadrant, which means the rectifier mode with positive dc-output voltage for the converter and the motoring clockwise mode for the machine. The firing angle, α_f, is kept below 90°.

Kinetic energy of a train running at full speed is enormous, and using only friction brakes to slow down or stop the train is ineffective and wasteful. Therefore, an electric braking scheme is implemented by transition to the second quadrant, in which the dc machine operates as a generator, and the converter works in the inverter mode. This situation is illustrated in Figure 4.69. The firing angle has been increased to over 90°,

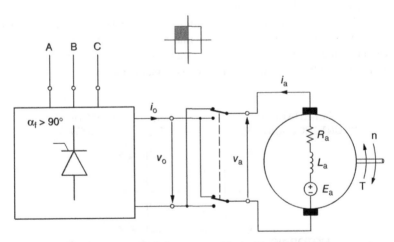

Figure 4.69   A dc motor supplied from a rectifier with cross-switch: second-quadrant operation.

and the electromechanical switch has been switched to the cross-connecting position, making $v_a = -v_o$ and $i_a = -i_o$. Although the direction of speed and the polarity of motor EMF have not changed, the EMF is now seen as negative from the rectifier terminals. The conditions for the inverter operation are thus satisfied, and the negative armature current produces a braking torque opposing the motion of the motor. The momentum of the train provides the driving torque for the dc machine, and the electric power generated is transferred via the rectifier to the ac supply line.

The third and fourth quadrants of operation represent the motoring and braking modes, respectively, when the locomotive runs backward and the motor rotates counterclockwise. Development of connection diagrams similar to those in Figures 4.68 and 4.69 is left to the reader.

It is worth noting that certain drives operate in the first and fourth quadrants only, so that the rectifier can directly be connected to the motor. Lift-type drives, such as those of elevators, are the typical example. No matter whether the load moves up or down, the motor torque must counter the torque imparted by the gravity. Therefore, the torque never changes its polarity, and the output current of the rectifier does not need to be reversed. When stopping from the upward motion, the gravity force, possibly supplemented with friction brakes, is sufficient to provide the braking torque.

High-voltage dc (HVDC) transmission lines link distant power systems, using dc current as a medium for energy exchange between the systems. Therefore, synchronous operation of the systems is not required (they may even operate with different frequencies), and the easy control of transferred power improves the stability of the systems. HVDC lines allow efficient transmission of electric power, because the inductance and capacitance of the line do not affect the voltage and the current.

A typical HVDC transmission system is illustrated in Figure 4.70. The link between the interconnected power systems consists of 12-pulse rectifiers RCT1 and RCT2, transformers TR1 and TR2, smoothing inductors, L1 through L4, and a transmission line. Each rectifier is composed of two six-pulse SCR-based rectifiers

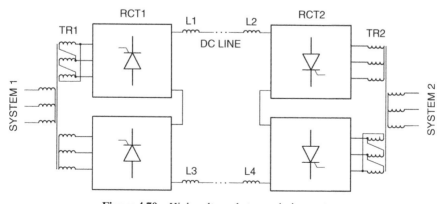

Figure 4.70   High-voltage dc transmission system.

connected in series, one supplied from a wye-connected secondary winding of the transformer and the other from a delta-connected secondary winding. Such 12-pulse rectifiers are characterized by a very smooth output voltage, with a minimal ripple factor.

The dc line is usually of the two-wire type, although single-wire lines, with the ground as a return path, are also feasible. If the power is transmitted from System 1 to System 2, rectifier RCT1 operates in the rectifier mode and rectifier RCT2 in the inverter mode. As a result, the three-phase ac power drawn from System 1 is converted into dc power, transmitted over the line to RCT2, and converted back into three-phase ac power in System 2. The rectifiers interchange the modes of operation if the power is transmitted in the opposite direction. Note that the dc current can only flow in one direction, and the reversal of power flow is accomplished by reversing polarity of the dc voltage.

A single switch of an HVDC rectifier is composed of dozens of high-rating SCRs connected in series and in parallel to withstand the high voltages and currents of the line. Individual SCRs are fired by photothyristors, to which the firing signals are delivered via fiber-optic cables to ensure simultaneous turn-on. The apparatus required affect the cost of HVDC lines, which are only economical when voltages of several hundred kilovolts, powers of several hundred to several thousand megawatts, and distances of several hundred miles are involved.

SUMMARY

The ac-to-dc power conversion is performed by rectifiers, which in practice are mostly of the six-pulse, three-phase bridge topology. Diode rectifiers produce a fixed dc voltage, whose value depends on the amplitude of the ac-input voltage. The output current in diode rectifiers is continuous, unless the load EMF has a value approaching the amplitude of the input voltage and the load inductance is low. Power flow is unidirectional, from the ac source, via the rectifier, to the load.

Controlled rectifiers are more flexible operationally, allowing control of the dc-output voltage and being capable of reversing the power flow. In phase-controlled, SCR-based rectifiers, the voltage control is realized by adjusting the firing angle of the SCRs. In the continuous conduction mode, the dc-output voltage is proportional to cosine of the firing angle. The inverter mode is possible only when the load EMF is negative and the firing angle is greater than 90°.

The load and source inductances affect the operation of rectifiers. Resonant input filters are recommended to reduce harmonic content of the input currents drawn from the supply system by uncontrolled and phase-controlled rectifiers.

An extension to four-quadrant operation of controlled rectifiers is possible by installing an electromechanical cross-switch at the output, which allows for a negative load current. Dual converters represent a more convenient and robust solution. A dual converter consists of two rectifiers connected in antiparallel. Circulating current-free dual converters are simpler but operationally inferior to the circulating current-conducting ones.

The quality of the input currents can greatly be enhanced using PWM rectifiers, based on fully controlled semiconductor power switches. The PWM rectifiers, either of the current-type (buck) or voltage-type (boost), allow maintaining a unity input power factor. The output voltage and current are also of high quality. One of these converters, the Vienna rectifier, employs only three fully controlled semiconductor switches. The maximum available volt-ampere ratings of PWM rectifiers are, however, lower than those of phase-controlled rectifiers with SCRs.

The selection of semiconductor power switches for rectifiers is primarily based on the amplitude of the ac supply voltage and the maximum average output current. Switches for PWM rectifiers must be sufficiently fast to realize the frequent switching required.

Dc motor control, HVDC transmission lines, battery charging, and supply of dc-input power electronic converters are the most common applications of rectifiers. They are also employed in various technological processes, such as electric arc welding or electrolysis.

EXAMPLES

Example 4.1   A three-pulse and a six-pulse diode rectifiers are supplied from a 460-V ac line. Compare the dc-output voltages available from these rectifiers.

Solution: According to Eq. (4.2), the dc voltage, $V_{o,dc(3)}$, of the three-pulse rectifier is $V_{o,dc(3)} = \frac{3\sqrt{3}}{2\pi} V_{LN,p}$ where the peak value, $V_{LN,p}$, of the input line-to-neutral voltage is

$$V_{LN,p} = \frac{\sqrt{2} \times 460}{\sqrt{3}} = 375.6 \text{ V}$$

(the rated voltage of any three-phase apparatus is always the rms line-to-line voltage). Thus,

$$V_{o,dc(3)} = \frac{3\sqrt{3}}{2\pi} \times 375.6 = 310.6 \text{ V}.$$

Following Eq. (4.4), the dc voltage of the six-pulse rectifier,

$$V_{o,dc(6)} = \frac{3}{\pi} V_{LL,p} = \frac{3}{\pi} \times \sqrt{2} \times 460 = 621.2 \text{ V},$$

that is, exactly twice as high as that of the three-pulse rectifier.

Example 4.2 A 270-V battery pack is charged from a six-pulse diode rectifier supplied from a 230-V ac line. The internal resistance of the battery pack is 0.72 Ω. Calculate the dc charging current.

Solution: The load EMF coefficient, ε, is

$$\varepsilon = \frac{270}{\sqrt{2} \times 230} = 0.83$$

and the load angle, φ, is zero, as no load inductance is assumed. Therefore, condition (4.19) for continuous conduction is

$$\varepsilon < \sin\left(\frac{\pi}{3}\right) = \frac{\sqrt{3}}{2} \approx 0.866$$

and it is satisfied. Consequently, the dc-output voltage, $V_{o,dc}$, of the rectifier can be calculated from Eq. (4.4) as

$$V_{o,dc} = \frac{3}{\pi} \times \sqrt{2} \times 230 = 310.6 \text{ V},$$

and, according to Equation (4.28), the dc-output current, $I_{o,dc}$, is

$$I_{o,dc} = \frac{310.6 - 270}{0.72} = 56.4 \text{ A}.$$

Example 4.3 A dc motor with the armature resistance, R_a, of 0.6 Ω and armature inductance, L_a, of 4 mH is supplied from a phase-controlled six-pulse rectifier fed from a 460-V, 60-Hz line. The motor rotates with such speed that the armature EMF, E_a, is 510 V. Find the dc-output voltage, $V_{o,dc}$, and current, $I_{o,dc}$, of the rectifier when the firing angle, α_f, is 30° and 60°.

Solution: First, the feasibility of the assumed firing angles must be checked. The load EMF coefficient, ε, is

$$\varepsilon = \frac{510}{\sqrt{2} \times 460} = 0.784$$

and condition (4.43) for the allowable values of firing angle gives

$$-0.146 \text{ rad} < \alpha_f < 1.193 \text{ rad},$$

which is satisfied by both $\alpha_f = 30°$ and $\alpha_f = 60°$. This is also confirmed by the diagram in Figure 4.21. Next, the conduction modes must be determined for proper selection of formulas for the dc-output voltage. The load angle, φ, is

$$\varphi = \tan^{-1}\left(\frac{\omega L_a}{R_a}\right) = \tan^{-1}\left(\frac{120\pi \times 4 \times 10^{-3}}{0.6}\right) = 1.192 \text{ rad} = 68.3°,$$

and when $\alpha_f = 30° = \pi/6$ rad, condition (4.45) for continuous output current is

$$\varepsilon < \left[\sin\left(\frac{\pi}{6} + \frac{\pi}{3} - 1.192\right) + \frac{\sin\left(1.192 - \frac{\pi}{6}\right)}{1 - e^{-\frac{\pi}{3\tan(1.192)}}}\right]\cos(1.192) = 0.81$$

which is satisfied. Similar calculations for $\alpha_f = 60°$ yield the requirement $\varepsilon < 0.447$, which is not satisfied, that is, the rectifier operates in the discontinuous conduction mode. The same conclusions could be reached by inspecting the diagram in Figure 4.22.

Using Eqs. (4.41) and (4.28), the dc-output voltage and current for $\alpha_f = 30°$ are calculated as

$$V_{o,dc} = \frac{3}{\pi} \times \sqrt{2} \times 460 \times \cos\left(\frac{\pi}{6}\right) = 538 \text{ V}$$

and

$$I_{o,dc} = \frac{538 - 510}{0.6} = 46.7 \text{ A}.$$

Calculation of $V_{o,dc}$ for $\alpha_f = 60°$, when the output current, i_o, is discontinuous, requires knowledge of the conduction angle, β. It can be determined by computing $i_o(\omega t)$ for sequential values of ωt, starting at $\omega t = \alpha_f$ and ending when the current drops to zero at $\omega t = \alpha_e = \alpha_f + \beta$. The impedance, Z, of the load is

$$Z = \sqrt{R_a^2 + (\omega L_a)^2} = \sqrt{0.6^2 + (120\pi \times 4 \times 10^{-3})^2} = 1.623\Omega,$$

and Eq. (4.46) gives

$$i_o(\omega t) = 400.8\left[\sin(\omega t - 0.145) - 2.12 + 1.335e^{\frac{\omega t - \pi/3}{2.513}}\right].$$

A computer should be employed to calculate the current. Beginning with $\omega t = 60°$ and proceeding with $0.1°$ increments (smaller increments could, of course, be used as well), the following printout is obtained:

Angle	Current (A)
60.0°	0.0000000
60.1°	0.0614110
60.2°	0.1221164
...	...
75.8°	0.1172305
75.9°	0.0513202
76.0°	−0.015876

Thus, the extinction angle, α_e, is about 76°, and the conduction angle, β, is $76°-60° = 16°$, that is, 0.279 rad. Using Eq. (4.47), the dc-output voltage is obtained as

$$V_{\text{o,dc}} = \frac{3}{\pi}\sqrt{2} \times 460 \left[2\sin\left(\frac{\pi}{3} + \frac{0.279}{2} + \frac{\pi}{3}\right) \sin\left(\frac{0.279}{2}\right) + 0.784 \left(\frac{\pi}{3} - 0.279\right) \right]$$
$$= 510.4 \text{ V}.$$

The dc-output current is

$$I_{\text{o,dc}} = \frac{510.4 - 510}{0.6} = 0.7 \text{ A},$$

that is, two orders of magnitude smaller than that in the continuous conduction mode with $\alpha_f = 30°$.

Note that if the discontinuous conduction mode had not been identified and Eq. (4.41) used for calculation of $V_{\text{o,dc}}$ instead of Eq. (4.47), the resultant value of 310.6 V would be lower than the load EMF, implying a negative output current, which is not possible.

Example 4.4   A phase-controlled six-pulse rectifier is supplied from a 460-V, 60-Hz power line, whose inductance, L_s, seen from the input terminals of the rectifier is 1 mH/phase. The rectifier operates with a firing angle, α_f, of 30°, producing an average output current, $I_{\text{o,dc}}$, of 140 A. Neglecting resistances of the supply circuit, find the overlap angle, μ, and average output voltage, $V_{\text{o,dc}}$, of the rectifier.

Solution: The source reactance, X_s, is given by

$$X_s = \omega L_s = 120\pi \times 10^{-3} = 0.377\Omega/\text{ph}$$

and the peak value of line-to-line input voltage, $V_{\text{LL,p}}$, is $\sqrt{2} \times 460 = 650.5\text{V}$. Substituting the values of α_f, X_s, $I_{\text{o,dc}}$, and $V_{\text{LL,p}}$ in Eq. (4.63) yields

$$\mu = \left| \cos^{-1} \left[\cos \left(\frac{\pi}{6} \right) - 2\frac{0.377 \times 140}{650.5} \right] - \frac{\pi}{6} \right| = 0.267 \text{ rad} = 15.3°.$$

If the source inductance were zero, the average output voltage, $V_{\text{o,dc(0)}}$, would be

$$V_{\text{o,dc(0)}} = \frac{3}{\pi} \times 650.5 \times \cos \left(\frac{\pi}{6} \right) = 538 \text{ V}.$$

However, as seen from Eq. (4.64), the load inductance causes reduction of this voltage by

$$\Delta V_{\text{o,dc}} = \frac{3}{\pi} \times 0.377 \times 140 = 50.4 \text{ V}$$

so that the actual dc-output voltage is 538 V–50.4 V $= 487.6$ V.

Example 4.5   A current-type PWM rectifier is supplied from a 460-V line and it is to produce 500 V of dc voltage at its output. At certain instant, the angle, β, of vector of the input voltages is 235°. Find the switching pattern of the rectifier if the switching frequency is 5 kHz.

Solution: The maximum dc voltage of the rectifier equals about 92% of the peak line-to-line supply voltage, that is,

$$V_{\text{o(max)}} = 0.92 \times \sqrt{2} \times 460 = 598 \text{ V}.$$

Thus, the modulation index, m, equals

$$m = \frac{V_o}{V_{\text{o,max}}} = \frac{500}{598} = 0.84.$$

For unity power factor, the space vector of input currents must be aligned with the input voltage vector, which is in sector V that extends from 210° to 270° (see Figure 4.46). Hence, the in-sector angle, α, of that vector is $235° - 210° = 25°$, and states X, Y, and Z are state 5, state 6, and state 9, respectively. With the switching period, $T_{\text{sw}} = 1/5\text{kHz} = 200$ µs, durations of these states are:

$$T_5 = 0.84 \times 200 \times \sin(60° - 25°) = 96.4 \text{ µs}$$

$$T_6 = 0.84 \times 200 \times \sin(25°) = 71 \text{ µs}$$

$$T_9 = 200 - 96.4 - 71 = 32.6 \text{ µs}.$$

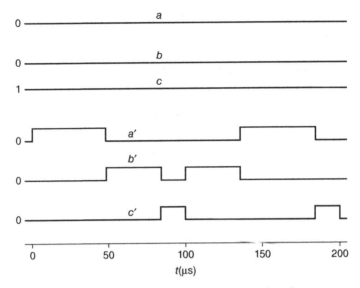

Figure 4.71   Switching pattern in Example 4.5.

The state sequence in sector V, as listed in Section 4.3.3, is $5 - 6 - 9 - 6 - 5 - 9 \ldots$, and the corresponding time intervals, in microseconds, are $48.2 - 35.5 - 16.3 - 35.5 - 48.2 - 16.3 \ldots$ Switches that turn on and off are those in the lower row, that is, SA', SB', and SC', while SA and SB are off and SC is on all the time. The described switching pattern is shown in Figure 4.71.

Example 4.6   A voltage-type PWM rectifier is so controlled that its input current vector lags the input voltage vector by 30°. At certain instant, the input line-to-neutral voltages v_{AN} and v_{BN} are 261.5 V and −90.8 V, while the input line currents i_A and i_B are 94.0 A and −76.6 A. After re-tuning the control system, the rectifier operates with the unity power factor. At the instant when the voltages are at the same levels as before, currents i_A and i_B are 98.5 A and −34.2 A, respectively. Calculate the instantaneous real, reactive, and apparent input power in both cases.

Solution: The line-to-neutral and line-to-line voltages are $v_{AN} = 261.5$ V, $v_{BN} = -90.8$ V, $v_{CN} = -v_{AN} - v_{BN} = -170.7$ V, $v_{AB} = v_{AN} - v_{BN} = 352.3$ V, $v_{BC} = v_{BN} - v_{CN} = 79.9$ V, and $v_{CA} = -v_{AB} - v_{BC} = -432.2$ V. In the first case, the line currents are $i_A = 94.0$ A, $i_B = -76.6$ A, and $i_C = -i_A - i_B = -17.4$ A. According to Eqs. (4.102) and (4.103), $p = 34,506$ W and $q = 19,911$ VAr. The instantaneous apparent power, s, which is a geometric sum (square root of sum of squares) of p and q, equals 39,839 VA.

In the second case, $i_A = 98.5$ A, $i_B = -34.2$ A, and $i_C = -i_A - i_B = -64.3$ A and the real, reactive, and apparent powers are $p = 39,839$ W, $q = 0$, $s = 39,839$ VA. It can be seen that the reactive power from the previous case has been converted into real power and added to the initial real power, so that now $s = p$.

PROBLEMS

P4.1 A three-pulse diode rectifier, fed from a 460-V ac line, supplies a 5-Ω resistive load. Calculate the average output voltage and current of the rectifier.

P4.2 The rectifier in Problem 4.1 charges a 360-V battery pack. Is the charging current continuous or discontinuous? Sketch the output voltage and current waveforms.

P4.3 A six-pulse diode rectifier, fed from a 230-V ac line, charges a 240-V battery pack. Is the charging current continuous or discontinuous? Sketch the output voltage and current waveforms.

P4.4 A smoothing inductor of 40 mH has been connected between the rectifier in Problem 4.3 and the battery pack. The internal resistance of the battery is 2.2 Ω. Determine the conduction mode and find the dc-output voltage and current of the rectifier.

P4.5 The rectifier in P4.3 charges a 300-V battery pack. Find the crossover, extinction, and conduction angles.

P4.6 Sketch the output voltage waveform of a phase-controlled three-pulse rectifier operating in the continuous conduction mode with the firing angle of 60°

P4.7 Assuming the continuous conduction mode, sketch the output voltage waveforms of a phase-controlled six-pulse rectifier corresponding to firing angles of 0°, 45°, 90°, 120°, and 180°.

P4.8 What firing angle is always feasible in an SCR-based rectifier if the load EMF does not exceed the peak value of line-to-line input voltage?

P4.9 Sketch the output voltage and current waveforms of a phase-controlled six-pulse rectifier supplying an RLE load with the load EMF equal 90% of the peak line-to-line input voltage, and operating in the discontinuous conduction mode with the firing angle of 60°.

P4.10 A phase-controlled six-pulse rectifier, fed from a 460-V, 60-Hz line, supplies a dc motor, whose armature resistance and inductance are 0.3Ω and 1.7 mH, respectively. The speed of the motor is such that the EMF induced in the armature is 520 V. Determine the conduction mode of the rectifier and calculate the dc armature voltage and current if the rectifier operates with the firing angle of 15°.

P4.11 Determine the conduction mode for the rectifier in P4.10 if it operates with the firing angle of 50°. Find the dc-output voltage and current.

P4.12 The rectifier in P4.10 operates in the inverter mode, with the firing angle of 125° and the motor working as a dc generator. What armature EMF is needed to maintain continuous armature current?

P4.13 A phase-controlled six-pulse rectifier operates in the continuous conduction mode with the firing angle of 40°. What is the input power factor of the rectifier?

P4.14 A phase-controlled six-pulse rectifier, fed from a 230-V, 60-Hz line, operates with a firing angle of 45°. The source inductance is such that the dc-output voltage of the rectifier is reduced by 12% in comparison to that with a zero-inductance source. Assuming the continuous conduction mode, find the overlap angle and dc-output voltage of the rectifier.

P4.15 A circulating current-conducting dual converter, fed from a 230-V line, operates in the continuous conduction mode, with the 50° firing angle in one of the constituent rectifiers. What is the firing angle of the other rectifier, and what is the average output voltage of the converter? Sketch the output voltage waveforms of both rectifiers of the converter and find the value of the differential voltage at $\omega t = \pi/9$ rad.

P4.16 At certain instant, voltages v_{AB} and v_{AC} in the three-wire power line supplying a three-phase rectifier are 498.4 V and 611.4 V, respectively, and currents i_A and i_C in the same line are 289.8 A and –212.2 A, respectively. Determine and sketch space vectors, $\vec{V}_{LL}$ and $\vec{I}_L$, of the line-to-line voltage and line current.

P4.17 A current-type PWM rectifier operates with the switching frequency of 4 kHz and the modulation index of 0.6. Determine and sketch the waveforms of switching variables within a single switching cycle when the reference current vector lies in the middle of sector III of the dq plane.

P4.18 Find the instantaneous real and reactive powers drawn by the rectifier in Problem 4.16 from the supply line.

P4.19 The SCRs in a phase-controlled six-pulse rectifier are rated at 3200 V and 1000 A. Assuming the voltage and current safety margins of 40% and 20%, respectively, find the maximum allowable voltage of the supply ac line and the minimum allowable resistance of the load. Remember that the voltage of a three-phase line is always meant as the rms value of line-to-line voltage.

P4.20 SCRs rated at 6.5 kV are used in a multipulse converter at one end of a 400-kV dc transmission line. The converter can be assumed to produce ideal dc voltage equal to the peak line-to-line supply voltage times the cosine of firing angle (see Eq. 4.42). When the interconnected systems operate with the rated ac voltage, the firing angle of the converter is 20° to provide a margin for control of the dc voltage when the ac voltage strains from its rated level. The ac voltage may vary from 95% to 105% of the rated value. Find the rated ac voltage at the input to the converter and, assuming the voltage safety margin of 20%, determine how many SCRs are connected in series in a single power switch of the converter.

COMPUTER ASSIGNMENTS

CA4.1* Run PSpice program *Contr_Rect_1P.cir* for a single-pulse phase-controlled rectifier. Observe voltage and current waveforms with various firing angles.

CA4.2 Use computer graphics to produce a background sheet for sketching voltage waveforms of six-pulse rectifiers. The sheet should contain waveforms of all six input line-to-line voltages drawn using broken lines as, for example, in Figure 4.8.

CA4.3* Run PSpice program *Diode_Rect_3P.cir* program for a three-pulse diode rectifier. Set appropriate values of the coefficient of load EMF and perform simulations for the continuous and discontinuous operation modes of the rectifier. In both cases, find for the output voltage and current:

(**a**) dc component

(**b**) rms value

(**c**) rms value of the ac component

(**d**) ripple factor

　　Observe oscillograms of the input voltages and currents.

CA4.4* Run PSpice program *Diode_Rect_6P.cir* for a six-pulse diode rectifier. Set appropriate values of the coefficient of load EMF and perform simulations for the continuous and discontinuous operation modes of the rectifier. In both cases, find for the output voltage and current:

(**a**) dc component

(**b**) rms value

(**c**) rms value of the ac component

(**d**) ripple factor

　　Observe oscillograms of the input voltages and currents.

CA4.5* Run PSpice program *Diode_Rect_6P_F.cir* for a six-pulse diode rectifier with an input filter. To evaluate the impact of the input filter, find for the current drawn from the supply source (before the filter) and current at the input terminals of the rectifier (after the filter):

(**a**) total harmonic distortion

(**b**) amplitudes of the 1st, 5th, 7th, 11th, and 13th harmonics

CA4.6* Run PSpice program *Contr_Rect_6P.cir* for a phase-controlled six-pulse rectifier. Find for both the output voltage and current:

(**A**) dc component

(**B**) rms value

(**C**) rms value of the ac component

(**D**) ripple factor

　　Observe oscillograms of the input voltages and currents.

CA4.7*   Run PSpice program *Rect_Source_Induct.cir* for a six-pulse controlled rectifier supplied from a source with inductance. For reference, run the *Contr_Rect_6P.cir* program for a similar rectifier fed from an ideal source. Compare the output voltage waveforms and their average values.

CA4.8   Develop a computer program for analysis of the phase-controlled six-pulse rectifier. For specified values of the supply voltage, parameters of the RLE load, and firing angle, the program should perform:

(a) determination of feasibility of the firing angle

(b) determination of the conduction mode

(c) calculation of the output voltage waveform

(d) calculation of the output current waveform

It is advised that the waveform calculations are done for one sub-cycle of the input voltage and then extended on the remaining sub-cycles. Storing the waveform data will allow making plots similar to Figures 4.8, 4.11, 4.17, 4.20, and 4.23.

CA4.9*   Run PSpice program *Dual_Conv.cir* for a six-pulse circulating current-conducting dual converter. Observe oscillograms of the output voltages of the constituent rectifiers, output voltage of the converter, differential output voltage, and circulating current.

CA4.10* Run PSpice program *PWM_Rect_CT.cir* for a current-type PWM rectifier. Determine:

(a) total harmonic distortion of the input current

(b) dc-output power

(c) real input power

(d) apparent input power

(e) input power factor

Observe oscillograms of the output voltage and current.

CA4.11* Run PSpice program *PWM_Rect_VT.cir* for a voltage-type PWM rectifier with an input filter. Determine:

(a) total harmonic distortion of the input current

(b) dc-output power

(c) real input power

(d) apparent input power

(e) input power factor

Observe oscillograms of the output voltage and current.

CA4.12   Develop a computer program emulating the SVPWM modulator for the six-pulse current-type PWM rectifier (see Figure 4.48). Based on information about the modulation index, switching frequency, and angle of the input voltage vector, the program should determine the switching pattern for the rectifier, that is, the waveforms of switching signals for the all six switches within a switching cycle.

CA4.13 Develop a computer program that will determine the minimum required voltage and current ratings of power switches in a six-pulse rectifier. The program should ask for the volt-ampere and voltage ratings of the rectifier, and produce the voltage and current ratings of the switches, taking into account specified safety margins.

FURTHER READING

[1] Kolar, J. W., Drofenik U., and Zach F. C., Vienna rectifier II—a novel high-frequency isolated three-phase PWM rectifier system, *IEEE Transactions on Industrial Electronics*, vol. 46, no. 4, pp. 674–691, 1999.

[2] Lee, K., Blasko, V., Jahns, T. M., and Lipo, T. A., Input harmonic estimation and control methods in active rectifiers, *IEEE Transactions on Power Delivery*, vol. 25, no. 2, pp. 953–960, 2010.

[3] Malinowski, M., Kazmierkowski, M, and Trzynadlowski, A. M., A comparative study of control techniques for PWM rectifiers in AC adjustable speed drives, *IEEE Transactions on Power Electronics*, vol. 18, no. 6, pp.1390–1396, 2003.

[4] Rodriguez, J. R., Dixon, J. W., Espinoza, J. R., Ponnt, J., and Lezana, P., PWM regenerative rectifiers: state of the art, *IEEE Transactions on Industrial Electronics*, vol. 52, no. 1, pp. 5–22, 2005.

[5] Singh, B., Singh, B. N., Chandra, A., Al-Haddad, K., Pandey, A, and Kothari, D. P., A review of three-phase improved power quality ac-dc converters, *IEEE Transactions on Industrial Electronics*, vol. 51, no. 3, pp. 641–660, 2004.

5 AC-to-AC Converters

In this chapter: power electronic converters supplied from an ac source and producing voltage of adjustable magnitude and frequency are presented; power circuits, control methods, and characteristics of ac voltage controllers, cycloconverters, and matrix converters are described; selection of power switches for ac-to-ac converters is explained; and typical applications of the ac-to-ac converters are reviewed.

5.1 AC VOLTAGE CONTROLLERS

If a fixed ac voltage needs to be adjusted, an ac voltage controller is used. No frequency control is possible, and the fundamental output frequency equals the input frequency. Single-phase controllers are supplied from a single-phase ac voltage source and have a single-phase output. In three-phase controllers, both the input and the output are of the three-phase type, usually in the three-wire version.

Ac voltage controllers are based on the pairs of antiparallel connected power switches. When SCRs or triacs used, the phase control is employed, and the controller belongs to the class of line-commutated power electronic converters. Fully controlled switches are used in the force-commutated, PWM ac voltage controllers, whose advantages are similar to those of PWM rectifiers.

5.1.1 Phase-Controlled Single-Phase AC Voltage Controller

The circuit diagram of a phase-controlled single-phase ac voltage controller is shown in Figure 5.1. In low-power converters, a triac is used instead of the two antiparallel connected SCRs shown. An RL load is assumed in the subsequent considerations. When one of the SCRs is conducting, the other SCR is reverse biased by the voltage drop across the conducting SCR. Only one path of the current exists, so the input current, i_i, equals the output current, i_o. With either SCR conducting, the input and output terminals of the converter are directly connected, and the output voltage, v_o, equals the input voltage, v_i.

Output Voltage and Current. Example voltage and current waveforms for a controller with an RL load are shown in Figure 5.2. When SCR T1 is forward biased and fired at certain firing angle, α_f, the conducted current initially increases thanks to

Introduction to Modern Power Electronics, Third Edition. Andrzej M. Trzynadlowski.
© 2016 John Wiley & Sons, Inc. Published 2016 by John Wiley & Sons, Inc.
Companion website: www.wiley.com/go/modernpowerelectronics3e

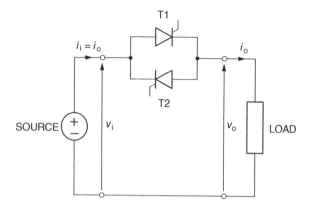

Figure 5.1   Single-phase ac voltage controller.

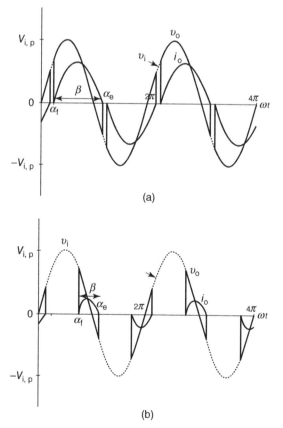

Figure 5.2   Waveforms of output voltage and current of a single-phase ac voltage controller ($\varphi = 30°$) : (a) $\alpha_f = 45°$, (b) $\alpha_f = 135°$.

the positive input voltage, but eventually drops to zero, somewhat later than does the voltage. Now the input voltage is negative, and it is SCR T2 that is forward biased and ready for firing. The firing occurs at $\omega t = \alpha_f + \pi$, where ω denotes the radian frequency, common for the input and the output. The waveform of current through T2 is a mirror image of that through T1, so both the output current and the voltage waveforms have the half-wave symmetry and no dc component. When a triac is used, the firing pulses are applied to its gate with the delay of α_f after each zero-crossing of the input voltage, that is, every half cycle of this voltage.

In Figure 5.2a, the firing angle is 45° and in Figure 5.2b it is 135°. An RL load with the load angle, φ, of 30° is assumed. This is relevant information since, as shown later, the *minimum feasible firing angle equals the load angle*. It can be seen that when the firing angle increases, the conduction angle, β, decreases and the current pulses become smaller and shorter. Consequently, the rms value of the current decreases, as does the rms value of output voltage. Note that the current is always discontinuous and waveforms of both the current and output voltages are strongly distorted in comparison with pure sinusoids, especially with large firing angles. Therefore, ac voltage controllers should be used sparingly in applications requiring high-quality ac currents, such as ac motors.

Accounting for the half-wave symmetry of the output voltage waveform, $v_0(\omega t)$, the rms value of this voltage can be found as

$$
V_0 = \sqrt{\frac{1}{\pi} \int_0^\pi v_0^2 (\omega t) \, d\omega t} = \sqrt{\frac{1}{\pi} \int_0^\pi \left[V_{i,p} \sin(\omega t) \right]^2 d\omega t}
$$

$$
= V_{i,p} \sqrt{\frac{1}{\pi} \left\{ \alpha_e - \alpha_f - \frac{1}{2} [\sin(2\alpha_e) - \sin(2\alpha_f)] \right\}}, \tag{5.1}
$$

where $V_{i,p}$ and V_i denote peak and rms values of the input voltage, respectively, and α_e is the extinction angle. The extinction angle depends on the firing angle, α_f, and the load angle, φ.

The expression for the load current waveform, $i_0(\omega t)$, within the α_f to α_e interval of ωt can be derived in a similar way to that used for a phase-controlled rectifier in the discontinuous conduction mode (see Section 4.2.1). The waveform is given by

$$
i_0(\omega t) = \frac{V_{i,p}}{Z} \left[\sin(\omega t - \varphi) - e^{-\frac{\omega t - \alpha_f}{\tan(\varphi)}} \sin(\alpha_f - \varphi) \right], \tag{5.2}
$$

where Z denotes the load impedance, as in Eq. (4.11). If $\omega t = \alpha_e$, the current has just reached zero, which allows numerical determination of the extinction angle. If $\alpha_f = \varphi$, the right-hand term in brackets disappears, and the current becomes purely sinusoidal, as if the load was directly connected to the supply source. This observation confirms the previously made statement about the minimum feasible firing angle. Expression for the negative pulse of the current differs from Eq. (5.2) by a minus sign only.

Voltage Control. The magnitude control ratio, M, for ac voltage controllers is defined with respect to the rms value, V_o, of output voltage, whose maximum available value equals the rms value, V_i, of input voltage. Since there is no closed-form expression for the extinction angle, α_e, which appears in Eq. (5.1), no closed-form expression for the voltage control characteristic can be derived. This is not a serious problem, since in the practical loads, the load angle, necessary for computation of the extinction angle, is usually unknown anyway. Therefore, only an envelope of control characteristics for load angles in the $0–\pi/2$ range can be determined. If $\varphi = 0$ (purely resistive load), Eq. (5.2) yields

$$i_o(\omega t) = \frac{V_{i,p}}{R} \sin(\omega t) \tag{5.3}$$

and the extinction angle is π rad. Conversely, if $\varphi = \pi/2$ (purely inductive load), then

$$i_o(\omega t) = \frac{V_{i,p}}{\omega L}[\cos(\alpha_f) - \cos(\omega t)] \tag{5.4}$$

and the extinction angle is $2\pi - \alpha_f$, as $\cos(\alpha_f) - \cos(2\pi - \alpha_f) = 0$.

Substituting the obtained values of α_e in Eq. (5.1), the envelope of control characteristics can be expressed as

$$V_{o(\varphi=0)}(\alpha_f) \le V_o(\alpha_f) \le V_{o(\varphi=\pi/2)}(\alpha_f), \tag{5.5}$$

where

$$V_{o(\varphi=0)}(\alpha_f) = V_i\sqrt{\frac{1}{\pi}\left[\pi - \alpha_f + \frac{1}{2}\sin(2\alpha_f)\right]} \tag{5.6}$$

and

$$V_{o(\varphi=\pi/2)}(\alpha_f) = \sqrt{2}V_{o(\varphi=0)}(\alpha_f). \tag{5.7}$$

Graphic representation of Eq. (5.5) is shown in Figure 5.3 in the $M = f(\alpha_f)$ form.

As already mentioned, the load angle is usually unknown. Interestingly, the behavior of an ac voltage controller when the firing angle is less than the load angle depends on the firing technique. As explained in Section 2.3.1, SCRs and triacs can be fired using a short single pulse of the gate current, i_g, or a long multipulse. Operation of the controller with these two firing schemes and $\alpha_f < \varphi$ is illustrated in Figure 5.4. In the case of single gate pulses (Figure 5.4a), only one SCR is fired, by gate pulse i_{g1}, since firing of the other SCR, by gate pulse i_{g2}, is attempted when it is reverse biased by the first, still conducting SCR. As a result, the converter operates faultily as a single-pulse rectifier. If multipulses, whose width should not be less than $\pi/2$ rad, are employed (Figure 5.4b), each SCRs is successfully fired by one of the constituent

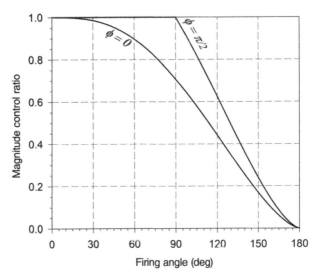

Figure 5.3 Envelope of control characteristics, $V_o = f(\alpha_f)$, of a single-phase ac voltage controller.

pulses of a multipulse immediately after the other SCR has ceased to conduct. Thus, the converter acts as a closed switch between the supply voltage and the load, and the actual firing angle equals the load angle. For this reason, multipulse gate signals are used in all phase-controlled ac voltage controllers, including the three-phase ones covered in the next section.

An ac voltage controller with the multipulse firing of SCRs operates correctly no matter what firing angle, measured from the zero crossing of the input voltage to the beginning of the multipulse, is applied. However, changes of the firing angle in the $0-\varphi$ range do not affect the output voltage and current, both of which stay at their maximum rms values. This dead zone, whose width varies with the load angle, is undesirable, especially in feedforward control schemes. Therefore, an alternate approach to the phase control has been devised, in which the firing angle is defined with respect to the instant when the current, not voltage, crosses zero. This technique requires current sensing, which actually is usually needed anyway for the control or protection purposes.

As seen in Figure 5.5, the angle interval, α'_f, between an end of a current pulse and the next firing instant is given by

$$\alpha'_f = \pi - \beta = \pi - \alpha_e + \alpha_f. \tag{5.8}$$

Angle α'_f, which represents a firing delay measured from the last zero crossing of current, will subsequently be called a *control angle*. Based on Eq. (5.8), the firing angle, α_f, can be expressed in terms of the control angle as

$$\alpha_f = \alpha'_f + \alpha_e - \pi \tag{5.9}$$

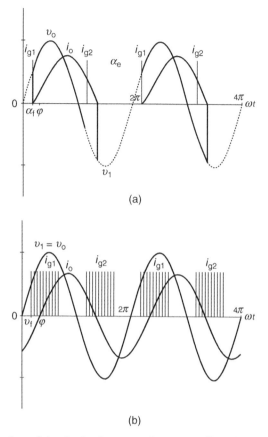

(a)

(b)

Figure 5.4   Operation of the single-phase ac voltage controller with: (a) single-pulse gate signal, (b) multipulse gate signal.

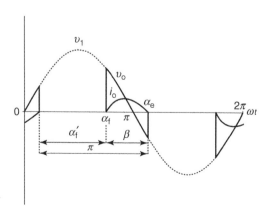

Figure 5.5   Definition of the control angle.

and substituted in Eq. (5.1), to yield

$$V_{\text{o}} = V_{\text{i}}\sqrt{\frac{1}{\pi}\left[\pi - \alpha_{\text{f}}' + \sin\left(\alpha_{\text{f}}'\right)\cos\left(2\alpha_{\text{e}} + \alpha_{\text{f}}'\right)\right]}. \tag{5.10}$$

It can be seen that with $\alpha_{\text{f}}' = 0$, $V_{\text{o}} = V_{\text{i}}$, and with $\alpha_{\text{f}}' = \pi$, $V_{\text{o}} = 0$, independently of the load angle. As before, an envelope of control characteristics corresponding to various load angles can be determined by substituting $\alpha_{\text{e}} = \pi$ in Eq. (5.10) for the $\varphi = 0$ case and $\alpha_{\text{e}} = 2\pi - \alpha_{\text{f}} = 3(\pi - \alpha_{\text{f}}')/2$ for the $\varphi = \pi/2$ case. Then,

$$V_{\text{o}(\varphi=\pi/2)}\left(\alpha_{\text{f}}'\right) \le V_{\text{o}}\left(\alpha_{\text{f}}'\right) \le V_{\text{o}(\varphi=0)}\left(\alpha_{\text{f}}'\right), \tag{5.11}$$

where

$$V_{\text{o}(\varphi=\pi/2)}\left(\alpha_{\text{f}}'\right) = V_{\text{i}}\sqrt{\frac{1}{\pi}\left[\pi - \alpha_{\text{f}}' - \sin\left(\alpha_{\text{f}}'\right)\right]} \tag{5.12}$$

and

$$V_{\text{o}(\varphi=0)}\left(\alpha_{\text{f}}'\right) = V_{\text{i}}\sqrt{\frac{1}{\pi}\left[\pi - \alpha_{\text{f}}' + \frac{1}{2}\sin\left(2\alpha_{\text{f}}'\right)\right]}. \tag{5.13}$$

Relation (5.11) is illustrated in Figure 5.6. No dead zone exists in the control characteristics and, particularly with highly inductive loads, the rms output voltage decreases almost linearly when the control angle increases.

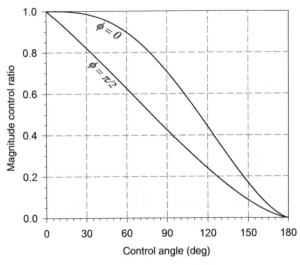

Figure 5.6 Envelope of control characteristics, $V_{\text{o}} = f\left(\alpha_{\text{f}}'\right)$, of a single-phase ac voltage controller.

Power Factor. The input power factor, PF, of a lossless ac voltage controller can be expressed as

$$\text{PF} = \frac{P_o}{S_i} = \frac{RI_o^2}{V_iI_i} = \frac{ZI_o}{V_i}\cos(\varphi). \tag{5.14}$$

When $\alpha_f \leq \varphi$ and, consequently, $I_o = V_i/Z$, the power factor equals that of the load, that is, $\cos(\varphi)$. When the firing angle increases beyond the load angle, the rms output current, I_o, decreases and so does the power factor. An ac voltage controller with a purely resistive load has the highest power factor, while a purely inductive load results in a power factor of zero for all values of α_f.

The relation between the input power factor, PF, and the firing angle, α_f, for an R load ($\varphi = 0$) can easily be derived substituting R for Z and V_o/R for I_o in Eq. (5.14). Then, PF $= V_o/V_i = M$ (magnitude control ratio). Thus, based on Eq. (5.6),

$$\text{PF}_{(\varphi=0)}(\alpha_f) = \sqrt{\frac{1}{\pi}\left[\pi - \alpha_f + \frac{1}{2}\sin(2\alpha_f)\right]}. \tag{5.15}$$

It can be seen that, similarly to phase-controlled rectifiers, phase-controlled ac voltage controllers are characterized by decreasing quality of the input current with the increase of firing angle. The input power factor decreases and the total harmonic distortion of the input current increases.

5.1.2 Phase-Controlled Three-Phase AC Voltage Controllers

Several configurations of three-phase ac voltage controllers are possible. Here, only the most common, *fully controlled* topology with a wye-connected load will be described in some detail. Detailed analysis of that converter requires lengthy considerations, omitted in this third edition. Interested readers are directed to the second edition of this text or to book [6]. Other types of three-phase controllers will briefly be described, with the stress on only their unique features.

Fully Controlled Three-Phase AC Voltage Controller. Operation of the fully controlled three-phase ac voltage controller, shown in Figure 5.7, is more complicated than that of the single-phase converter. Note, for instance, that to get the controller started, two triacs must be fired simultaneously to provide the path for the current necessary to maintain their on-state.

Operation with a single triac conducting a current is impossible. Therefore, the controller can operate with two or three triacs conducting or none at all. If no triac is conducting, the load is cut off from the supply and all currents and output voltages are zero. The two- and three-triac conduction cases are illustrated in Figure 5.8. A balanced load is assumed and, for better visualization, the triacs are represented by generic switches. It can be seen in Figure 5.8a that with two triacs, TA and TB, conducting, the line-to-line voltage, v_{AB}, is equally split between the load impedances in phases A and B. Consequently, $v_a = v_{AB}/2$, $v_b = -v_{AB}/2 = v_{BA}/2$, and $v_c = 0$. When all three triacs are conducting, as in Figure 5.8b, the output voltages, v_a, v_b, and v_c, equal their input counterparts, v_A, v_B, and v_C.

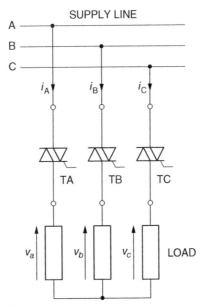

Figure 5.7   Fully controlled three-phase ac voltage controller.

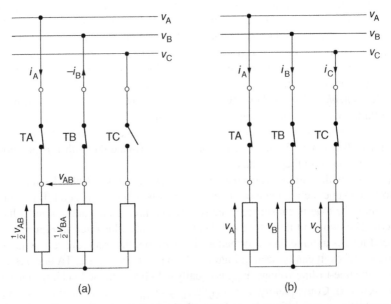

Figure 5.8   Voltage and current distribution in a fully controlled three-phase ac voltage controller: (a) two triacs conducting, (b) three triacs conducting.

Switching variables, a, b, and c, can be introduced for triacs TA, TB, and TC, respectively and are equal to 1 when a given triac is conducting and is equal 0 otherwise. It can be shown that the output voltages of the controller are given by

$$\begin{bmatrix} v_a \\ v_b \\ v_c \end{bmatrix} = \frac{1}{2} \begin{bmatrix} a & -b & -c \\ -a & b & -c \\ -a & -b & c \end{bmatrix} \begin{bmatrix} v_A \\ v_B \\ v_C \end{bmatrix}. \tag{5.16}$$

Indeed, if no triac is conducting, then $a = b = c = 0$ and $v_a = v_b = v_c = 0$. When, for example, triacs TA and TB are conducting, then $a = b = 1$, $c = 0$ and $v_a = (v_A - v_B)/2 = v_{AB}/2$, $v_b = (-v_A + v_B)/2 = v_{BA}/2$, and $v_c = (-v_A - v_B)/2 = v_c/2$. The last equation is only true when $v_c = 0$. Finally, if all triacs are conducting, then $a = b = c = 1$ and $v_a = (v_A - v_B - v_C)/2 = [2v_A - (v_A + v_B + v_C)]/2 = (2v_A - 0)/2 = v_A$, and, similarly, $v_b = v_B$, and $v_c = v_C$.

For simplicity, a purely resistive load is assumed, and only the waveform of phase-A output voltage, v_a, will be considered. It can be seen from Eq. (5.16) that, in general, the waveform in question is composed of segments of the v_A, $v_{AB}/2$, and $v_{AC}/2$ waveforms. For example, in Figure 5.9a, the firing angle, α_f, is zero and all three triacs conduct all the time connecting the source with the load. With the firing angle of $30°$, in Figure 5.9b, the triacs cease to conduct when their currents reach zero. As a result, either two or three triacs are simultaneously conducting and the output voltage, v_a, sequentially equals v_A, $v_{AB}/2$, v_A, $v_{AC}/2$, and zero.

Operation of the controller with large firing angles is illustrated in Figure 5.10. An example waveform of the output voltage at $\alpha_f = 75°$ is shown in Figure 5.10a and that at $\alpha_f = 120°$ in Figure 5.10b.

For completeness, formulas for the rms output voltage, V_o, of the fully controlled ac voltage controller with purely resistive and purely inductive loads are provided below without derivation.

Resistive load:

$$V_o = V_i \sqrt{\frac{1}{\pi} \left[\pi - \frac{3}{2}\alpha_f + \frac{3}{4}\sin(2\alpha_f) \right]} \tag{5.17}$$

for $0 \le \alpha_f < 60°$,

$$V_o = V_i \sqrt{\frac{1}{\pi} \left[\frac{\pi}{2} + \frac{3\sqrt{3}}{4}\sin\left(2\alpha_f + \frac{\pi}{6}\right) \right]} \tag{5.18}$$

for $60° \le \alpha_f < 90°$, and

$$V_o = V_i \sqrt{\frac{1}{\pi} \left[\frac{5}{4}\pi - \frac{3}{2}\alpha_f + \frac{3}{4}\sin\left(2\alpha_f + \frac{\pi}{3}\right) \right]} \tag{5.19}$$

for $90° \le \alpha_f < 150°$.

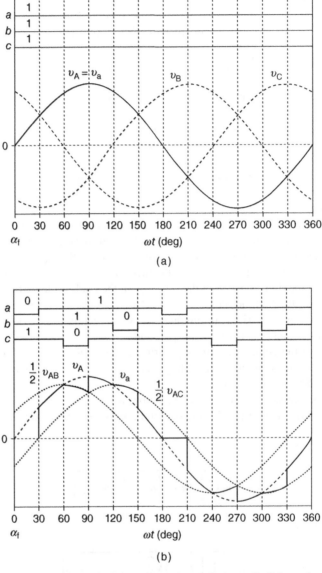

Figure 5.9 Example output voltage waveforms in a fully controlled three-phase ac voltage controller with R load: (a) $\alpha_f = 0°$, (b) $\alpha_f = 30°$.

Inductive load:

$$V_o = V_i \sqrt{\frac{1}{\pi} \left[\frac{5}{2}\pi - 3\alpha_f + \frac{3}{2} \sin(2\alpha_f) \right]} \qquad (5.20)$$

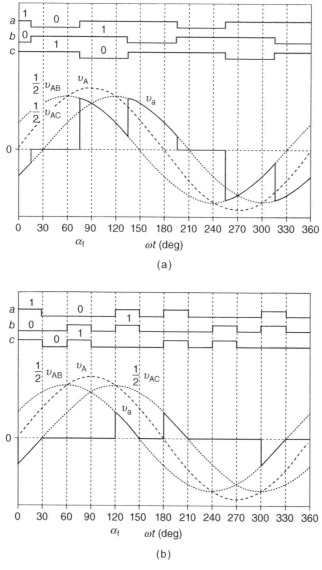

Figure 5.10   Output voltage waveforms in a fully controlled three-phase ac voltage controller with R load: (a) $\alpha_f = 75°$, (b) $\alpha_f = 120°$.

for $90° \le \alpha_f < 120°$, and

$$V_o = V_i \sqrt{\frac{1}{\pi} \left[\frac{5}{2}\pi - 3\alpha_f + \frac{3}{2}\sin(2\alpha_f+) \right]} \tag{5.21}$$

for $120° \le \alpha_f < 150°$.

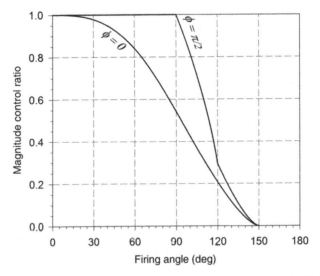

Figure 5.11   Envelope of control characteristics, $V_o = f(\alpha_f)$, of a fully controlled three-phase ac voltage controller.

The envelope of control characteristics given by Eqs. (5.17) through (5.21) is shown in Figure 5.11. In order to avoid the dead zone, as in the single-phase controller, the control angle measured from the last zero crossing of current in the given phase of a controller can be employed in place of the firing angle. Relations between the control angle, α_f', and firing angle, α_f, are given by

$$\alpha_f' = \begin{cases} \alpha_f & \text{for} & 0° \leq \alpha_f < 60° \\ 60° & \text{for} & 60° \leq \alpha_f < 90° \\ \alpha_f - 30° & \text{for} & 90° \leq \alpha_f < 150° \end{cases} \qquad (5.22)$$

for a resistive load and

$$\alpha_f' = 2(\alpha_f - 90°) \qquad (5.23)$$

for an inductive load. The envelope of voltage control characteristics when the control angle is used is shown in Figure 5.12. With a purely resistive load, the control characteristic is discontinuous at $\alpha = 60°$. Therefore this control method should only be employed when the load contains substantial inductance. The sensitivity of the rms output voltage to changes of the control angle is higher than that in a single-phase ac voltage controller, since the upper limit of control angle is 120° instead of 180°.

When a three-phase load of the fully controlled ac voltage controller is connected in delta, instead of wye, the control characteristics remain unchanged although different waveforms of the output voltage are generated. The input power factor of all ac voltage controllers is given by the general equation (5.14), where V_i represents

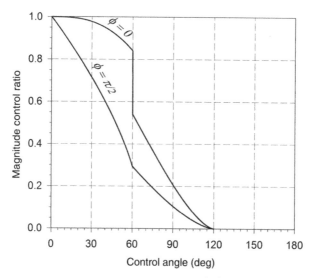

Figure 5.12   Envelope of control characteristics, $V_o = f\left(\alpha_f'\right)$, of a fully controlled three-phase ac voltage controller.

the rms line-to-neutral voltage, V_{LN} for wye-connected loads, and the rms line-to-line voltage, V_{LL}, for delta-connected loads. With a purely resistive load, the power factor, PF, equals the magnitude control ratio, M.

Other Types of Three-Phase AC Voltage Controllers. Some other types of phase-controlled three-phase ac voltage controllers are shown in Figures 5.13 and 5.14. In high-power three-phase ac voltage controllers, SCRs must be used because of the low

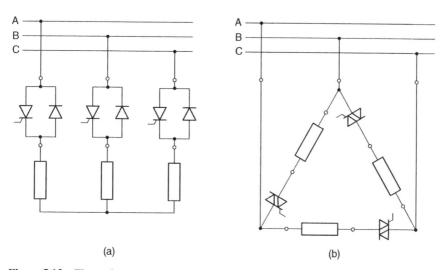

(a)   (b)

Figure 5.13   Three-phase ac voltage controllers connected before the load: (a) half controlled, (b) delta connected.

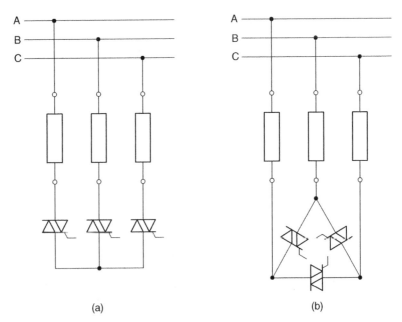

Figure 5.14   Three-phase ac voltage controllers connected after the load: (a) wye connected, (b) delta connected.

current ratings of the available triacs. In such controllers, the *half-controlled* topology, depicted in Figure 5.13a, can be employed. Instead of another SCR, each of the three SCRs has a diode connected in antiparallel. When a current flows through a diode in one phase, it is indirectly controlled by the conducting SCRs in another phase. Interestingly, the maximum firing angle in the half-controlled ac voltage controller is 210°.

Three triacs and phase loads can be connected in delta, as shown in Figure 5.13b. Characteristics of each phase controller are the same as those of the single-phase controller, but triple harmonics are absent in the line currents drawn from the ac supply line. Controllers shown in Figure 5.14 are connected *after* the load, which must have all the six terminals accessible. On the other hand, the controllers have only three terminals. The triacs in the wye-connected controller in Figure 5.14a have a common node, which simplifies the control circuitry. A half-controlled configuration as in Figure 5.13a is also possible. Such configuration is, however, infeasible in the delta-connected controller in Figure 5.14b, since the diodes connecting the load terminals at the controller side would make the control impossible. The advantage of the delta topology consists of low ampere ratings of the triacs, which carry lower currents than those in the load.

The circuit diagram of a four-wire ac voltage controller is shown in Figure 5.15. The load is connected in wye, and the neutrals of the load and supply line are connected. As such, the converter operates as three independent single-phase ac voltage controllers, whose operating properties have already been described in Section 5.1.1.

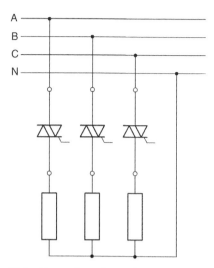

Figure 5.15   Three-phase four-wire ac voltage controller.

In practice, four-wire controllers are used only when there is a need to control individual phase loads independently, and when the load has the neutral point accessible. Otherwise, even with unbalanced loads, the arrangement in Figure 5.13b is preferable.

5.1.3   PWM AC Voltage Controllers

The circuit diagram of a single-phase PWM ac voltage controller, also known as an *ac chopper*, is shown in Figure 5.16. An input filter such as that used for PWM rectifiers is required to attenuate the high-frequency harmonic currents drawn from the source, usually the grid. The inductance provided by the source is often sufficient, so that only the capacitor is installed. Ac choppers have been developed to improve the input power factor, control characteristics, and quality of the output current.

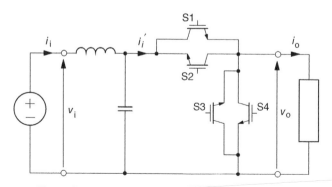

Figure 5.16   Single-phase ac chopper with input filter.

Fully controlled main switches S1 and S2 connected in antiparallel allow controlling rms values of the output voltage and current. The freewheeling switches S3 and S4 provide a path for this current when both S1 and S2 are off. The switches are turned on and off many times within a cycle of the input voltage. Assigning switching variables x_1 through x_4 to switches S1 through S4, respectively, the duty ratios for switches S1 and S2 in the nth switching interval, $d_{1,n}$ and $d_{2,n}$, are given by

$$d_{1,n} = \left\{ \begin{array}{ll} F(m, \alpha_n) & \text{for } 0 < \alpha_n \leq \pi \\ 0 & \text{otherwise} \end{array} \right\}$$

$$d_{2,n} = \left\{ \begin{array}{ll} F(m, \alpha_n) & \text{for } \pi < \alpha_n \leq 2\pi \\ 0 & \text{otherwise} \end{array} \right\} \quad (5.24)$$

and for switches S3 and S4

$$x_3 = \bar{x}_1$$
$$x_4 = \bar{x}_2. \quad (5.25)$$

Symbol $F(m, \alpha_n)$ in Eq. (5.24) denotes the value of the so-called *modulating function* $F(m, \omega t)$ at $\omega t = \alpha_n$, where m is the modulation index and α_n is the central angle of the nth switching interval. The modulating function determines the magnitude of output voltage and improves the quality of output current. It can be as simple as $F(\omega t) = m$, or more complex, for example, $F(\omega t) = m |\sin(\omega t)|$.

In practice, it is convenient to control the switches by continuously applying the on and off switching signals to all switches, so that the *intended* duty ratios and switching variables are

$$d^*_{1,n} = d^*_{2,n} = F(m, \alpha_n)$$
$$x^*_3 = x^*_4 = \bar{x}^*_1. \quad (5.26)$$

Because of the bias of individual switches during the periods of positive and negative input voltage, the *actual* duty ratios and switching variables are such as described by Eqs. (5.24) and (5.25). Unsuccessful attempts at firing reverse-biased switches do not affect the operation of the controller. In contrast with controlled rectifiers, this control scheme does not require synchronization of firing signals with the input voltage.

Operation of an ac chopper with 12 switching intervals per cycle, modulation index of 0.8, and the simplest modulating function $F(m, \omega t) = m$, is illustrated in Figure 5.17. Clearly, the output current is of higher quality than that in phase-controlled ac voltage controllers. The output voltage waveform is identical with that in Figure 1.21b for the generic PWM controller (see also Figure 1.23 for a typical harmonic spectrum). Pulsed current, i_i, is drawn from the filter. The fundamental, i_{i1}, of this current lags the input voltage, v_i, by angle θ_1 that is practically equal to the load angle φ. Thanks to the input capacitor, the power factor at the terminals of the filter is higher than $\cos(\theta_1)$. The input current, i_i, delivered by the supply system is sinusoidal, with only minor ripple.

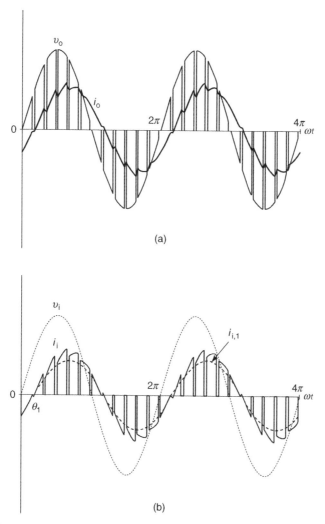

Figure 5.17 Waveforms of voltages and currents in a single-phase ac chopper: (a) output voltage and current, (b) input voltage and current (after the input filter) and the fundamental output current.

It can be proved that the magnitude control ratio, M, equals $\sqrt{m}$, that is,

$$V_o = \sqrt{m}\,V_i \qquad (5.27)$$

which results in the control characteristic shown in Figure 5.18. However, the ratio, $V_{o,1}/V_{i,1}$, of fundamentals of the output and input voltages equals the modulation index m.

The simple modulation technique described is most popular, but not exclusive. Several advanced PWM techniques with variable duty ratios have been developed, for

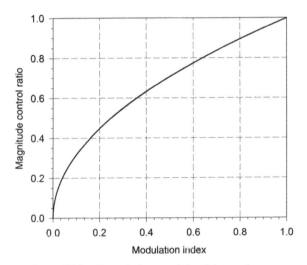

Figure 5.18 Control characteristic of the ac chopper.

the purpose of reducing the filter size and improving the quality of input and output currents. Three-phase PWM ac voltage controllers in the wye and delta configurations are shown in Figures 5.19 and 5.20, respectively.

The pulse width modulation mode cannot extend over the whole theoretical 0–1 range of the modulation index, because the minimum on- and off-times of switches

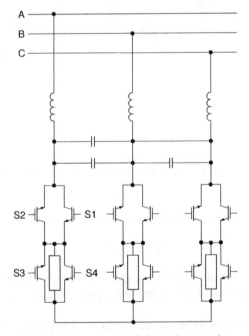

Figure 5.19 Wye-connected three-phase ac chopper.

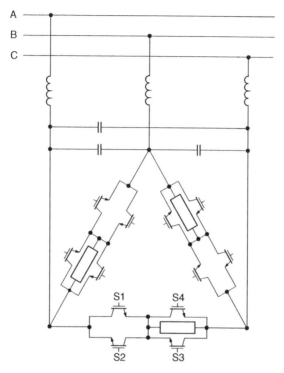

Figure 5.20   Delta-connected three-phase ac chopper.

would have to approach zero. Therefore, the actual modulation is performed with the values of the modulation index between $m_{min} > 0$ and $m_{max} < 1$. If the desired modulation index falls below m_{min}, it is rounded up to m_{min}, limiting the minimum available value of the magnitude control ratio to $\sqrt{m_{min}}$. When the desired value of m is greater than m_{max}, the modulation index is set to unity, causing the controller to operate as a closed ac switch.

5.2   CYCLOCONVERTERS

Cycloconverters are ac-to-ac power converters in which the output frequency is a fraction of the input frequency. A single-phase two-pulse generic cycloconverter in the mode of operation resulting in an integer ratio of the input frequency to output frequency has been shown in Examples 1.1 and 1.2. Practical cycloconverters have a three-phase output although a hypothetical single-phase six-pulse cycloconverter will be used for the illustration of the operating principles. It is simply the same circulating current-conducting dual converter described in Section 4.2.2, that is, an inverse-parallel connection of two phase-controlled rectifiers. Analogously, three-phase cycloconverters are made of three pairs of such rectifiers. The rectifiers are of the three-pulse or, more often, the six-pulse type.

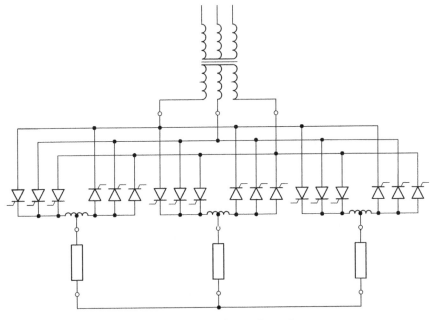

Figure 5.21 Three-phase three-pulse cycloconverter.

Most cycloconverters are high-power converters supplied from dedicated transformers. The transformer establishes the maximum available output voltage of the converter and, for certain types of cycloconverters, provides isolation of the supply sources for individual phases.

The circuit diagram of a three-phase three-pulse cycloconverter is shown in Figure 5.21, and two types of three-phase six-pulse cycloconverters are depicted in Figures 5.22 and 5.23. Phase loads of the cycloconverter in Figure 5.22 are isolated from each other, and the converter is supplied from a single three-phase source. When the phase loads are connected, as in Figure 5.23, the interphase isolation is provided by a supply transformer with three secondary windings. All three cycloconverters

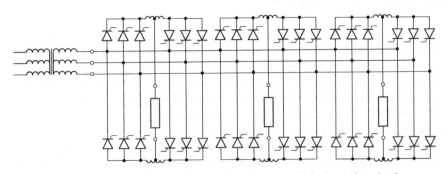

Figure 5.22 Three-phase six-pulse cycloconverter with isolated phase loads.

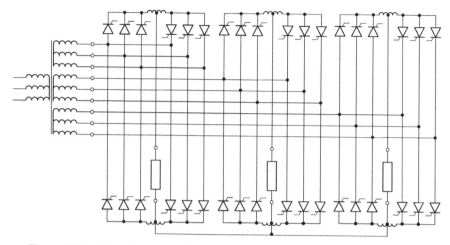

Figure 5.23   Three-phase six-pulse cycloconverter with interconnected phase loads.

shown are of the circulating current-conducting type, necessitating the use of separating inductors.

As the dc output voltage of a phase-controlled rectifier depends on the firing angle, α_f, it is possible to vary that angle in such a way that the magnitude of that voltage changes along a positive half wave of a given reference sinusoid, from zero to a peak value and back to zero. Now, the inverse-parallel complementary rectifier can be so controlled that its dc-output voltage changes along a negative half wave, from zero to the negative peak value and back to zero. The radian frequency, ω_o, of the reference sinusoids is the output frequency of the converter and, inherently, it cannot be higher than the input frequency, ω. To maintain reasonable quality of the output current, it is recommended that the ω/ω_o ratio be at least 3.

As the dc-output voltage of a dual converter can be expressed as

$$V_{o,dc} = V_{o,dc(max)} \cos(\alpha_f) \qquad (5.28)$$

(see Eq. 4.42), where $V_{o,dc(max)}$ denotes the maximum available value of this voltage, corresponding to the firing angle of zero. If the dual converter operates as a single-phase cycloconverter, the dc-output voltage should vary according to the equation

$$V_{o,dc}(t) = V_{o,1,p} \sin(\omega_o t), \qquad (5.29)$$

where $V_{o,1,p}$ is the desired peak value of the fundamental output voltage of the cycloconverter.

Based on Eqs. (5.28) and (5.29), the required variations of the firing angle can be expressed as

$$\cos[\alpha_f(\omega_o t)] = \frac{V_{o,1,p}}{V_{o,dc(max)}}, \qquad (5.30)$$

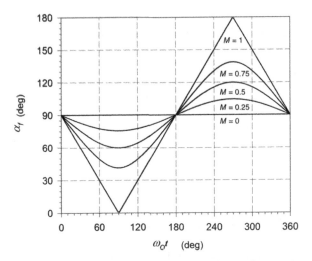

Figure 5.24   Waveforms of the firing angle in a cycloconverter.

where the $V_{o,1,p}/V_{o,dc(max)}$ ratio represents the magnitude control ratio, M. Thus, the firing angle as a function of $\omega_o t$ is given by

$$\alpha_f(\omega_o t) = \cos^{-1}[M \sin(\omega_o t)] \tag{5.31}$$

and illustrated in Figure 5.24 for various values of M.

Example waveforms of the output voltage, v_{o1}, of one of the constituent rectifier pairs of a six-pulse cycloconverter are shown in Figure 5.25 for the frequency ratio, ω_o/ω, of 0.2. The magnitude control ratio is 1 in Figure 5.25a and 0.5 in Figure 5.25b. It is assumed that the cycloconverter is of the circulating current-conducting type. Otherwise, the changeover from one constituent rectifier to another, when the output current crosses zero, would have to be accompanied by a short delay period to allow for a complete turnoff of the SCRs in the outgoing rectifier. Circulating current-conducting cycloconverters require separating inductors, but they make the output voltage waveforms smoother and closer to ideal sinusoids. Consequently, the output currents are also of high quality.

Neglecting the voltage drops across the separating inductors, the rms value, $V_{o,LN,1}$, of fundamental line-to-neutral output voltage in the three-pulse cycloconverter in Figure 5.21 is given by

$$V_{o,LN,1} = \frac{3\sqrt{3}}{2\pi}MV_{i,LN} \approx 0.827\,MV_{i,LN} \tag{5.32}$$

while the same value in the six-pulse cycloconverters in Figures 5.22 and 5.23 is

$$V_{o,LN,1} = \frac{3}{\pi}MV_{i,LL} \approx 0.955\,MV_{i,LL}, \tag{5.33}$$

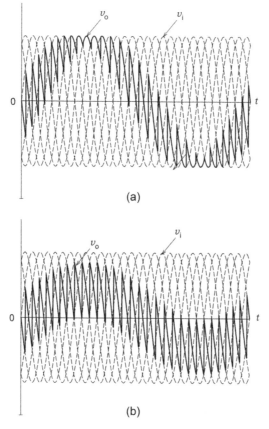

Figure 5.25 Output voltage waveforms in a six-pulse cycloconverter ($\omega_o/\omega = 0.2$): (a) $M = 1$, (b) $M = 0.5$.

where $V_{i,LN}$ and $V_{i,LL}$ denote rms line-to-neutral and line-to-line input voltages, respectively.

Cycloconverters are perfectly suited for high-power low-frequency applications, because the quality of the output current increases with the decrease in output frequency. In contrast with the commonly used rectifier–inverter cascade, they provide direct ac-to-ac power conversion. They are inherently capable of four-quadrant operation and are quite reliable, since even with a failure of one SCR a cycloconverter may still be functioning, albeit with somewhat increased distortion of the output voltage.

Besides the reduced range of the output frequency, disadvantages of cycloconverters include the high number of SCRs and the associated complexity of control. Also, as in phase-controlled rectifiers, the input power factor is generally low, particularly when a cycloconverter operates with a low magnitude control ratio. Correction of the power factor is inconvenient and expensive, because a high-power cycloconverter requires a proportionally sized large input filter.

5.3   MATRIX CONVERTERS

Although first proposed four decades ago, matrix converters still struggle for a significant share of the power electronic market. The concept of matrix converter represents an extension of the principle of generic converter, introduced in Chapter 1. Each terminal of a K-phase ac voltage source is connected with each terminal of an L-phase load by a *bidirectional* fully controlled switch. Thus, the voltage at any input terminal can be made to appear at any output terminal or terminals, while the current in any phase of the load can be drawn from any phase or phases of the supply source.

5.3.1   Classic Matrix Converters

Circuit diagram of the classic three-phase to three-phase (3Φ-3Φ) matrix converter is shown in Figure 5.26. The converter consists of nine switches, denoted by S_{Aa} through S_{Cc}. An input filter is employed to screen the supply system from harmonic currents generated in the converter, and the load is assumed to contain inductance that ensures continuity of the output currents. Clearly, at any time, one and only one switch in each row must be closed. Otherwise, either the supply lines would be shorted or one or more of the output currents would be interrupted. It is easy to demonstrate that out of the theoretically possible 512 (2^9) states of the converter, only 27 are permitted.

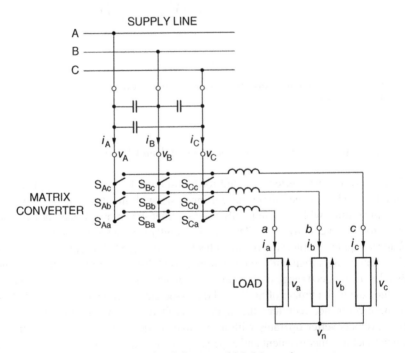

Figure 5.26   Functional diagram of 3Φ-3Φ matrix converter.

The voltages, v_a, v_b, and v_c, at the output terminals are given by

$$\begin{bmatrix} v_a \\ v_b \\ v_c \end{bmatrix} = \begin{bmatrix} x_{Aa} & x_{Ba} & x_{Ca} \\ x_{Ab} & x_{Bb} & x_{Cb} \\ x_{Ac} & x_{Bc} & x_{Cc} \end{bmatrix} \begin{bmatrix} v_A \\ v_B \\ v_C \end{bmatrix}, \tag{5.34}$$

where x_{Aa} through x_{Cc} denote switching variables of switches S_{Aa} through S_{Cc}, and v_A, v_B, and v_C are the voltages at the input terminals. It can be proved that with a balanced, linear, wye-connected load, the voltage, v_n, at the common point of the load is given by

$$v_n = \frac{1}{3}(v_a + v_b + v_c). \tag{5.35}$$

Consequently, the line-to-neutral output voltages, v_{an}, v_{bn}, and v_{cn}, can be expressed as

$$\begin{bmatrix} v_{an} \\ v_{bn} \\ v_{cn} \end{bmatrix} = \frac{1}{3} \begin{bmatrix} 2 & -1 & -1 \\ -1 & 2 & -1 \\ -1 & -1 & 2 \end{bmatrix} \begin{bmatrix} v_a \\ v_b \\ v_c \end{bmatrix}. \tag{5.36}$$

The input currents, i_A, i_B, and i_C, are related to the output currents, i_a, i_b, and i_c, as

$$\begin{bmatrix} i_A \\ i_B \\ i_C \end{bmatrix} = \begin{bmatrix} x_{Aa} & x_{Ab} & x_{Ac} \\ x_{Ba} & x_{Bb} & x_{Bc} \\ x_{Ca} & x_{Cb} & x_{Cc} \end{bmatrix} \begin{bmatrix} i_a \\ i_b \\ i_c \end{bmatrix}. \tag{5.37}$$

It can be seen that the matrix of switching variables in Eq. (5.37) is a transpose of the respective matrix in Eq. (5.34). Based on those equations, fundamentals of both the output voltages and the input currents can be controlled. It is done by employing appropriately timed sequences of either the individual switching variables or, as in the case of space-vector-controlled PWM rectifiers, whole states of the converter (see Sections 4.3.2–4.3.4). As a result of such control, the fundamental output voltages should be balanced and having the desired frequency and amplitude, while the fundamental input currents should also be balanced and having the required phase shift, usually zero, with respect to the corresponding input voltages.

There exist a number of control methods for matrix converters, some quite complex, and a variety of solutions, including closed-loop control of voltages and currents, have been proposed. The feasibility of independent control of output voltages and input currents has been demonstrated by several researchers. It has been found for the 3Φ-3Φ matrix converter that the maximum obtainable voltage gain, defined here as the ratio of the peak fundamental output voltage to peak input voltage, is $\sqrt{3}/2 \approx 0.866$. It is so because the set of six input line-to-line voltages (see Figure 4.8), which constitutes "raw material" for the output voltages, dips every 60°

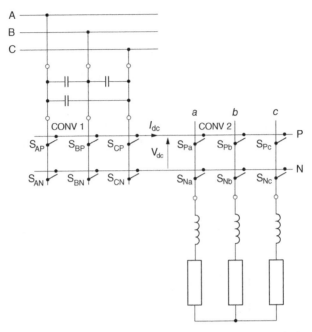

Figure 5.27   Arrangement of 3Φ-1Φ and 1Φ-3Φ matrix converters equivalent to a 3Φ-3Φ matrix converter.

to $\sqrt{3}/2$ of peak value of the input voltages. This is considered a disadvantage, since other means of direct or indirect (with an intermediate dc stage) ac-to-ac power conversion yield voltage gains close to unity. On the other hand, a credit must be given to matrix converter for the inherent capability of the bidirectional power flow, lacking in certain popular indirect ac-to-ac conversion schemes.

Space vector PWM technique, already outlined in Section 4.3, is most often used in control of matrix converters. It is based on representation of the three-phase input currents and output voltages as space vectors. An arrangement of two six-switch matrix converters equivalent to the nine-switch converter under consideration is shown in Figure 5.27. The six-switch converters are placed between the source and the load, and connected via lines P ("positive") and N ("negative"), which constitute a "virtual dc link." Converter CONV 1 is of the 3Φ-1Φ type and converter CONV 2 of the 1Φ-3Φ type. Topology of CONV 1 is identical with that of a six-pulse current-type PWM rectifier, thus it can be so operated that the voltage between lines P and N has a specified dc component, V_{dc}, while input currents to the converter are sinusoidal and satisfying the unity power factor requirement. Conversely, CONV 2 can be operated as a generic PWM three-phase inverter, producing balanced ac currents in the load. Subsequently, converters CONV 1 and CONV 2 are called a "virtual rectifier" and a "virtual inverter," respectively.

One and only one switch in each *row* of the virtual rectifier and one and only one switch in each *column* of the virtual inverter must be conducting at any time. Thus, six different states are permitted for the rectifier and eight for the inverter, and the total number of allowable states of the combined 12-switch converter is 48. However, it can be shown that connections between the source and load terminals corresponding to each of these states can be realized using the original, nine-switch, 27-state 3Φ-3Φ matrix converter in Figure 5.26. Consider, for example, a situation when switches S_{AP}, S_{BN}, S_{Pa}, S_{Pb}, and S_{Nc} are closed. It can be seen that phase A of the source is connected, through line P, to phases a and b of the load, and phase B, through line N, to phase c. The same connections can directly be made in the original matrix converter by closing switches S_{Aa}, S_{Ab}, and S_{Bc}.

The same example can be analyzed in a more general way. In the subsequent considerations each of the allowed states of the matrix converter is assigned a three-letter code, specifying which input terminals, A, B, or C, are connected to the output terminals a, b, and c, respectively. The state described above can thus be designated AAB. Note that the virtual rectifier connects terminal A to line P and terminal B to line N, while terminal C remains unconnected. Therefore, the rectifier state can be called PN0. The virtual inverter connects terminals a and b to line P and terminal c to line N, so its state can be named PPN. Examining the rectifier state indicates an association of P with A and N with B. Thus, the inverter state PPN can be translated into state AAB of the whole matrix converter, and to connect the input terminals AAB with the output terminals abc, switches S_{Aa}, S_{Ab}, and S_{Bc} must be activated. Realization of the described state is illustrated in Figure 5.28.

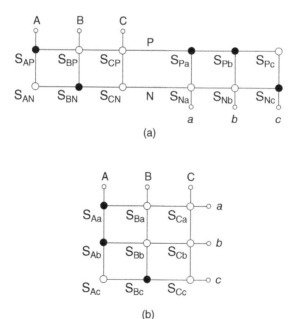

Figure 5.28   State AAB as realized by activation of switches in: (a) virtual rectifier and inverter, (b) matrix converter.

If the P and N lines are shorted by the switches in either of the input phases, A, B, or C, then $V_{dc} = 0$. These zero states of the virtual rectifier are subsequently called Z00, 0Z0, and 00Z. State 0Z0, for example, represents the situation when only switches S_{BP} and S_{BN} are closed. Respectively, zero states of the virtual inverter cause its output voltages to be zero, which happens when all the three output terminals, a, b, and c, are clamped to either the P or N line. Consequently, these zero states of the virtual inverter are denoted by PPP and NNN, the latter state, for instance, meaning that those are switches S_{Na}, S_{Nb}, and S_{Nc} that are closed.

In the example state PN0, input currents i_A, i_B, and i_C are equal to I_{dc}, $-I_{dc}$, and 0, respectively, and the corresponding space vector of input currents is

$$\vec{I}_{PN0} = \frac{3}{2}I_{dc} + j\frac{\sqrt{3}}{2}I_{dc}, \tag{5.38}$$

where I_{dc} denotes the dc current in the virtual dc link (see Figure 5.27). Current vectors for the remaining five states of the virtual rectifier can be determined similarly. All six stationary current vectors and the revolving reference vector, $\vec{i}^*$, are shown in Figure 5.29 which, unsurprisingly, is practically identical with Figure 4.46 for the real current-type PWM rectifier. As in that rectifier, in order to satisfy the unity power factor condition in the matrix converter under consideration, the reference current vector should be aligned with the space vector of input voltages.

Considering the example state PPN of the virtual inverter, it can be seen that $v_a = v_b = v_P$ and $v_c = v_N$, where v_P and v_N denote potentials of lines P and N, respectively. According to Eq. (5.36), $v_{an} = v_{bn} = (v_P - v_N)/3 = V_{dc}/3$ and

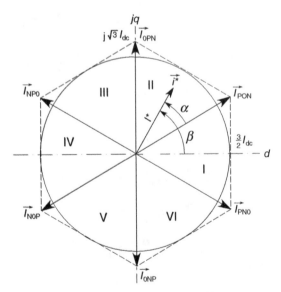

Figure 5.29   Reference current vector in the vector space of input currents of the virtual rectifier.

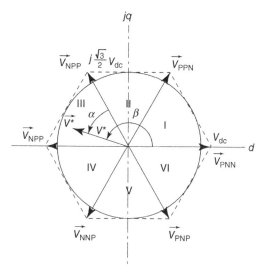

Figure 5.30 Reference voltage vector in the vector space of line-to-neutral output voltages of the virtual inverter.

$v_{cn} = -2V_{dc}/3$. Employing Eq. (4.74), the space vector, $\vec{V}_{PPN}$, of the line-to-neutral output voltages is found to be

$$\vec{V}_{PPN} = \frac{1}{2}V_{dc} + j\frac{\sqrt{3}}{2}V_{dc}. \tag{5.39}$$

The remaining five active voltage vectors are shown in Figure 5.30 with the reference voltage vector $\vec{v}^*$, which represents the desired line-to-line voltages of the matrix converter. Figure 5.30 is almost identical with Figure 4.56 for the voltage-type rectifier, which, as subsequently explained in Section 7.1.3, has the same topology as the so-called voltage-source inverter.

Eqs. (4.77), (4.78), and (4.80) for duty ratios of states of a converter controlled by space vector PWM can be adapted to the virtual rectifier and inverter. The question is how to specify the modulation indexes for these converters. Let the rectifier modulation index, m_{rec}, be defined as

$$m_{rec} \equiv \frac{I_{i,p}}{I_{dc}}, \tag{5.40}$$

where $I_{i,p}$ denotes the peak value of the input currents (a balanced load is assumed). Neglecting losses in the virtual rectifier, the input power, P_i, equals the power, P_{DC}, in the virtual dc link, that is,

$$\sqrt{3}V_iI_i\cos(\varphi_i) = V_{dc}I_{dc}, \tag{5.41}$$

where V_i and I_i denote rms values of the input line-to-line voltage and current, respectively, and $\cos(\varphi_i)$ is the power factor of the rectifier. Consequently,

$$V_{dc} = \frac{\sqrt{3}}{2} m_{rec} V_{i,p} \cos(\varphi_i), \qquad (5.42)$$

where $V_{i,p}$ denotes the peak value of the input line-to-line voltages.

Now, let the inverter modulation index, m_{inv}, be defined as

$$m_{inv} \equiv \frac{V_{o,p}}{V_{dc}}, \qquad (5.43)$$

where $V_{o,p}$ denotes the peak value of fundamental line-to-line output voltage. Substituting Eq. (5.42) in Eq. (5.43) gives

$$m_{inv} = \frac{2V_{o,p}}{\sqrt{3}V_{i,p}m_{rec}\cos(\varphi_i)} = \frac{2m}{\sqrt{3}m_{rec}\cos(\varphi_i)}, \qquad (5.44)$$

where

$$m \equiv \frac{V_o}{V_i} \qquad (5.45)$$

is the modulation index of the whole matrix converter. Equation (5.44) can be rearranged to

$$m = \frac{\sqrt{3}}{2} m_{rec} m_{inv} \cos(\varphi_i). \qquad (5.46)$$

Eq. (4.46) confirms the already mentioned $\sqrt{3}/2$ limit on the voltage gain of the matrix converter. To maximize the control range of the matrix converter, m_{rec} is set to 1. Note that the magnitude of vector of the input currents depends on the load of the converter, and it is only the phase shift of that vector with respect to the vector of input voltages that is adjusted for the unity power factor.

Let the states that produce vectors framing the reference current vector in Figure 5.29 be denoted by X_I and Y_I, and those that produce vectors framing the reference voltage vector in Figure 5.30 by X_V and Y_V. The corresponding zero states are designated by Z_X and Z_V. One of the commonly used switching patterns, in which the switching cycle is divided into nine sub-cycles, is presented in Table 5.1. The sequential number of a sub-cycle is denoted by n and its duration, relative to the switching period T_{sw}, by t_n. Duty ratios of individual states are designated by d with an appropriate subscript, for example, d_{XV} for State X_V.

Typical waveforms of the output voltage and current of a matrix converter supplying an RL load are illustrated in Figure 5.31. For reference, waveforms of the six

Table 5.1   Switching Pattern for 3Φ-3Φ Matrix Converter with Space Vector PWM

Switching Subcycle	Rectifier State	Inverter State	t_n/T_{sw}
1	X_I	X_V	$d_{XI}d_{XV}/2$
2	X_I	Y_V	$d_{XI}d_{YV}/2$
3	Y_I	Y_V	$d_{YI}d_{YV}/2$
4	Y_I	X_V	$d_{YI}d_{XV}/2$
5	Z_I	Z_V	$1 - (d_{XI} + d_{YI})$ $(d_{XV} + d_{YV})$
6	Y_I	X_V	$d_{YI}d_{XV}/2$
7	Y_I	Y_V	$d_{YI}d_{YV}/2$
8	X_I	Y_V	$d_{XI}d_{YV}/2$
9	X_I	X_V	$d_{XI}d_{XV}/2$

line-to-line input voltages are also shown. The switching frequency is 48 times higher than the input frequency, ω. The output frequency, ω_o, in Figure 5.31a is 2.8 times higher than the input frequency and the modulation index, m, is 0.75. The respective parameters in Figure 5.31b are 0.7 and 0.35. The input current waveforms are similar to those in the PWM rectifiers (see Figure 4.60).

Practical implementation of matrix converters was hampered by the unavailability of bidirectional fully controlled semiconductor power switches. Recently, several small companies started producing such switches, whose scheme is based on two IGBTs, in the common-emitter connection, and two diodes, as shown in Figure 5.32a. Another solution involves a combination of one IGBT and four diodes, depicted in Figure 5.32b and already mentioned in Section 4.3.5. In each case, the circuit diagram of a classic 3Φ-3Φ matrix converter contains a relatively large number of semiconductor devices (18 IGBTs and 18 diodes, or 9 IGBTs and 36 diodes) and the associated drivers. Therefore, integration of the devices in a single case per switch represents valuable progress. Circuit diagram of the classic 3Φ-3Φ matrix converter based on the two-transistor two-diode switches is shown in Figure 5.33.

Typically, thanks to the high switching frequency, the filter capacitors and inductors are small. The output inductors shown here as distinct devices may in practice represent an internal inductance of the load, such as the stator inductance of an induction motor. Because of the absence of large energy storage components, operation of matrix converters is sensitive to fluctuations of the grid voltage. Therefore, in practical applications, the converter must be equipped with fast-acting protection systems.

5.3.2 Sparse Matrix Converters

The classic matrix converter is often termed a *direct matric converter* because, as seen in Figure 5.33, only one set of switches separates the output from the input (the generic converter in Chapter 1 has the same property). Consequently, a single-stage power conversion is realized.

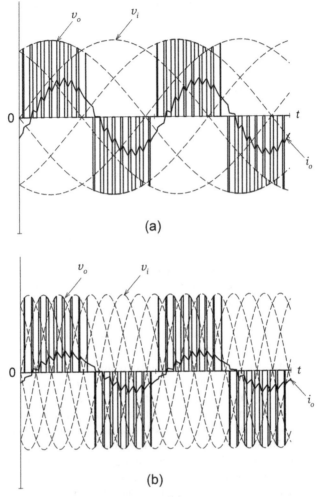

Figure 5.31 Output voltage and current waveforms in a 3Φ-3Φ matrix converter: (a) $m = 0.75$, $\omega_o/\omega = 2.8$, (b) $m = 0.35$, $\omega_o/\omega = 0.7$.

Increasing the number of conversion stages while maintaining the functionality results in an *indirect matrix converter* shown in Figure 5.34. The bidirectional switches on the left-hand side constitute a rectifier stage and on the right-hand side circuit is an inverter stage. Buses connecting these stages constitute a dc link marked in Figure 5.34 by "P" and "N". Note that the topology of indirect matrix converter corresponds to the arrangement of switches shown in Figure 5.27 and equivalent to the classic matrix converter. The inverter circuit comprising six transistors and six diodes is widely available as an integrated power module (see Section 2.6). As a result, manufacturing of an indirect matrix converter requires less effort than that of the direct one.

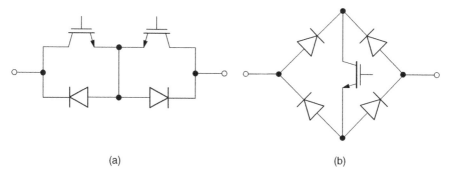

(a)   (b)

Figure 5.32   Bidirectional semiconductor power switches: (a) two IGBTs and two diodes, (b) one IGBT and four diodes.

The rectifier stage allows the indirect matrix converter to operate with both positive and negative dc link. However, most of the practical applications of matrix converters, such as control of ac machines, require only the positive polarity. In that case, the rectifier stage can be simplified as shown in Figure 5.35. With the unchanged number of diodes, the resultant *sparse matrix converter* contains 15 transistors instead of the 18 ones of the classic matrix converter.

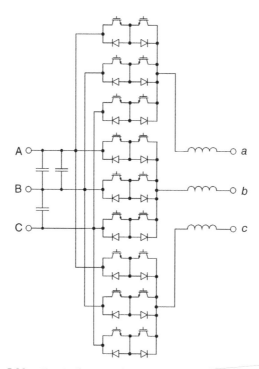

Figure 5.33   Circuit diagram of the classic 3Φ-3Φ matrix converter.

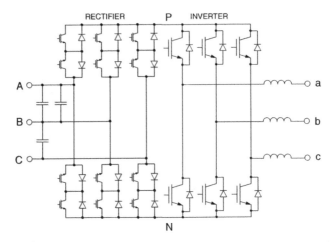

Figure 5.34   Indirect matrix converter.

Certain ac loads experience only a positive power flow. For example, an ac motor driving a mixer cannot operate in the generator mode because the mixer is incapable of reciprocating the driving action. In that case, the ultra-sparse matrix converter shown in Figure 5.36 can be employed. The number of transistors is now reduced to 9. Due to control constraints, the load current cannot be shifted by more than 30° from the fundamental output voltage. Several other topologies of indirect matrix converters, developed for various cost, reliability, and efficiency reasons, have also been proposed.

5.3.3   Z-Source Matrix Converters

Sparse matrix converters are cheaper and more reliable than the classic ones, but the disadvantage of low voltage gain remains. One of the promising means

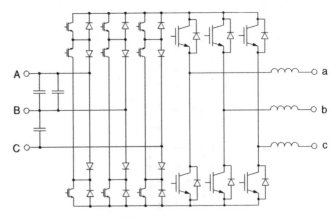

Figure 5.35   Sparse matrix converter.

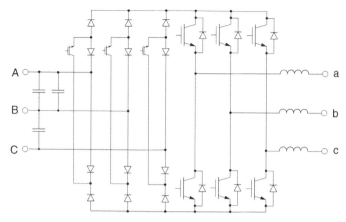

Figure 5.36   Ultra-sparse matrix converter.

for boosting the output voltage of an indirect matrix converter is the so-called
Z-*source* (impedance source) recently gaining significant popularity in a variety of
applications.

The Z-source used as a dc link between a dc source and a dc-input PWM power
electronic converter is shown in Figure 5.37. It consists of two identical inductors,
L_1 and L_2, and two cross-connected identical capacitors, C_1 and C_2. It is assumed
that the source, such as a diode rectifier, can only conduct current in one direction,
hence the diode D in the scheme. The supplied converter, for example, an inverter,
is represented by a current source, due to the fact that the load inductance prevents
the load current from significant changes within the short time intervals considered.
The shunt switch S symbolizes the so-called shoot-through state of the converter, in
which the dc buses of the converter are shorted by turning on switches connecting
these buses.

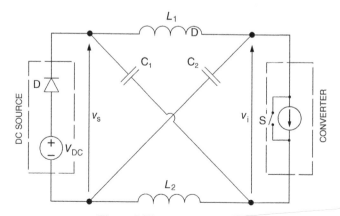

Figure 5.37   Z-source dc link.

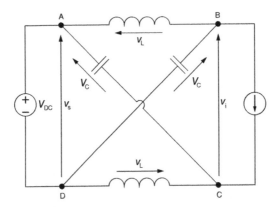

Figure 5.38 Z-source during the nonshoot-through converter state.

Most common in practice is a capacitive dc link, in which a high-capacitance electrolytic capacitor is connected between the dc buses. It stabilizes the bus voltage and absorbs sporadic currents flowing back from the converter. The shoot-through is hazardous to the capacitor and the switches of the converter, which can be damaged by the high short-circuit current. Therefore, in converters with a capacitive dc link the shoot-through must be avoided. The Z-source dc link is not only impervious to the shoot-through thanks to the inductors in the path of the current, but the shoot-through is systematically employed to boost the voltage supplying the converter.

A nonshoot-through state of the converter results in the circuit shown in Figure 5.38. The Kirchhoff voltage law equations are

$$\begin{cases} v_s - V_C - v_L = 0 & \text{(loop DACD)} \\ V_C - v_i - v_L = 0 & \text{(loop DCBD)} \end{cases} \tag{5.47}$$

and when solved they yield $v_L = V_{DC} - V_C$ and $v_i = 2V_C - V_{DC}$. The last equation indicates that $2V_C > V_{DC}$, which will be taken into account when analyzing the shoot-through state.

The shoot-through situation is illustrated in Figure 5.39. Identical voltages, V_C, across the capacitors are practically unchanged within the involved time intervals. The input voltage, v_i, to the converter is zero and, as shown later, the voltage, v_s, across the source terminals is higher than the source EMF, V_{DC}. Consequently, diode D is reverse-biased and the dc source is cut off from the link. Employing the Kirchhoff voltage law, the following set of equations is obtained:

$$\begin{cases} v_L + V_C - v_s = 0 & \text{(loop DCAD)} \\ v_L - V_C = 0 & \text{(loop DCBD)} \end{cases} \tag{5.48}$$

When solved, the equations yield $v_L = V_C$ and $v_s = 2V_C$, which, as stated before, is higher than V_{DC}, proving the reverse bias of diode D.

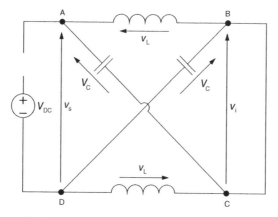

Figure 5.39 Z-source during the shoot-through converter state.

As a dc voltage, the average voltage, V_L, across inductor L equals zero. Thus,

$$V_L = \frac{t_{NS}(V_{DC} - V_C) + t_{ST}V_C}{t_{NS} + t_{ST}} = 0, \tag{5.49}$$

from which

$$V_C = \frac{t_{NS}}{t_{NS} - t_{ST}} V_{DC}, \tag{5.50}$$

where t_{NS} and t_{ST} denote durations of the adjacent nonshoot-through and shoot-through states, respectively. Consequently, in the nonshoot-through state,

$$v_i = 2V_C - V_{DC} = 2\frac{t_{NS}}{t_{NS} - t_{ST}} V_{DC} - V_{DC} = \frac{t_{NS} + t_{ST}}{t_{NS} - t_{ST}} V_{DC}. \tag{5.51}$$

Equation (5.51) can be re-written as

$$v_i = BV_{DC}, \tag{5.52}$$

where

$$B = \frac{t_{NS} + t_{ST}}{t_{NS} - t_{ST}} \tag{5.53}$$

denotes the so-called *boost factor*, that is, the voltage gain of the Z-source dc link. Clearly, if $0 < t_{ST} < t_{NS}$, then $B > 1$, which means that using that link in an indirect matrix converter allows boosting the converter gain to unity and beyond. As an example, the Z-source sparse matrix converter is shown in Figure 5.40. Other variants of Z-source matrix converters have also been proposed. See [1] for more details.

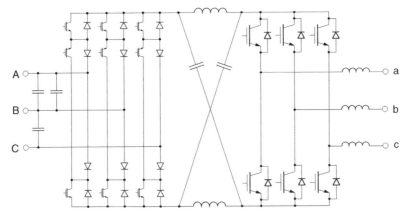

Figure 5.40 Z-source sparse matrix converter.

It must be pointed out that Eq. (5.53) is imprecise, because the circuit resistances, especially those of the inductors, have been neglected in its derivation. In practice, the boost factors are limited to single digits. Finally, note that the Z-source introduces energy storage components to the topology of an indirect matrix converter. The resultant desensitizing of the converter to grid voltage disturbances is an additional advantage, but it could be disputable if the Z-source matrix converters truly deserve the "matrix" designation.

5.4 DEVICE SELECTION FOR AC-TO-AC CONVERTERS

The highest instantaneous voltage appearing in an ac voltage controller is the peak value of the input voltage. Thus, the selected switches must satisfy condition (4.107) in Section 4.4. For three-phase ac voltage controllers, the peak value, $V_{LL,p}$, of the line-to-line supply voltage is substituted for $V_{i,p}$.

The rules of device selection with respect to the rated current depend on whether triacs or separate, anti-parallel connected devices are used in the designed controller. For triacs, the rated current is meant as the maximum allowable *rms value* of a sustained sinusoidal ac current, while in the other semiconductor power switches the current ratings involve the allowable *average* current. Therefore, for triac-based ac voltage controllers, the condition to be satisfied is

$$I_{rat} \geq (1 + s_I)I_{o(rat)}, \tag{5.54}$$

where s_I denotes the safety margin for the current and $I_{o(rat)}$ is the rated rms output current of the controller.

The maximum average current in a single power switch of an ac voltage controller occurs at the full-wave conduction, that is, when the output current is sinusoidal and has the maximum rms value representing the rated current, $I_{o(rat)}$, of the controller.

The corresponding maximum average value, $I_{ave(max)}$, of the current conducted by a single switch is given by

$$I_{ave(max)} = \frac{1}{2\pi} \int_0^\pi \sqrt{2} I_{o(rat)} \sin(\omega t)\,d\omega t = \frac{\sqrt{2}}{\pi} I_{o(rat)} \approx 0.45 I_{o(rat)}. \quad (5.55)$$

Consequently, the condition for the rated current of switches is

$$I_{rat} \geq \frac{\sqrt{2}}{\pi}(1 + s_I) I_{o(rat)}. \quad (5.56)$$

If the switches are connected in delta after the load, as in Figure 5.14b, the rms current conducted by them is lowered from the rms load current by the factor of $\sqrt{3}$. Therefore, the rated current of these switches can be reduced by $\sqrt{3}$ in comparison with that given by condition (5.56).

Usually, the freewheeling switches in PWM ac voltage controllers carry much lower currents than do the main switches. If the application of a given controller is restricted to a passive load, the current rating of the freewheeling switches can be half of that of the main switches. However, in general, it is prudent to use identical semiconductor devices for both the main and freewheeling switches. This avoids the danger of overloading the latter switches when the controller supplies an active, power-generating load, for example, an ac motor that can operate as a generator.

For ac choppers, the minimum on-time and off-time, $t_{ON(min)}$ and $t_{OFF(min)}$, of the main switches are

$$t_{ON(min)} = m_{min} T_{sw} \quad (5.57)$$

and

$$t_{OFF(min)} = (1 - m_{max}) T_{sw}, \quad (5.58)$$

where m_{min} and m_{max} denote the minimum and maximum values of the modulation index. The same rules apply to switches in matrix converters. Respective times for the freewheeling switches in ac choppers are obtained from Eqs. (5.57) and (5.58) by interchanging the "ON" and "OFF" subscripts.

Cycloconverters are composed of controlled rectifiers. Therefore, the selection rules for power switches are the same as for rectifiers (see Section 4.4).

5.5   COMMON APPLICATIONS OF AC-TO-AC CONVERTERS

Ac voltage controllers can be used for the adjustment of the rms value of ac voltages and currents, and as *static ac switches*. In the latter case, a controller connects the supply and the load for a continuous operation, passing an uncontrolled, sinusoidal current. Depending on the transferred power, ac switches are based on SCRs or triacs. In the on-state, an ac switch operates with the minimum firing angle, while the turn-off is accomplished by removing the gate signals.

Ac voltage controllers are primarily used in the lighting and heating control, and in the so-called *soft starters* for induction motors. A soft starter, connected between a supply line and a motor, reduces the starting current that otherwise would be excessive. Ac switches are employed as transformer tap changers, allowing voltage control in power systems. Another typical application of ac switches involves speed control of high-inertia induction-motor drives, such as large centrifuges. The driving motor is switched on when the speed of the centrifuge drops below the minimum allowable level, and it is turned off when the speed reaches the maximum allowable value. With the de-energized motor, the centrifuge is driven by its momentum, and the high mechanical inertia and low friction result in a low deceleration rate. In this simple way, the average speed is maintained at a constant level, and the instantaneous speed does not stray from the assigned tolerance band. A similar mode of control can be used for temperature control, with an ac switch intermittently turning on and off an electric heater or air conditioner.

Cycloconverters are mostly used for the control of large ac motors, usually the synchronous ones, in the low-speed, high-power drive systems, such as those of rolling mills in metal industries or kilns in cement factories. Speed of ac motors is proportional to the supply frequency, and the low output frequency of a cycloconverter translates into a low speed of the motor, allowing direct, gearless drive of the load. Cycloconverter-fed ac drives are capable of rapid acceleration and deceleration, and the regenerative operation mode (with the reversed power flow) is available over the complete speed range. Recently, cycloconverters started to appear in certain renewable energy systems.

Matrix converters have been increasingly employed in applications in which wide-range frequency/magnitude control and bidirectional power flow are required. The low voltage gain is one of the problems preventing matrix converters from extensive use in the control of ac motors. Voltage ratings of mass-produced motors correspond to the common voltage levels in the existing electrical infrastructure. The torque of an induction motor is proportional to the supply voltage squared, so the 15% reduction in the voltage translates into a 28% reduction in the developed torque. Still, in certain "custom" applications, such as dedicated high-performance ac drives, matrix converters seem to have found their market niche.

Another growing application of matrix converters is in low- and medium-power wind-turbine systems, in which the converter serves an interface between a variable-speed ac generator and a grid. The issue of low voltage gain is not very important here, as the whole system is designed for the optimum match between individual components. Use of matrix converters in electric and hybrid vehicles is under serious consideration.

SUMMARY

Ac voltage controllers allow adjustment of the rms values of output voltage and current by means of phase control or pulse width modulation. In phase-controlled controllers, SCRs or triacs are used, while fully controlled switches are employed in PWM controllers (ac choppers). Several types of three-phase ac voltage controllers are available. Ac voltage controllers are often operated as static ac switches.

Three-phase cycloconverters, which are composed of three dual converters, perform direct ac-to-ac conversion with adjustable output frequency and voltage. However, the output frequency is limited to a fraction of the input frequency. Cycloconverters have several advantages, but the device count is high and the control system is complex. Typically, cycloconverters are high-power converters, supplied from dedicated transformers.

Matrix converters, based on bidirectional fully controlled switches that provide direct connections between each input terminal and each output terminal, allow independent sinusoidal modulation of output voltages and input current. Many solutions to reduce the number of switches and increase the inherently low voltage gain have been recently proposed.

Typical applications of ac voltage controllers and static ac switches include the lighting and heating control, transformer tap changing, soft starters for induction motors, and integral-cycle control of ac motors. Cycloconverters are mostly used in high-power, low-speed, ac adjustable-speed drives. The still relatively rare applications of matrix converters include adjustable-speed industrial ac drives, renewable energy systems, and electric vehicles.

EXAMPLES

Example 5.1   A single-phase ac voltage controller is used in a movie theater to control the incandescent lighting. Neglecting the temperature-related changes of resistance of the lamps, find the reduction in power consumed by the lighting when the firing angle is 60°. What firing angle would result in a 50% reduction of that power?

Solution: The incandescent lighting constitutes a resistive load. Consequently, the ratio of rms values of the output and input voltages, that is, the magnitude control ratio, can be found from Eq. (5.6) as

$$\frac{V_o}{V_i} = \sqrt{\frac{1}{\pi}\left[\pi - \frac{1}{3}\pi + \frac{1}{2}\sin(120°)\right]} = 0.897.$$

With a constant resistance, the power is proportional to the squared rms value of voltage, so the resultant power constitutes $0.897^2 = 0.805$ of the full power. Thus, the consumed power is reduced by 19.5%.

The magnitude control ratio for the 50% power reduction is $\sqrt{0.5} = 0.707$. From the control characteristic in Figure 5.3 for $\varphi = 0$, the corresponding value of the firing angle is determined as 90°. Notice that at this firing angle, exactly half of the sinusoidal waveform of the input voltage is passed to the load.

Example 5.2 A single-phase ac chopper is supplied from 120-V line and operates with a modulation index of 0.75. The load impedance is $2\,\Omega$. Determine the rms value of output voltage and, neglecting the ripple of output current, the rms value of this current.

Solution: The rms value, V_o, of the chopped output voltage is given by Eq. (5.27) as $V_o = \sqrt{0.75} \times 120 = 103.9\,\text{V}$. However, the rms value, $V_{o,1}$, of the fundamental of this voltage is $V_{o,1} = 0.75 \times 120 = 90\,\text{V}$ and the output current, which is assumed to be purely sinusoidal, has the rms value, I_o, of

$$I_o = I_{o,1} = \frac{V_{o,1}}{Z} = \frac{90}{2} = 45\,\text{A}.$$

Example 5.3 A three-phase, six-pulse cycloconverter is supplied from a 460-V, 60-Hz line. The load is connected in wye, with the resistance of $0.9\,\Omega/\text{ph}$ and inductance of 15 mH/ph. The cycloconverter operates with the output frequency of 10 Hz and magnitude control ratio of 0.7. Neglecting the ripple, find the rms value of output currents.

Solution: Since the load is connected in wye, the rms value, I_o, of output currents is given by

$$I_o = \frac{V_{o,LN,1}}{Z},$$

where $V_{o,LN,1}$ denotes the rms value of the fundamental line-to-neutral output voltage and Z is the load impedance, which at the frequency of 10 Hz is $Z = \sqrt{0.9^2 + (2\pi \times 10 \times 0.015)^2} = 1.303\,\Omega/\text{ph}$. Using Eq. (5.33), $V_{o,LN,1}$ can be determined as $V_{o,LN,1} = \frac{3}{\pi} \times 0.7 \times 460 = 439.3\,\text{V/ph}$, which yields

$$I_o = \frac{439.3}{1.303} = 337.1\,\text{A/ph}.$$

Example 5.4 A 3Φ-3Φ matrix converter, supplied from a 460-V line, operates with the switching frequency of 5 kHz, the modulation index of 0.5, and a unity power factor. Consider a switching cycle in which the phase angles of the reference input current and output voltage are $135°$ and $100°$, respectively, and determine the switching pattern in that cycle.

Solution: Assuming $m_{rec} = 1$, the modulation index, m_{inv}, of the virtual inverter can be found from Eq. (5.46) as $m_{inv} = \dfrac{m}{\frac{\sqrt{3}}{2} m_{rec} \cos(\varphi_i)} = \dfrac{2}{\sqrt{3}} 0.5 = 0.577$. The phase angles, β_I and β_V, of the reference vectors indicate that the current reference vector,

$\vec{\imath}^*$, is in sector III and the voltage reference vector, $\vec{v}^*$, is in sector II (see Figures 5.29 and 5.30). Thus, $X_{\mathrm{I}} = 0\mathrm{PN}$, $Y_{\mathrm{I}} = \mathrm{NP0}$, $X_{\mathrm{V}} = \mathrm{PPN}$, and $Y_{\mathrm{V}} = \mathrm{NPN}$. The in-sector (local) angles, α_{I} and α_{V}, are $15°$ and $10°$, respectively. Based on Eqs. (4.77), (4.79), and (4.80),

$$d_{\mathrm{XI}} = 1 \times \sin(60° - 15°) = 0.707$$
$$d_{\mathrm{YI}} = 1 \times \sin(15°) = 0.259$$
$$d_{\mathrm{ZI}} = 1 - d_{\mathrm{XI}} - d_{\mathrm{YI}} = 0.034$$

and

$$d_{\mathrm{XV}} = 0.577 \times \sin(60° - 10°) = 0.442$$
$$d_{\mathrm{YV}} = 0.577 \times \sin(10°) = 0.100$$
$$d_{\mathrm{ZV}} = 1 - d_{\mathrm{XV}} - d_{\mathrm{YV}} = 0.458.$$

The switching period, T_{sw}, which is a reciprocal of the switching frequency, f_{sw}, equals 200 μs. States of the virtual rectifier and inverter and their relative durations are listed in Table 5.2, which is an example-specific version of Table 5.1. Based on the information in Table 5.2, activation of individual switches of the matrix converter can be determined as shown in Table 5.3 and Figure 5.41. Notice that within the whole switching cycle considered, switches S_{Ab} and S_{Cb} are off and switch S_{Bb} is on. The zero state BBB in the fifth switching subcycle is preferable over the other two zero states, AAA or CCC, as the preceding and following state is BBA, so that only two switches, S_{Ac} and S_{Bc}, must change their states.

The input and output frequencies of the matrix converter have not been specified. This information is not needed, as in the space vector PWM methods the required voltage and current vectors are independently synthesized in each switching cycle. Thus, the speed of a vector can only be assessed by comparing its position in two consecutive switching cycles. Here, for example, if the vector of output voltages has

Table 5.2 Switching Pattern for the Example Matrix Converter

Switching Subcycle	Rectifier State	Inverter State	t_n/T_{sw}
1	0PN	PPN	0.156
2	0PN	NPN	0.035
3	NP0	NPN	0.009
4	NP0	PPN	0.057
5	Z00 or 0Z0	PPP	0.486
6	NP0	PPN	0.057
7	NP0	NPN	0.009
8	0PN	NPN	0.035
9	0PN	PPN	0.156

Table 5.3 Activation of Switches in the Example Matrix Converter

Switching Subcycle	State of Matrix Converter	Activated Switches	Duration (μs)
1	BBC	S_{Ba}, S_{Bb}, S_{Cc}	31.2
2	CBC	S_{Ca}, S_{Bb}, S_{Cc}	7.0
3	ABA	S_{Aa}, S_{Bb}, S_{Ac}	1.8
4	BBA	S_{Ba}, S_{Bb}, S_{Ac}	11.4
5	BBB	S_{Ba}, S_{Bb}, S_{Bc}	97.2
6	BBA	S_{Ba}, S_{Bb}, S_{Ac}	11.4
7	ABA	S_{Aa}, S_{Bb}, S_{Ac}	1.8
8	CBC	S_{Ca}, S_{Bb}, S_{Cc}	7.0
9	BBC	S_{Ba}, S_{Bb}, S_{Cc}	31.2

moved by 3.6°, then the output frequency is 50 Hz, because 5000 switching cycles would produce 18,000°, that is, 50 cycles of output voltage.

Example 5.5 A three-phase, delta-connected, PWM ac voltage controller is rated at 10 kVA and 460 V. The supply frequency is 60 Hz, and the modulation index is limited to the 0.05–0.95 range. Determine the minimum voltage and current ratings of switches, assuming safety margins of 0.4 and 0.2 for the rated voltage and current, respectively. What is the maximum allowable switching frequency for the minimum on-time and off-time to be at least 20 μs long?

Solution: The rated voltage, V_{rat}, of switches of the controller is given by Eq. (4.107) as $V_{rat} \geq (1 + 0.4) \times \sqrt{2} \times 460 = 911$ V. The rated rms value, $I_{o,rat}$, of output

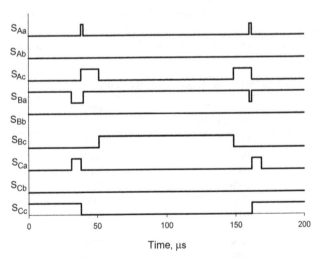

Figure 5.41 Switching signals for individual switches in the example matrix converter.

currents can be calculated as $I_{o,rat} = \frac{10 \times 10^3}{3 \times 460} = 7.25$ A and, consequently, the rated current, I_{rat}, of the switches, as required by condition (5.56), must be $I_{rat} \geq \frac{\sqrt{2}}{\pi} (1 + 0.2) \times 7.25 = 3.92$ A. From Eq. (5.57), $T_{sw} \geq \frac{t_{ON(min)}}{m_{min}} = \frac{20}{0.05} = 400$ μs, that is, $f_{sw} \leq \frac{1}{400} = 0.0025$ MHz = 2.5 kHz. Equation (5.58) would yield the same result, since in the considered case, $m_{max} = 1 - m_{min}$.

PROBLEMS

P5.1 For a single-phase ac voltage controller, sketch the waveforms of output voltage and current for the following values of the firing and extinction angles:

(a) 60° and 225°

(b) 90° and 210°

(c) 120° and 205°

Assume that the load angle is less than 60°.

P5.2 A single-phase ac voltage controller supplies a load consisting of a 2.5-Ω resistance and a 6-mH inductance. The controller is fed from a 120-V, 60-Hz line, and multipulse gate signals are used to activate controller's triacs. Estimate the rms output voltage for firing angles of 45°, 90°, and 120°.

P5.3 A single-phase ac voltage controller supplies a resistive load. Use the characteristic in Figure 5.3 to determine the firing angle resulting in a 35% reduction of output voltage.

P5.4 A three-phase ac controller in Figure 5.7 is supplied from a 230-V line, and the load is purely resistive. Find the rms output voltage for firing angles of 30°, 90°, and 120°.

P5.5 A single-phase ac chopper supplied from a 120-V line operates with the modulation index of 0.75. Find the rms output voltage.

P5.6 A single-phase ac chopper is supplied from a 120-V, 60-Hz line and operates with a constant modulation index of 0.6 and 20 switching intervals per cycle. Find the fundamental output voltage and durations of the pulses and notches of the output voltage.

P5.7 A three-phase, delta-connected ac chopper supplied from a 460-V line operates with the modulation index of 0.6. Find the rms output voltage and fundamental output voltage.

P5.8 A three-phase six-pulse cycloconverter is supplied from a 460-V line and operates with the magnitude control ratio of 0.75. Find the rms values of the line-to-neutral and line-to-line output voltages of the cycloconverter.

P5.9 List all allowable states of the 3Φ-3Φ matrix converter using the three-letter designation, for example, AAB.

P5.10 A 3Φ-3Φ matrix converter is supplied from a 230-V line with the positive phase sequence. The 5-Ω resistive load is connected in wye. Neglecting the voltage drop across the filter inductances, the phase-A input voltage can be expressed as $v_A = V_{i,p} \cos(\omega t)$. Find the instantaneous line-to-line output voltages and input currents at $\omega t = 135°$ if switches S_{Ab}, S_{Ac}, and S_{Ca} are closed.

P5.11 Repeat Example 5.4 with the following data: $f_{sw} = 4$ kHz, $m = 0.45$, $\beta_I = 235°$, $\beta_V = 15°$. What is the output line-to-neutral voltage if the matrix converter in question is supplied from a 230-V line?

P5.12 Determine the minimum required voltage and current ratings of SCRs in a 20-kVA, 460-V three-phase fully controlled ac voltage controller. Assume safety margins of 0.4 for the voltage rating and 0.25 for the current rating.

P5.13 An ac voltage controller with the same ratings as those in Problem 5.12 has the SCRs connected in delta after the load. Assuming the same safety margins as in P5.12, determine the minimum required voltage and current ratings of the SCRs.

P5.14 A 10-kVA, 120-V, 60-Hz single-phase ac chopper operates with 50 switching intervals per cycle. The minimum and maximum values of the modulation index for the PWM operation of the chopper are 0.05 and 0.95, respectively. Assuming the same safety margins as in Problem 5.12, determine the minimum required voltage and current ratings of the main and freewheeling switches. Find the minimum on-time and off-time of the switches.

COMPUTER ASSIGNMENTS

CA5.1* Run PSpice program *AC_Volt_Contr_1ph.cir* for a single-phase ac voltage controller. For the firing angles of 40° and 80°, find for the output voltage and current:
   (a) rms value
   (b) rms value of the fundamental
   (c) total harmonic distortion

CA5.2* Run PSpice program *AC_Volt_Contr_1ph.cir* for a single-phase ac voltage controller. For firing angles of 40° and 110°, obtain the harmonic spectra of the output voltage and current. For the 10 most prominent harmonics, determine the harmonic number and amplitude as a fraction of amplitude of the fundamental.

CA5.3* Run PSpice program *AC_Volt_Contr_3ph.cir* for a three-phase ac voltage controller. For firing angles of 45° and 80°, find for the output voltage and current:
   (a) rms value

(b) rms value of the fundamental

(c) total harmonic distortion

CA5.4* Run PSpice program *AC_Volt_Contr_3ph.cir* for a three-phase ac voltage controller. For firing angles of 40° and 100°, obtain the harmonic spectra of the output voltage and current. For the 10 most prominent harmonics, determine the harmonic number and amplitude as a fraction of amplitude of the fundamental.

CA5.5* Run PSpice program *AC_Chopp.cir* for a single-phase ac chopper with an input filter. For $N = 20$ switching intervals per cycle and the magnitude control ratio, M, of 0.75, find for the output voltage and current:

(a) rms value

(b) rms value of the fundamental

(c) total harmonic distortion

Repeat the assignment for $N = 20$ and $M = 0.6$.

CA5.6* Run PSpice program *AC_Chopp.cir* for a single-phase ac chopper with an input filter. For $N = 10$ switching intervals per cycle and the magnitude control ratio, M, of 0.4, find:

(a) total harmonic distortion of the current drawn from the supply source

(b) real input power to the filter (not equal to the output power, because of the nonideal switches employed)

(c) apparent input power to the filter,

(d) input power factor.

Repeat the assignment for $N = 20$ and $M = 0.6$. Observe oscillograms of the current drawn from the supply source and current drawn by the chopper from the filter. Also, compare the voltage waveform at the input terminals of the chopper (after the filter) with that of the supply source.

CA5.7* Run PSpice program *Cyclocon.cir* for a single-phase six-pulse cyclocon-verter. Observe oscillograms of the output voltages of the cycloconverter and constituent rectifiers, output current, and currents in the separating inductors.

FURTHER READING

[1] Baoming, G., Qin, L., Wei, Q., and Peng, F. Z., A family of Z-source matrix converters, *IEEE Transactions on Industrial Electronics*, vol. 59, no. 1, pp. 35–46, 2012.

[2] Empringham, L., Kolar, J. W., Rodriguez, J., Wheeler, P. W., and Clare, J. C., Technological issues and industrial application of matrix converters: a review, *IEEE Transactions on Industrial Electronics*, vol. 60, no. 10, pp. 4260–4271, 2013.

[3] Kolar, J. W., Friedli, T, Rodriguez, J., and Wheeler, P. W., Review of three-phase PWM ac-ac converter topologies, *IEEE Transactions on Industrial Electronics*, vol. 58, no. 11, pp. 4988–5006, 2011.

[4] Kolar, J. W., Schafmeister, F., Round, S., and Ertl, H., Novel three-phase AC-AC sparse matrix converters, *IEEE Transactions on Power Electronics*, vol. 22, no. 5, pp. 1649–1661, 2007.

[5] Rashid, M. H., *Power Electronics Handbook*, 2nd ed., Academic Press, Boston, MA, 2007, Chapter 18.

[6] Rombaut, C., Seguier, G., and Bausiere, R., *Power Electronic Converters—AC/AC Conversion*, McGraw-Hill, New York, 1987.

6 DC-to-DC Converters

In this chapter: power electronic converters for dc-to-dc conversion are described; static dc switches with fully controlled semiconductor power switches and forced-commutated SCRs are explained; single-, two-, and four-quadrant step-down choppers and a step-up chopper are analyzed; guidelines for device selection are provided; and applications of dc-to-dc converters are presented.

6.1 STATIC DC SWITCHES

The simplest type of dc-to-dc power conversion involves connection and disconnection of a dc supply source to a load, with no control of the supplied voltage. Power electronic *static dc switches*, although incapable of physical isolation of the source from the load, find a variety of applications in industrial and consumer electronics. The term "static" implies that the switch changes its state infrequently. In this respect, static dc switches replace the traditional electromechanical switches. The latter have a limited life span due to the wear of contacts and other mechanical parts, and they usually require substantial activating power. In contrast, power electronic switches can, theoretically, operate for an infinite amount of time and their *power gain*, that is, the ratio of connected power to activating power, is much higher than that in electromechanical switches.

Circuit diagram of a static dc switch based on a fully controlled semiconductor power switch is shown in Figure 6.1. The semiconductor switch, S, is connected in series with the load. The freewheeling diode, D, parallel to the load, provides a path for the lingering load current, when the switch is off. In the on-state of the switch, the output voltage, v_o, equals the fixed dc input voltage, V_i, and the output current, i_o, equals the input current, i_i. The diode is reverse biased and its current, i_D, is zero.

In all the subsequent considerations, an RLE load is assumed, making the freewheeling diode necessary. When the switch is turned off, the output current is forced to decrease. The negative rate of change of the current produces a negative voltage across the load inductance. The diode becomes forward biased and starts conducting the output current. Now, $i_o = i_D$, $v_o \approx 0$, and $i_i = 0$. After a short period of time the load current dies out. The voltage drop across switch S equals the input voltage, so the switch is forward biased and ready for the next turn-on. Waveforms of the gate signal,

Introduction to Modern Power Electronics, Third Edition. Andrzej M. Trzynadlowski.
© 2016 John Wiley & Sons, Inc. Published 2016 by John Wiley & Sons, Inc.
Companion website: www.wiley.com/go/modernpowerelectronics3e

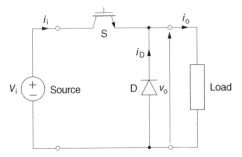

Figure 6.1   Static dc switch based on a fully controlled semiconductor power switch.

g (current or voltage, depending on the type of switch), output voltage and current, input current and diode current are shown in Figure 6.2. For simplicity, instantaneous turn-on and turn-off are assumed.

A fully controlled semiconductor power switch can be turned off by an appropriate gate signal, but an SCR operating in a dc circuit requires an auxiliary *commutating circuit* for forced turn-off of the device. Various such circuits have been developed. However, the currently available fully controlled switches have phased out SCRs from most of the dc-input converters. Still, to give the reader an idea about the issue, an example of SCR-based static dc switch employing a resonant commutating circuit is described below.

The switch is shown in Figure 6.3. The commutating circuit comprises an auxiliary SCR, T2, diode, D2, inductor, L, and capacitor, C. Polarity of the voltage, v_c, across the capacitor is shown for the on-state of the switch, that is, when the main SCR, T1, is conducting and T2 is off.

To turn T1 off, the forward-biased T2 is fired. To better explain the mechanism of forced commutation, the T1-T2-C sub-circuit is shown in Figure 6.4 at the instant when T2 starts conducting and, therefore, it can be represented by a closed switch. It

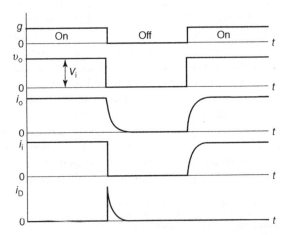

Figure 6.2   Voltage and current waveforms in the static dc switch.

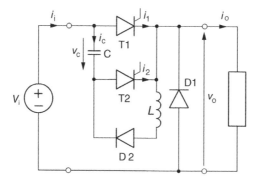

Figure 6.3   SCR-based static dc switch with a resonant commutating circuit.

can be seen that the capacitor has been connected across T1 and the capacitor voltage imposes reverse bias on that SCR. Since T1 is still conducting, the capacitor is shorted and its discharge current, i_c, flows into the cathode of T1. As a result, a reverse recovery current in T1 is enforced and the SCR turns off, which is tantamount to turning off the whole dc switch considered.

The same mechanism is employed to turn off the auxiliary SCR. When T1 is off, T2 closes the source-C-T2-load loop. Thus, following turn-off of T1, capacitor C is charged, via the load, to the voltage equal $-V_i$. Note that the forced commutation of T2 may not be needed if the off-time of T1 is sufficiently long. Then, T2 turns off by itself because of the decay of current in the dc-supplied capacitor circuit in question.

When the capacitor is fully charged, T1 can be turned on again, closing the T1-L-D2-C loop. This loop is a resonant circuit in which, if T1 and D2 were absent, the capacitor voltage and current would have the oscillating, sinusoidal waveforms (assuming insignificant resistance of the circuit). However, the diode blocks a positive current and the resonant process is interrupted after a half period of the oscillation. At this instant, the capacitor has been recharged to $v_c = V_i$, that is, to the voltage required for a successful turn-on of the main SCR. Pertinent waveforms are shown in Figure 6.5, where g_1 and g_2 denote gate signals of SCRs T1 and T2, respectively,

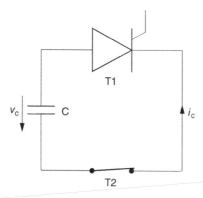

Figure 6.4   Equivalent sub-circuit of the SCR-based static dc switch.

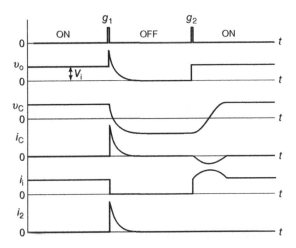

Figure 6.5   Voltage and current waveforms in the SCR-based static dc switch.

while i_1 and i_2 are currents conducted by these SCRs. For simplicity, instantaneous switching transients of the SCRs and a purely resistive load have been assumed.

A commutating circuit increases the size, weight, and cost of a static dc switch and adversely affects its reliability. Moreover, the forced commutation reduces the maximum available operating frequency of the switch because of the substantial duration of the transient conditions illustrated in Figure 6.5. The output voltage spike following the turn-on of the auxiliary SCR constitutes an additional disadvantage as it may damage the load.

6.2   STEP-DOWN CHOPPERS

Power electronic choppers are dc-to-dc converters with adjustable average output voltage. As already explained in Section 1.5, the principle of operation of a chopper consists in high-frequency on–off switching. The dc supply source is alternatively connected with the load and disconnected from it. As a result, the output voltage waveform constitutes a train of short pulses interspersed with short notches. The output current in the assumed inductive load does not have enough time to change significantly within the duration of a single pulse or notch, which makes for low-current ripple.

Most practical choppers are of the *step-down* type. Assuming negligible voltage drops across conducting switches, the magnitude of pulses of the output voltage equals the input voltage, V_i. The average output voltage, $V_{o,dc}$, depends linearly on the duty ratios of chopper switches and, generally, can be adjusted in the $-V_i$ to $+V_i$ range. Consequently, the magnitude control ratio, M, of a chopper can be taken as

$$M = \frac{V_{o,dc}}{V_i}. \qquad (6.1)$$

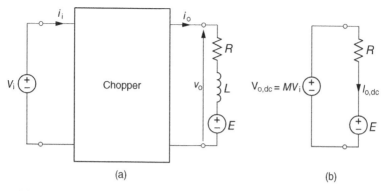

(a)   (b)

Figure 6.6   Step-down chopper: (a) block diagram, (b) equivalent circuit for the dc-output voltage and current.

General block diagram of a chopper and the corresponding equivalent circuit for dc components of the output voltage and current are shown in Figure 6.6. The RLE load may represent a dc motor or a battery charged through an inductive filter. It must be stressed that in this and the subsequent figures, the arrows of the voltages and currents indicate the *assumed positive polarities* of these quantities. As is the case of rectifiers covered in Chapter 4, the polarity of the load EMF is the same as that of the average output voltage. In the first and third quadrants of operation, the EMF can be zero since the power flows from the source to the load.

The differential equation of the load is

$$L\frac{di_o}{dt} + Ri_o + E = v_o, \tag{6.2}$$

where the output voltage, v_o, can assume values of zero depending on the type of chopper, $-V_i$ and/or $+V_i$. Therefore, analyzing a given state of chopper, v_o can be treated as a constant, which makes the solution of Eq. (6.2) to be

$$i_o(t) = \frac{v_o - E}{R} + \left[\frac{E - v_o}{R} + i_o(t_0)\right] e^{-\frac{R}{L}(t-t_0)}, \tag{6.3}$$

where $i_o(t_0)$ is the output current at the initial instant, t_0, of the state considered. The average output current, I_o, can be found from the circuit in Figure 6.6a as

$$I_{o,dc} = \frac{V_{o,dc} - E}{R} = \frac{MV_i - E}{R}. \tag{6.4}$$

Depending on the circuit configuration, step-down choppers can operate in a single, two, or four quadrants of the operation plane. Subsequent sections provide detailed description of those converters.

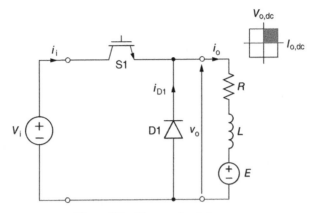

Figure 6.7   First-quadrant chopper.

6.2.1   First-Quadrant Chopper

The first-quadrant chopper can only produce positive dc-output voltage and current, and the average power flows always from the source to the load. Shown in Figure 6.7, the circuit diagram of the chopper is identical with that of the static dc switch in Figure 6.1. A switching variable, x_1, can be assigned to the fully controlled switch S, and the output voltage of the chopper can be expressed as

$$v_o = x_1 V_i. \tag{6.5}$$

The value of the switching variable indicates the state of chopper. Equivalent circuits of the chopper in state 1 and state 0 are shown in Figures 6.8a and 6.8b, respectively. In this and the subsequent equivalent circuits, the switch and diode currents are shown with their *actual polarities* while, as already mentioned, the shown polarities of the input and output quantities indicate the assumed positive directions of these quantities.

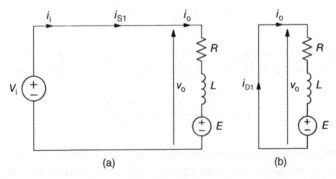

Figure 6.8   Equivalent circuits of the first-quadrant chopper: (a) state 1, (b) state 0.

In state 1, the switch is on, the input voltage appears across the load, and the load inductance is charged with electromagnetic energy. The output current waveform follows a growth function expressed by Eq. (6.3) with $v_o = V_i$. In state 0, the switch is off, the freewheeling diode D1 shorts the output terminals, and the discharge of energy stored in the load inductance results in the output current decaying according to Eq. (6.3) with $v_o = 0$.

Denoting by t_{ON} and t_{OFF} the times during which switch S is on and off, that is, the chopper is in state 1 and state 0, respectively, the dc-output voltage, $V_{o,dc}$, can be calculated as a time-weighted average of output voltages in the equivalent circuits in Figure 6.8. Specifically,

$$V_{o,dc} = \frac{V_i \times t_{ON} + 0 \times t_{OFF}}{t_{ON} + t_{OFF}} = \frac{t_{ON}}{t_{ON} + t_{OFF}} V_i = d_1 V_i, \tag{6.6}$$

where d_1 is the duty ratio of switch S1. Note that Eq. (6.6) could be derived by averaging both sides of Eq. (6.5) and taking into account that the average value of a switching variable equals the duty ratio of the corresponding switch.

From Eqs. (6.1) and (6.6),

$$M = d_1 \tag{6.7}$$

and, as the average output current given by Eq. (6.4) cannot be negative, the available range of magnitude control of the output voltage is

$$\frac{E}{V_i} < M \leq 1. \tag{6.8}$$

According to Eq. (6.7), the E/V_i ratio in condition (6.8) represents the minimum duty ratio of the switch allowable for continuous conduction. If operation with a lower value of d is attempted, the output current becomes discontinuous which, as in rectifiers, should possibly be avoided.

Example waveforms of the output voltage and current of the first-quadrant chopper are shown in Figure 6.9 with the E/V_i ratio of 0.25. The magnitude control ratio, M, had initially been 0.50, then changed to 0.75. The output voltage responds instantaneously to the change in M, while the current response is inertial, with the time

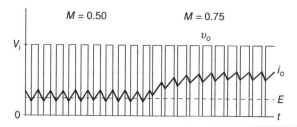

Figure 6.9 Example waveforms of output voltage and current in a first-quadrant chopper.

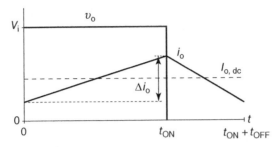

Figure 6.10   Single cycle of the output voltage and current in the first-quadrant chopper.

constant, τ, imposed by the load. Note that waveforms of other voltages and currents in the chopper are easy to determine from the output current and voltage. For instance, voltage across the switch is $V_i - v_o$, while the switch current equals $x_1 i_o$.

In practice, output current waveforms in choppers can be assumed piecewise linear, thanks to the high switching frequencies employed. This approximation greatly simplifies the analysis of choppers, as demonstrated by the following derivation of formulas for the ac and dc components of the output current.

A single cycle of the output voltage and current in the steady state of a first-quadrant chopper is shown in Figure 6.10. The output current, i_o, increases by Δi_o during the t_{ON} time and decreases by the same amount during the t_{OFF} time. The current increment, Δi_o, can be thought of as doubled amplitude of the ac component (ripple) of the output current. The waveform of this component is triangular, and it can easily be shown that the rms value, $I_{o,ac}$, of such waveform is $\sqrt{3}$ times lower than the amplitude. Thus,

$$I_{o,ac} = \frac{\Delta i_o}{2\sqrt{3}}. \tag{6.9}$$

From Eq. (6.2),

$$di_o = \frac{1}{L}(v_o - E - Ri_o)dt. \tag{6.10}$$

During the on-time, $di_o \approx \Delta i_o$, $v_o = V_i$, $i_o \approx I_{o,dc}$, and $dt \approx t_{ON}$. Consequently,

$$\Delta i_o = \frac{1}{L}(V_i - E - RI_{o,dc})t_{ON}. \tag{6.11}$$

Analogously, during the off-time, $di_o \approx -\Delta i_o$, $v_o = 0$, $i_o \approx I_{o,dc}$, $dt \approx t_{OFF}$, and

$$\Delta i_o = \frac{1}{L}(E - RI_{o,dc})t_{OFF}. \tag{6.12}$$

Comparing Eqs. (6.11) and (6.12), and solving for $I_{o,dc}$, yields

$$I_{o,dc} = \frac{1}{R} \left(\frac{t_{ON}}{t_{ON} + t_{OFF}} V_i - E \right) = \frac{d_1 V_i - E}{R} = \frac{M V_i - E}{R},$$ (6.13)

that is, Eq. (6.4), already determined in the preceding section.
  Substituting Eq. (6.13) in Eq. (6.12) gives

$$\Delta i_o = \frac{M V_i}{L} t_{OFF}.$$ (6.14)

Note that t_{OFF} can be expressed as

$$t_{OFF} = (1 - d)(t_{ON} + t_{OFF}) = \frac{1 - M}{f_{sw}},$$ (6.15)

where $f_{sw} \equiv 1/(t_{ON} + t_{OFF})$ denotes the switching frequency of a chopper. Also, $L = \tau R$, where $\tau \equiv L/R$ is the time constant of the load. Thus, Eq. (6.14) can be rearranged to

$$\Delta i_o = \frac{V_i}{R} \frac{M(1 - M)}{\tau f_{sw}}.$$ (6.16)

  For generality, to accommodate choppers operating in the third and fourth quadrants, that is, with negative magnitude control ratios, M in Eq. (6.16) can be replaced with $|M|$. Then, Eqs. (6.9) and (6.16) yield

$$I_{o,ac(pu)} = \frac{|M|(1 - |M|)}{2\sqrt{3} f_{sw(pu)}},$$ (6.17)

where $I_{o,ac(pu)}$ denotes a per-unit rms value of the output ripple current defined as

$$I_{o,ac(pu)} \equiv \frac{I_{o,ac}}{\frac{V_i}{R}},$$ (6.18)

while $f_{sw(pu)}$ is the per-unit switching frequency, defined as

$$f_{sw(pu)} \equiv \tau f_{sw}.$$ (6.19)

  Relation (6.17) is illustrated in Figure 6.11 in the form of a three-dimensional graph. It can be seen that the maximum ripple occurs at $|M| = 0.5$. The switching frequency has a strong impact on the ripple current, particularly at low values of this frequency, when the switching times of the chopper are comparable with the time constant of the load. Excessively high switching frequencies are impractical as small

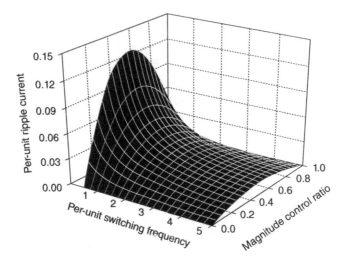

Figure 6.11 Current ripple in a chopper as a function of the magnitude control ratio and switching frequency.

improvements in the quality of output current are obtained at the expense of high switching losses. The per-unit switching frequency of about 3 offers the best tradeoff between the quality and the efficiency of chopper operation.

6.2.2 Second-Quadrant Chopper

The second-quadrant chopper, depicted in Figure 6.12, operates with a positive output voltage and a negative average output current. The average power flows always from the load to the source. Equivalent circuits of the chopper in states 1 and 0 are shown in Figure 6.13. Comparing these circuits with those in Figure 6.8 for a first-quadrant

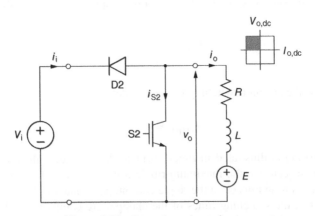

Figure 6.12 Second-quadrant chopper.

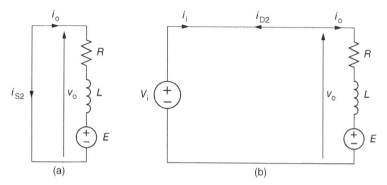

Figure 6.13 Equivalent circuits of a second-quadrant chopper: (a) state 1, (b) state 0.

chopper, it can be seen that the circuits have simply been interchanged. Clearly, configuration of the chopper prevents the input and output currents from flowing in the positive direction.

In state 1, switch S2 shorts the load and the load EMF, E, supplies the resultant circuit that includes the load inductance, L. In state 0, the energy stored in this inductance maintains the current, which is now forced to flow through diode D2 to the supply source. The instantaneous output voltage is given by

$$v_o = (1 - x_2)V_i, \tag{6.20}$$

and the average output voltage can be determined as

$$V_{o,dc} = \frac{0 \times t_{ON} + V_i \times t_{OFF}}{t_{ON} + t_{OFF}} = \frac{t_{OFF}}{t_{ON} + t_{OFF}} V_i = (1 - d_2)V_i, \tag{6.21}$$

that is,

$$M = 1 - d_2, \tag{6.22}$$

where d_2 denotes the duty ratio of switch S2. For the output current to be continuous and its dc component, $I_{o,dc}$, to be negative, the control range of the chopper must be limited to

$$0 \leq M \leq \frac{E}{V_i} \tag{6.23}$$

if $E < V_i$. Otherwise, M can be controlled in the full 0–1 range. However, allowing a discontinuous current, energy can be transferred from an active load to the input source even if the input voltage is higher than the load EMF.

Operation of the second-quadrant chopper is illustrated in Figure 6.14. The E/V_i ratio is 0.75, while the magnitude control ratio has initially been 0.50, and then

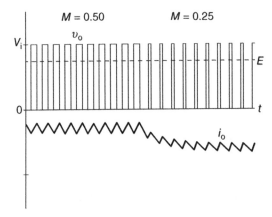

Figure 6.14 Example waveforms of output voltage and current in a second-quadrant chopper.

changed to 0.25. This results in the output current waveform being a mirror image of that in Figure 6.9.

6.2.3 First-and-Second-Quadrant Chopper

While the single-quadrant choppers described in Sections 6.2.1 and 6.2.2 can only transmit energy in one direction, two-quadrant choppers are capable of bidirectional power flow. Circuit diagram of a first-and-second quadrant chopper, that is, one operating with a positive output voltage and an output current of either polarity, is shown in Figure 6.15. Topologically, the chopper is a combination of the first- and second-quadrant choppers. Indeed, if branch S2-D2 is removed, the remaining circuit is identical with that of the first-quadrant chopper and, vice-versa, removal of branch S1-D1 would result in a second-quadrant chopper.

Denoting by x_1 and x_2 switching variables of switches S1 and S2, respectively, the state of the chopper can be designated as $(x_2 x_1)_2$. If, for instance, switch S1 is on, that is, $x_1 = 1$, and S2 is off, that is, $x_2 = 0$, then the chopper is said to be in state 2 since $10_2 = 2$. Theoretically, a total of four states are possible. However, state 3 is forbidden, as it would short the supply source.

In the first quadrant of operation, only switch S1 is controlled, while S2 is turned off. Thus, the chopper alternates between states 0 and 2. Diode D2 is permanently reverse biased, so the whole branch S2-D2 is inactive. Conversely, when the chopper operates in the second quadrant, it is only switch S2 that performs the chopping, the chopper alternates between states 0 and 1, and branch S1-D1 is inactive. Therefore, the equations derived in the preceding two sections for the first- and second-quadrant choppers can easily be adapted here. In particular, in the first quadrant of operation,

$$v_o = x_1 V_i, \quad x_2 = 0 \tag{6.24}$$

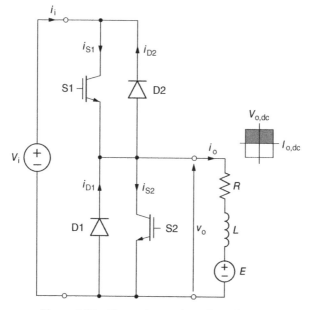

Figure 6.15   First-and-second-quadrant chopper.

and

$$M = d_1, \tag{6.25}$$

where d_1 denotes the duty ratio of switch S1. In the second quadrant,

$$v_o = (1 - x_2)V_i, \quad x_1 = 0 \tag{6.26}$$

and

$$M = 1 - d_2, \tag{6.27}$$

where d_2 is the duty ratio of switch S2. Note that the designations of switches agree with the quadrants of operation, that is, switch S1 is associated with the first quadrant and S2 with the second quadrant. This convention is observed in the description of all multi-quadrant choppers covered in this chapter. The available ranges of the magnitude control of the output voltage are given by Eq. (6.8) for the first quadrant and Eq. (6.23) for the second quadrant.

An example of operation of the chopper is illustrated in Figure 6.16. The E/V_i ratio is 0.5. At the beginning, the chopper operates in the first quadrant with the magnitude control ratio of 0.75, and later in the second quadrant with $M = 0.25$. The average output voltage is always positive while, according to Eq. (6.4), the output current,

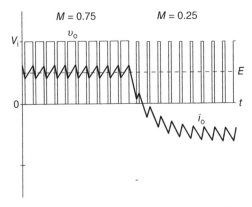

Figure 6.16 Example waveforms of output voltage and current in a first-and-second-quadrant chopper.

initially positive, changes its polarity to negative but with the same absolute value of the dc component.

6.2.4 First-and-Fourth-Quadrant Chopper

The first-and-fourth-quadrant chopper shown in Figure 6.17 allows bidirectional power flow with a positive output current. The load EMF must be positive for the first-quadrant operation and negative when the chopper is to operate in the fourth quadrant. The state of the chopper is designated as $(x_1 x_4)_2$, where x_1 and x_4 denote switching variables of switches S1 and S4, respectively.

For the first-quadrant operation, switch S4 must be turned on permanently to provide a path for the output current. Switch S1 performs the chopping, with the duty ratio d_1, so the chopper operates alternatively in states 2 and 3. In the fourth quadrant, switch S1 is off, S4 operates with the duty ratio d_4, and the chopper alternates between states 0 and 1.

Operation in the first quadrant has already been covered in Section 6.2.3 although Eq. (6.24) must be modified to

$$v_o = x_1 V_i, \ x_4 = 1. \tag{6.28}$$

Equivalent circuits of the chopper in the fourth quadrant of operation are shown in Figure 6.18. Inspection of these circuits yields

$$v_o = (x_4 - 1)V_i, \ x_1 = 0 \tag{6.29}$$

and

$$V_{o,dc} = \frac{0 \times t_{ON,4} - V_i \times t_{OFF,4}}{t_{ON,4} + t_{OFF,4}} = -\frac{t_{OFF,4}}{t_{ON,4} + t_{OFF,4}} V_i = (d_4 - 1)V_i, \tag{6.30}$$

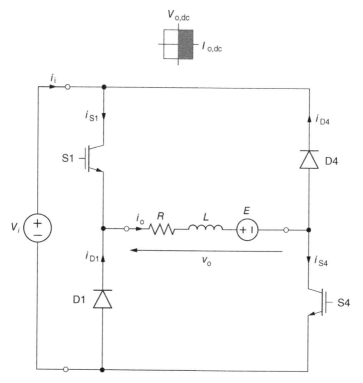

Figure 6.17   First-and-fourth-quadrant chopper.

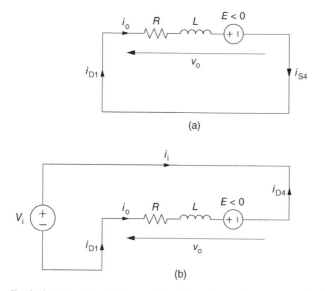

Figure 6.18   Equivalent circuits of a first-and-fourth-quadrant chopper operating in the fourth quadrant: (a) state 2, (b) state 0.

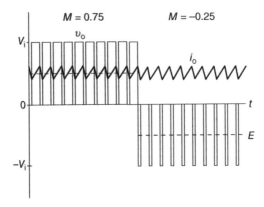

Figure 6.19   Example waveforms of output voltage and current in a first-and-fourth-quadrant chopper.

where $t_{ON,4}$ and $t_{OFF,4}$ denote the on-time and off-time of switch S4, respectively. Consequently,

$$M = d_4 - 1. \qquad (6.31)$$

For the output current to be continuous and positive, the allowable control range is

$$\frac{E}{V_i} < M \le 0. \qquad (6.32)$$

Operation of the first-and-fourth-quadrant chopper is illustrated in Figure 6.19, where initially, for the first-quadrant operation, $E/V_i = 0.5$ and $M = 0.75$ (see Figure 6.16). The dc-output current is maintained constant when the load EMF changes its polarity, so that $E/V_i = -0.5$. It can easily be found from Eq. (6.4) that the magnitude control ratio in the fourth quadrant must be -0.25.

It is worth pointing out that the first-quadrant operation could also be performed using switch S4 for chopping, with S1 continuously on. Then, when S4 is off, the output current would close through diode D4. However, this operating mode is not recommended as it violates the control symmetry, which requires that in each quadrant of operation a different switch–diode pair is employed.

6.2.5   Four-Quadrant Chopper

The four-quadrant chopper is the most versatile of all step-down choppers. Its power circuit, shown in Figure 6.20, comprises four power switches, S1 through S4, and four freewheeling diodes, D1 through D4. Designating a state of the chopper as $(x_1 x_2 x_3 x_4)_2$, where x_1 through x_4 are switching variables of the switches, a theoretical total of 16 states is obtained. However, only states 0, 1, 4, 6, and 9 are utilized. The remaining states would either short-circuit the supply source or spoil the symmetry of control.

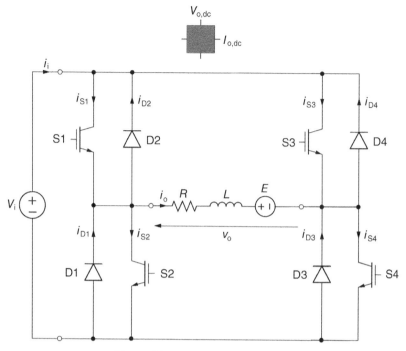

Figure 6.20 Four-quadrant chopper.

Instead of describing operation of the chopper separately for each quadrant, a summary of operating conditions is provided in Table 6.1. The "ON-state" and "OFF-state" denote states of the chopper corresponding to states of the chopping switch. The "ON-circuit" and "OFF-circuit" indicate switches and diodes carrying the output current. Duty ratios of the individual switches are denoted by d_1 through d_4. As already mentioned, designations of the switches correspond to respective operating

Table 6.1 Operating Features of the Four-Quadrant Chopper

Quadrant:	I	II	III	IV
E	≥ 0	> 0	≤ 0	< 0
x_1	0, 1, 0, 1, …	0	0	0
x_2	0	0, 1, 0, 1, …	1	0
x_3	0	0	0, 1, 0, 1, …	0
x_4	1	0	0	0, 1, 0, 1, …
ON state	9	4	6	1
OFF state	1	0	4	0
ON circuit	S1-D4	S2-D3	S2-S3	S4-D1
OFF circuit	S4-D1	D2-D3	S2-D3	D1-D4
v_o	$x_1 V_i$	$(1-x_1)V_i$	$-x_3 V_i$	$(x_4-1)V_i$
M	d_1	$1-d_2$	$-d_3$	d_4-1
M range	E/V_i to 1	0 to E/V_i	-1 to E/V_i	E/V_i to 0

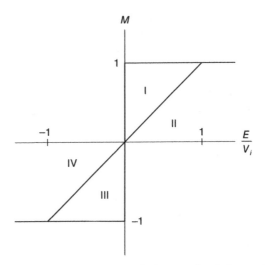

Figure 6.21 Allowable ranges of the magnitude control ratio in step-down choppers.

quadrants of the chopper. For instance, in the third quadrant, it is switch S3 that acts as the modulating (chopping) switch. The M ranges listed are those for a continuous output current. The same magnitude control ranges, valid for all step-down choppers, are illustrated in Figure 6.21.

6.3 STEP-UP CHOPPER

The pulsed output voltage in step-down choppers described in Section 6.2 has a fixed amplitude, practically equal to the input voltage, and an adjustable average value which, taking into account voltage drops in the chopper, is *lower* than the input voltage. In contrast, the step-up chopper produces an output voltage with a fixed average value equal to the input voltage, and with adjustable amplitude of the pulses, which is *higher* than the input voltage.

Shown in Figure 6.22, the step-up chopper consists of a fully controlled switch, a diode, and an inductor. The diode prevents reversal of the output current due to the load EMF when the switch is closed. This chopper cannot be emulated by the generic power converter since the inductor constitutes a crucial element for the power conversion performed. Notice a topological similarity of the step-up chopper with the second-quadrant chopper in Figure 6.12, with the load EMF considered a supply source.

When the switch is on for the time t_{ON}, the output voltage, v_o, is zero and the voltage drop, v_L, across the chopper's inductor, conducting the input current, i_i, equals the input voltage, V_i. Thus,

$$v_L = L_c \frac{di_i}{dt} = V_i, \tag{6.33}$$

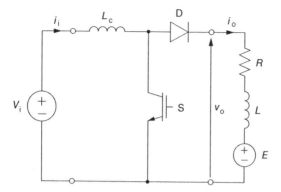

Figure 6.22 Step-up chopper.

where L_c denotes coefficient of inductance of the inductor. Consequently,

$$\frac{di_i}{dt} = \frac{V_i}{L_c},\qquad(6.34)$$

that is, the current in the inductor increases linearly in time. The increment, Δi_i, of the current by the end of the interval is

$$\Delta i_i = \frac{V_i}{L_c}t_{ON}.\qquad(6.35)$$

During the following time interval, t_{OFF}, the switch is off and the input current decreases as the load is no longer shorted by the switch. In the steady-state of operation of the chopper, the average input current is constant, which implies that the current drops by the same amount, Δi_i, as it has increased during the preceding time interval, t_{ON}. Assuming a linear change of the decreasing current, its derivative can be expressed as

$$\frac{di_i}{dt} = -\frac{\Delta i_i}{t_{OFF}} = -\frac{V_i}{L_c}\frac{t_{ON}}{t_{OFF}}.\qquad(6.36)$$

Consequently, the output voltage during the off-time of the switch is given by

$$v_o = V_i - v_L = V_i - L_c\frac{di_i}{dt} = V_i + L_c\frac{V_i}{L_c}\frac{t_{ON}}{t_{OFF}} = V_i\left(1 + \frac{t_{ON}}{t_{OFF}}\right).\qquad(6.37)$$

According to Eq. (6.37), the output voltage pulse, appearing during the off-time, has the peak value, $V_{o,p}$, equals

$$V_{o,p} = V_i\left(1 + \frac{t_{ON}}{t_{OFF}}\right) = \frac{V_i}{1-d},\qquad(6.38)$$

where d denotes the duty ratio of switch S.

The average output voltage, $V_{o,dc}$, can be determined as

$$V_{o,dc} = \frac{0 \times t_{ON} + \frac{V_i}{1-d} \times t_{OFF}}{t_{ON} + t_{OFF}} = \frac{V_i}{1-d} \frac{t_{OFF}}{t_{ON} + t_{OFF}}$$

$$= \frac{V_i}{1-d}(1-d) = V_i, \tag{6.39}$$

that is, as mentioned before, it is equal to the input voltage. Since $0 \leq d \leq 1$, the amplitude of pulses of the output voltage is higher than the input voltage. The magnitude control ratio, M, as defined by Eq. (6.1), is unity, independently on the duty ratio, d.

Operation of a step-up chopper is illustrated in Figure 6.23. For simplicity, a purely resistive load has been assumed. The duty ratio is 0.75, so that the pulses of the output voltage have amplitude four times higher than the input voltage. The pulses are not exactly rectangular because, in reality, the current in the chopper inductance decays exponentially, not linearly.

To obtain continuous output voltage, a capacitor must be connected in parallel with the load. The diode, D, protects the capacitor from discharging through switch S. If the capacitance is sufficiently high, the voltage is maintained at an average level equal to the input voltage. Then,

$$M = \frac{1}{1-d}. \tag{6.40}$$

Here, it must be pointed out that the parasitic resistances, mostly those of the source and inductor, affect the performance of the chopper. In particular, Eq. (6.38) remains valid only for low values of the duty ratio, d, that is, for low voltage gains $V_{o,p}/V_i$. The maximum available voltage gain, theoretically approaching infinity when d approaches unity, is actually quite limited (see Section 8.2.2). Therefore, in the design of step-up choppers, particularly those with smoothing capacitors and/or when a truly high output voltage is desired, circuit simulation programs, such as PSpice, should be used, taking the parasitic resistances into account.

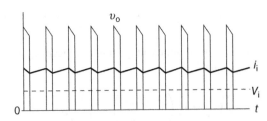

Figure 6.23 Waveforms of the output voltage and input current in a step-up chopper ($d = 0.75$).

6.4   CURRENT CONTROL IN CHOPPERS

Control of the output voltage by means of pulse width modulation is most commonly employed in choppers. However, in certain applications, such as the torque control of dc motors, it is the output current that needs direct control. This requires a closed-loop control system that imposes such switching patterns in the chopper that the actual current follows the desired reference signal. If the current is to change the polarity, the control system must cause a transition to an appropriate operating quadrant.

In practice, current-controlled inverters are more common than current-controlled choppers, and their current control systems are very similar to those of choppers. Both these converter types are supplied from dc sources. Therefore, detailed coverage of the current control is delayed until Section 7.2.2.

6.5   DEVICE SELECTION FOR CHOPPERS

Clearly, a static dc switch represents a first-quadrant chopper operating with a low switching frequency. Thus, with regard to selection of the semiconductor devices, there is no need for different treatment of the dc switches and step-down choppers. On the other hand, the rules for device selection for step-down choppers differ from those for the step-up choppers. Therefore, these two types of choppers will be considered separately.

Determination of the required rated voltage of semiconductor devices (switches and diodes) used in static dc switches and step-down choppers is straightforward, since the highest voltage across a device equals, at worst, the peak value, $V_{i,p}$, of the input voltage. Although assumed constant (ideal dc), the input voltage may in practice contain certain ripple, so that the peak value of this voltage is somewhat greater than the average value. Consequently, taking into account the voltage safety margin, s_V, the rated voltage, V_{rat}, of the devices must satisfy the condition

$$V_{rat} \geq (1 + s_V)V_{i,p}. \tag{6.41}$$

If, for example, a chopper is fed by a rectifier supplied from a 460-V line, the peak input voltage, $V_{i,p}$, should be taken as $\sqrt{2} \times 460 = 651\text{V}$.

The contribution of a switch operating with the duty ratio d to the output current, $I_{o,dc}$, is $I_{o,dc}d$, and that of the freewheeling diode is $I_{o,dc}(1 - d)$. As d can vary from 0 to 1, then the rated current, I_{rat}, of the semiconductor devices must at least be as high as the rated dc-output current, $I_{o,dc(rat)}$, of the chopper. Thus, employing the current safety margin, s_I, the condition to be met is

$$I_{rat} \geq (1 + s_I)I_{o,dc(rat)}. \tag{6.42}$$

In a step-up chopper, the rated voltage of the switch and diode must exceed the peak output voltage, $V_{o,p}$. The maximum allowed value, $V_{o,p(max)}$, of the instantaneous output voltage must be specified as a basis for device selection. Then, the maximum allowable value, d_{max}, of the duty ratio of the switch is

$$d_{max} = 1 - \frac{V_i}{V_{o,p(max)}}, \tag{6.43}$$

and the rated voltage, V_{rat}, of the switch and diode must satisfy the condition

$$V_{rat} \geq (1 + s_V)V_{o,p(max)}. \tag{6.44}$$

The diode carries the whole output current. Therefore, its current rating, $I_{D(rat)}$, must at least equal the rated average output current, $I_{o,dc(rat)}$, of the chopper. Thus,

$$I_{D(rat)} \geq (1 + s_I)I_{o,dc(rat)}. \tag{6.45}$$

To determine the required rated current, $I_{S(rat)}$, of the switch, a chopper with a resistive load and no smoothing capacitor is considered. If the switching frequency is sufficiently high, the output current consists of rectangular pulses whose amplitude is denoted by $I_{o,p}$, while the input current is practically constant and equals the same $I_{o,p}$. The switch carries the input current during the on-time, so that the average value, $I_{S,dc}$, of the switch current is

$$I_{S,dc} = I_{o,p}d. \tag{6.46}$$

The average output current, $I_{o,dc}$, which flows within the off-time, is given by

$$I_{o,dc} = I_{o,p}(1 - d). \tag{6.47}$$

Solving Eqs. (6.46) and (6.47) for $I_{S,dc}$ yields

$$I_{S,dc} = \frac{d}{1 - d}I_{o,dc}, \tag{6.48}$$

which leads to the condition,

$$I_{S(rat)} \geq \frac{d_{max}}{1 - d_{max}}(1 + s_I)I_{o,dc(rat)}. \tag{6.49}$$

The same condition is valid for a chopper with a smoothing capacitor. Assuming that the input and output voltages and currents have the dc quality, that is, $v_o = V_{o,dc}$, $i_o = I_{o,dc}$, and neglecting power losses in the chopper,

$$V_i I_i = V_{o,dc}I_{o,dc} = \frac{V_i}{1 - d}I_{o,dc}. \tag{6.50}$$

Thus,

$$I_i = \frac{I_{o,dc}}{1-d} \tag{6.51}$$

and, as

$$I_{S,dc} = I_i d = \frac{d}{1-d} I_{o,dc}, \tag{6.52}$$

the same relation as Eq. (6.48) has been obtained.

In any PWM converter, the pulse width modulation cannot be performed within the whole 0–1 range of the duty ratio, d, of switches, since at d close to zero or unity, the required on-time, t_{ON}, or off-time, t_{OFF}, would be too short in comparison with the feasible switching times of the switches. These times are related to the minimum and maximum values, d_{min} and d_{max}, of the duty ratio as

$$t_{ON(min)} = \frac{d_{min}}{f_{sw}} \tag{6.53}$$

and

$$t_{OFF(min)} = \frac{1-d_{max}}{f_{sw}}. \tag{6.54}$$

The last two equations facilitate selection of the switching frequency, f_{sw}, that is appropriate for a given type of semiconductor power switches.

6.6   COMMON APPLICATIONS OF CHOPPERS

Step-down choppers produce high-quality output currents that can be adjusted within a wide range. Voltage control characteristics of choppers are linear. Therefore, step-down choppers are mostly employed in high-performance dc-drive systems, such as those in machine tools and electrical traction.

In chopper systems fed from an ac line, a diode rectifier with the so-called *dc link* is to be used, as shown in Figure 6.24. The inductor, which prevents the chopper from drawing a high-frequency current from the ac line, is needed when the line inductance alone does not suffice. The capacitor is necessary to absorb the negative input current, which cannot close through the rectifier. In choppers supplied from a battery, the dc link is often used too, in order to avoid extra losses in the battery due to the undesirable ac component of the input current.

Sustained operation of the chopper in the second and fourth quadrants with the supply arrangement as in Figure 6.24 is not feasible, because the rectifier is incapable of transferring the power to the supply line. If operation of the chopper with a negative power flow was attempted, the resultant charging up of the filter capacitor would

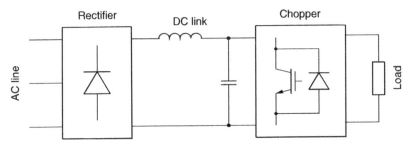

Figure 6.24 Typical supply system of a chopper.

make the input voltage to the chopper increase. This could cause an overvoltage damage to the semiconductor devices or the capacitor. One way to solve this problem is to replace the diode rectifier with a controlled rectifier. This solution is preferable when a number of choppers are supplied from a single dc network. A dc mass transit system is a typical example.

In the case of a single chopper feeding a dc motor and supplied through a diode rectifier, a *braking resistor* operating in the PWM mode is usually employed. It dissipates the energy generated by the motor in the second and fourth quadrants of operation. This arrangement is shown in Figure 6.25. When an overvoltage across the dc link capacitor is detected, the semiconductor switch in series with the braking resistor turns on, bleeding the capacitor of the excessive charge.

Second-quadrant choppers are utilized in autonomous power supply systems based on solar arrays charging a battery pack. The battery pack is connected to the input terminals while the array, whose voltage is fluctuating, is connected through a reactor to the output terminals of the chopper. Even when the source's voltage is lower than that of the battery pack, the pack can be charged.

Step-up choppers are primarily used as high-voltage sources in certain radar and ignition systems. As explained in Chapter 8, the operating principles of step-down and step-up choppers are utilized in switching power supplies.

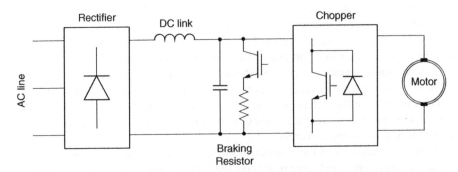

Figure 6.25 Chopper with breaking resistor in a dc-drive system.

SUMMARY

Dc-to-dc power conversion is performed using choppers. In step-down choppers, the dc input voltage is "chopped" into a train of rectangular pulses whose average value is adjustable by means of the width modulation. The load inductance attenuates the ac component of the output current, so that the current is practically of the dc quality. The higher the switching frequency is, the lower the current ripple becomes.

Step-down choppers can be structured to operate in a single, two, or all four quadrants. The number of the switch-diode pairs equals that of the operating quadrants. A first-quadrant chopper, when operated intermittently with the duty ratio of zero or unity, serves as a static dc switch connecting a load to a dc source.

A step-up chopper generates a train of voltage pulses whose average value is equal to the input voltage but the amplitude is higher than that voltage. A smoothing capacitor connected in parallel with the load allows obtaining a stepped-up dc-output voltage. The voltage gain of a step-up chopper is limited by parasitic resistances in the chopper.

Step-down choppers are primarily employed in high-performance dc-drive systems. Step-up choppers find application in radar and ignition systems.

EXAMPLES

Example 6.1   A step-down chopper fed from a 200-V source operates a dc motor whose armature EMF is 170 V and armature resistance is 0.5 Ω. With the magnitude control ratio of 0.4, find the average output voltage and current of the chopper and determine the quadrant of operation.

Solution: From Eq. (6.1), the average output voltage of the chopper is

$$V_{o,dc} = 0.4 \times 200 = 80 \text{ V},$$

and from Eq. (6.4), the average output current is

$$I_{o,dc} = \frac{80 - 170}{0.5} = -180 \text{ A}.$$

As $V_{o,dc} > 0$ and $I_{o,dc} < 0$, the chopper operates in the second quadrant.

Example 6.2   The same system as in Example 6.1 is considered. The chopper operates with a switching frequency of 1 kHz and the armature inductance of the motor is 20 mH. Find the ac component of the armature current and ripple factor of the current.

Solution: The time constant, τ, of the load of the chopper is

$$\tau = \frac{20 \times 10^{-3}}{0.5} = 0.04 \text{ s}$$

and, from Eqs. (6.17) and (6.19),

$$I_{\text{o,ac(pu)}} = \frac{0.4(1 - 0.4)}{2\sqrt{3} \times 0.04 \times 1 \times 10^3} = 0.0017.$$

Now, based on Eq. (6.18), the ac component, $I_{\text{o,ac}}$, of the armature current of the motor supplied by the chopper can be calculated as

$$I_{\text{o,ac}} = \frac{200}{0.5} \times 0.0017 = 0.68 \text{ A}.$$

In Example 6.1, the absolute value of the dc component of the output current of the chopper was found to be 180 A. Hence, the ripple factor, RF_I, equals

$$\text{RF}_I = \frac{0.68}{180} = 0.0038 = 0.38\%.$$

Example 6.3   A four-quadrant chopper operates with a switching frequency of 1 kHz and supplies power to a dc motor whose armature EMF is –216 V. The input voltage of the chopper is 240 V. Find:

(a) voltage control ranges for the third and fourth quadrants of operation of the chopper
(b) on-time and off-time of the modulating switch when the chopper operates in the third quadrant with the magnitude control ratio of –0.95
(c) on-time and off-time of the modulating switch when the chopper operates in the fourth quadrant with the magnitude control ratio of –0.75.

Solution:

(a) The negative load EMF implies operation in the third or fourth quadrant. The input voltage ratio, E/V_i, is $-216/240 = -0.9$. Therefore, according to Table 6.1, the magnitude control ratio, M, when the chopper operates in the third quadrant must be within the -1 to -0.9 range. The allowable range of M for the fourth quadrant is from -0.9 to 0.

(b) Considering the circuit diagram of the chopper in Figure 6.19, the modulating switch in the third quadrant is S3. From Table 6.1, the duty ratio, d_3, of this switch is

$$d_3 = -M = 0.95.$$

Consequently, the on-time, $t_{ON,3}$, of the switch is

$$t_{ON,3} = d_3 T_{sw} = 0.95 \times 1 \text{ ms} = 0.95 \text{ ms},$$

and the off-time, $t_{OFF,3}$, is

$$t_{OFF,3} = T_{sw} - t_{ON,3} = 1 \text{ ms} - 0.95 \text{ ms} = 0.05 \text{ ms},$$

where T_{sw} is the switching period, equal 1 ms as a reciprocal of 1 kHz.

(c) In this case, it is the switch S4 that performs the modulation. Since its duty ratio, d_4, is

$$d_4 = M + 1 = -0.75 + 1 = 0.25$$

then

$$t_{ON,4} = 0.25 \times 1 \text{ ms} = 0.25 \text{ ms}$$

and

$$t_{OFF,4} = 1 \text{ ms} - 0.25 \text{ ms} = 0.75 \text{ ms}$$

Example 6.4   A step-up chopper is supplied from a 200-V source. Find the duty ratio of the chopper switch such that the output voltage has a peak value of 500 V. What is the duration of output voltage pulses if the switching frequency is 5 kHz?

Solution: From Eq. (6.38), the required duty ratio, d, is

$$d = 1 - \frac{V_i}{V_{o,p}} = 1 - \frac{200}{500} = 0.6.$$

As the pulses of the output voltage appear when the switch is off, then their duration, t_{OFF}, is

$$t_{OFF} = \frac{1 - 0.6}{5 \times 10^3} = 8 \times 10^{-5} \text{ s} = 80 \text{ μs}.$$

Example 6.5   A two-quadrant chopper, rated at 15 kVA and 200 V, is supplied from a diode rectifier with a low-pass filter such that the peak supply voltage does not exceed 250 V. Find the minimum voltage and current ratings of switches and diodes of the chopper. Use safety margins of 0.4 for the rated voltage and 0.2 for the rated current.

Solution: According to Eq. (6.41), the rated voltage of the two semiconductor power switches and two freewheeling diodes of the chopper must be at least 1.4 × 250 V = 350 V. The rated current, $I_{o,dc(rat)}$, of the chopper is (15,000 VA)/(200 V) = 75 A. From Eq. (6.42), the rated current of the switches and diodes may not be less than 1.2 × 75 A = 90 A.

PROBLEMS

P6.1   A chopper is supplied from a 240-V dc source. The load of the chopper consists of a 20-Ω resistance in series with a 10-mH inductance. The chopper operates with the switching frequency of 0.8 kHz and produces the load current of 4.8 A. Find:

   (a) quadrant of operation of the chopper

   (b) duty ratio of the chopping switch

   (c) average output current

P6.2   Can the chopper in Problem 6.1 operate in the second quadrant? Justify the answer.

P6.3   For the chopper in Problem 6.1, find:

   (a) switching period

   (b) on-time and off-time of the chopping switch

   (c) output ripple current

   (d) ripple factor of the output current

P6.4   For the chopper in Problem 6.1, find:

   (a) per-unit switching frequency

   (b) per-unit output ripple current

P6.5   A two-quadrant chopper is supplied from a 400-V dc source and operates with the magnitude control ratio of –0.7 and switching frequency of 1.2 kHz. The chopper supplies power to a dc motor whose armature resistance is 0.4 Ω and armature inductance is 1.6 mH. The motor rotates with a constant speed such that the armature EMF is –320 V. Find:

   (a) operating quadrant of the chopper

   (b) average armature voltage of the motor

   (c) average armature current

P6.6   For the chopper in Problem 6.5, find:

   (a) duty ratio of the modulating switch

   (b) on-time of the switch

   (c) off-time of the switch

   (d) ac component of the output current

P6.7 The same chopper and motor as in Problem 6.5 are considered, but the armature voltage and EMF are 250 V and 200 V, respectively, and the armature current is positive. Find:

(a) operating quadrant of the chopper

(b) average armature current of the motor

(c) magnitude control ratio of the chopper

(d) duty ratio of the modulating switch

P6.8 A two-quadrant chopper, fed from a 400-V dc source, supplies power to a dc motor whose armature resistance is 0.2 Ω. The armature current is to be maintained constant at 250 A, independently of the motor speed that affects the armature EMF. Find the magnitude control ratio of the chopper and duty ratio of the modulating switch when the armature EMF is

(a) 320 V

(b) –320 V

P6.9 For the chopper in Problem 6.8, find the average output current when

(a) the armature EMF is 200 V and magnitude control ratio is 0.6

(b) the armature EMF is –200 V and magnitude control ratio is –0.6

P6.10 The armature inductance of the motor in Problem 6.9 is 0.75 mH. Find the switching frequency of the chopper such that the per-unit ripple current does not exceed 0.03 pu.

P6.11 A four-quadrant chopper is supplied from a 300-V dc source. The absolute value of the load EMF is 220 V. For each operating quadrant of the chopper, find the magnitude control range and the range of duty ratio of the modulating switch.

P6.12 A step-up chopper is supplied from a 120-V battery pack and operates with the switching frequency of 4 kHz. Find the amplitude of pulses of the output voltage and their duration when the duty ratio of the chopper switch is 0.8.

P6.13 A single-quadrant chopper rated at 420 VA is to be supplied from a 12-V battery. Determine the minimum required voltage and current ratings of the semiconductor devices of the chopper.

P6.14 A two-quadrant chopper is rated at 6 kVA and 600 V. A six-pulse diode rectifier fed from a three-phase 460-V line is to be used as the supply source. Determine the minimum required voltage and current ratings of the semiconductor devices of the chopper.

P6.15 Repeat Problem 6.14 for a chopper rated at 9 kVA and 300 V, and the rectifier fed from a three-phase 230-V line.

P6.16 A step-up chopper is supplied from a 6-V battery and rated at 300 VA. The highest allowable duty ratio of the chopper switch is 0.9. Determine the minimum required rated currents of the switch and diode of the chopper.

COMPUTER ASSIGNMENTS

CA6.1* Run PSpice program *DC_Switch.cir* for an SCR-based static dc switch. Observe oscillograms of the voltages and currents illustrated in Figure 6.5 (this simulation may need a long time to run).

CA6.2* Run PSpice program *Chopp_1Q.cir* for a first-quadrant chopper. For the magnitude control ratio of 0.8, find for both the output voltage and current of the chopper:

(a) dc component

(b) rms value

(c) rms value of the ac component

(d) ripple factor

CA6.3* Run PSpice program *Chopp_2Q.cir* for a second-quadrant chopper. For the magnitude control ratio of 0.4, find for both the output voltage and current of the chopper:

(a) dc component

(b) rms value

(c) rms value of the ac component

(d) ripple factor

CA6.4* Run PSpice program *Chopp_12Q.cir* for a first-and-second-quadrant chopper. With an arbitrarily set value of the magnitude control ratio, observe oscillograms of the voltages and the currents in both quadrants of operation of the chopper.

CA6.5 Develop a PSpice circuit file for a first-and-fourth-quadrant chopper. With the arbitrarily set magnitude control ratio, observe oscillograms of the voltages and the currents in both quadrants of operation of the chopper.

CA6.6 Develop a PSpice circuit file for a four-quadrant chopper. For each quadrant of operation, write separate set of lines of program for switching signals of individual switches of the chopper (when running the program for a specific quadrant, lines for the other three quadrants will be "commented out" using asterisks).

CA6.7* Run PSpice program *Step_Up_Chopp.cir* for a step-up chopper. Set the duty ratio of the switch to 0.75. Compare the voltage and current waveforms with those in Figure 6.22. Repeat the simulation for the duty ratio of 0.5.

CA6.8 Based on Eqs. (6.5) through (6.7), develop a computer program for computation of the ripple current and ripple factor in a chopper. Use the program to verify the results of Assignments 6.2(c) and 6.3(c).

FURTHER READING

[1] Bausiere, R., Labrique, F., and Sequier, G., *Power Electronic Converters: DC-DC Conversion*, Electric Energy Systems and Engineering Series, Springer-Verlag, Berlin Heidelberg, 2014.

[2] Rashid, M. H., *Power Electronics Handbook*, 3rd Ed., Chapters 13 and 14, Butterworth-Heinemann, Waltham, MA, 2011, Chapter 13.

7 DC-to-AC Converters

In this chapter: principles of dc-to-ac power conversion using power electronic inverters are introduced; voltage-source inverters (VSIs) and current-source inverters (CSIs) in the square-wave and pulse width modulation (PWM) operation modes are discussed, and voltage and current control techniques are reviewed; multilevel and soft-switching inverters are presented; device selection for inverters is outlined, and major inverter applications are described.

7.1 VOLTAGE-SOURCE INVERTERS

As already mentioned in Chapter 1, the dc-to-ac power conversion is performed by inverters. An inverter is supplied from a dc source while the ac-output voltage has a strong fundamental component with adjustable frequency and amplitude. Depending on the type of the source, *voltage-source inverters* (VSIs) and *current-source inverters* (CSIs) can be distinguished. Besides rectifiers, VSIs are the most common power electronic converters.

The dc-input voltage for a VSI can be obtained from an uncontrolled or controlled rectifier, or from another dc source, such as a battery or a photovoltaic array. As shown in Figure 7.1, if a rectifier is used, the inverter is supplied via an LC dc link, similar to that used for choppers (see Figure 6.24). The link capacitor can be thought of as a voltage source, since voltage across it cannot change instantly. Importantly, it serves as a tank for electric charges carried by the sporadic negative input currents to the inverter bridge. The main task of the inductor is to isolate the supplying rectifier and power system from the high-frequency component of the inverter input current. In contrast to the capacitor, the link inductor is not inherently necessary. Indeed, in some practical inverters, the inductor is disposed of to reduce the size and cost of the converter.

Inverters can be built with any number of output phases. In practice, single-phase and three-phase inverters are most common. However, construction of ac motors with more than three phases has recently been proposed, to increase reliability of certain critical applications. Such motors must be supplied from inverters with the same numbers of phases.

Introduction to Modern Power Electronics, Third Edition. Andrzej M. Trzynadlowski.
© 2016 John Wiley & Sons, Inc. Published 2016 by John Wiley & Sons, Inc.
Companion website: www.wiley.com/go/modernpowerelectronics3e

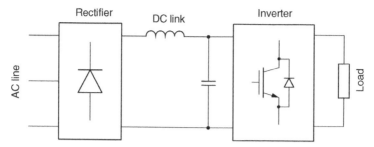

Figure 7.1   VSI supplied from a diode rectifier.

In the past, SCRs were used in high- and medium-power inverters. SCR-based inverters required commutating circuits, similar to that described in Section 6.1, to turn the SCRs off. The commutating circuits increased the inverter size and cost and reduced reliability and switching frequency. Presently, fully controlled semiconductor power switches, mostly IGBTs (in medium-power inverters) and GTOs or IGCTs (in high-power inverters), are employed. Therefore, fully controlled switches are assumed in all the subsequent considerations in this chapter.

7.1.1   Single-Phase VSI

The identity of block diagrams in Figures 6.24 and 7.1 is not coincidental. Indeed, a four-quadrant chopper, when appropriately controlled, becomes a single-phase inverter (analogously to a dual converter, which can also serve as a single-phase cyclo-converter). The inverter is so controlled that the time-averaged output voltage changes sinusoidally in time. As the output current is usually shifted in phase with respect to the fundamental output voltage, the converter operates in all four quadrants within each cycle of operation.

Circuit diagram of a single-phase VSI based on fully controlled semiconductor power switches is shown in Figure 7.2. As already mentioned, the bridge topology of the inverter is the same as that of the four-quadrant chopper in Figure 6.20 although drawn here with a slightly different arrangement and designation of components. An ideal voltage source, producing a dc-input voltage, V_i, is assumed.

Switches in any leg of the inverter may not be simultaneously on, as they would short the supply source. This condition, called a shoot-through, has already been covered in Section 3.3. Semiconductor switches, even the fast ones, require finite transition times from one conduction state to the other. Therefore, in practice, to avoid the shoot-through, a switch is turned off shortly before the turn-on of the other switch in the same leg. The interval between the turn-off and turn-on signals is called a *dead time* or *blanking time*.

Considering the issue of dead time, it is worth recalling the Z-source introduced in Section 5.3.3. The power circuit of a Z-source inverter is identical to that of the VSI, but the dc-link capacitor is replaced with the inductive-capacitive circuit shown in

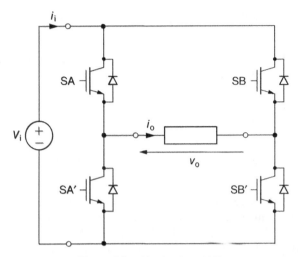

Figure 7.2   Single-phase VSI.

Figure 7.3. Other configurations are also possible, as is the application of the idea to other power electronic converters. Thanks to the presence of inductors in the dc link, the need for the dead time is eliminated, as turning on both switches in a given leg of the inverter is no longer dangerous. Even more importantly, as already explained in Chapter 5, the Z-source circuit allows both stepping down and stepping up (boosting) the link voltage.

For simplicity, in the subsequent considerations, a zero dead time will be assumed. If so, each leg of the inverter can assume two states only: either the upper (common-anode) switch is on and the lower (common-cathode) switch is off, or the other way

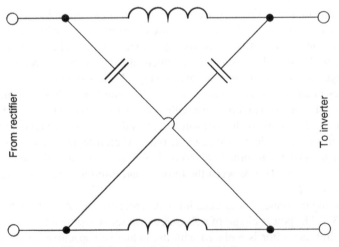

Figure 7.3   Circuit diagram of the Z-source.

around. Thus, two switching variables a and b can be assigned to the inverter legs and defined as

$$a = \begin{cases} 0 & \text{if SA is OFF and SA' is ON} \\ 1 & \text{if SA is ON and SA' is OFF} \end{cases}$$
$$b = \begin{cases} 0 & \text{if SB is OFF and SB' is ON} \\ 1 & \text{if SB is ON and SB' is OFF} \end{cases} \tag{7.1}$$

An inverter state is designated as ab_2. If, for instance, $a = 1$ and $b = 1$, the inverter is said to be in state 3 as $11_2 = 3$. Clearly, four states are possible, from state 0 through state 3.

When a given switching variable assumes a value of 1, the positive terminal of the supply source is connected to the corresponding output terminal of the inverter. Vice-versa, a value of 0 indicates connection of the negative terminal of the source to the output terminal of the inverter. Consequently, the output voltage, v_o, of the inverter can be expressed as

$$v_o = V_i (a - b) \tag{7.2}$$

and the voltage can assume three values only: V_i, 0, and $-V_i$, corresponding to state 2, states 0 or 3, and state 1, respectively.

When the output current, i_o, is of such polarity that it cannot flow through a switch that is turned on, the freewheeling diode parallel to the switch provides a path for the current. If, for instance, the output current is positive, as shown in Figure 7.2, and switch SB is on, the current is forced to close through diode DB because switch SB' must be off. In a similar way, it is easy to determine conditions under which each of the other three diodes becomes necessary for continuous conduction of the output current. The freewheeling diodes would not be needed if the load was purely resistive. Otherwise, interrupting a current in the load inductance would cause a dangerous overvoltage.

The basic version of the so-called *square-wave* operation mode of the inverter is described by the following control law:

$$a = \begin{cases} 1 & \text{for } 0 < \omega t \leq \pi \\ 0 & \text{otherwise} \end{cases} \quad b = \begin{cases} 1 & \text{for } \pi < \omega t \leq 2\pi \\ 0 & \text{otherwise} \end{cases}, \tag{7.3}$$

where ω denotes the fundamental output radian frequency of the inverter. Only states 1 and 2 are used. Waveforms of the output voltage and current in an RL load of an inverter in the basic square-wave mode are shown in Figure 7.4. The rms fundamental output voltage, $V_{o,1}$, and harmonic content, $V_{o,h}$, of this voltage equal $0.9V_i$ and $0.435V_i$, respectively, so the total harmonic distortion, THD, is 0.483.

The total harmonic distortion of the output voltage can be minimized by interspersing states 1 and 2 with states 0 and 3 lasting in the ωt domain 0.81 rad (46.5°) each, as shown in Figure 7.5. Then, in comparison with the basic square-wave operation,

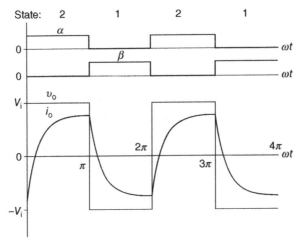

Figure 7.4 States, switching variables, and waveforms of output voltage and current in a single-phase VSI in the basic square-wave mode.

the fundamental output voltage decreases by 8%, to $0.828V_i$, but the total harmonic distortion is reduced by as much as 40%, to 0.29. The control law yielding the optimal square-wave mode is

$$a = \begin{cases} 1 & \text{for } \alpha_d < \omega t \leq \pi + \alpha_d \\ 0 & \text{otherwise} \end{cases} \qquad b = \begin{cases} 1 & \text{for } \pi + \alpha_d < \omega t \leq 2\pi - \alpha_d \\ 0 & \text{otherwise} \end{cases}, \quad (7.4)$$

where $\alpha_d = 0.405$ rad (23.2°).

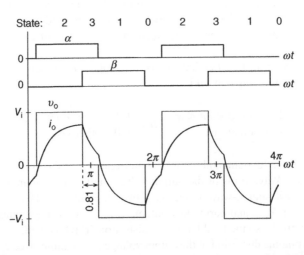

Figure 7.5 States, switching variables, and waveforms of output voltage and current in a single-phase VSI in the optimal square-wave mode.

The quality of operation of the inverter can be improved further by PWM. The single-phase inverter has only one output voltage, so the space vector PWM technique is not applicable. Here, for illustration purposes, a simple PWM strategy based on a sinusoidal modulating function, $F(m,\omega t) = m\sin(\omega t)$, is assumed. The concept of modulating function has been introduced in Section 5.1.3.

As in all PWM converters, the operating time represents a sequence of short switching cycles. Denoting the duty ratios (average values) of switching variables a and b in the nth switching cycle by d_{an} and d_{bn}, the control law for the inverter is

$$d_{an} = \frac{1}{2}\left[1 + F(m, \alpha_n)\right]$$
$$d_{bn} = \frac{1}{2}\left[1 - F(m, \alpha_n)\right] \tag{7.5}$$

where m denotes the modulation index and α_n is the phase angle of the output voltage in the center of the switching cycle.

Waveforms of the output voltage and current in an RL load, when the inverter operates in the PWM mode, are shown in Figure 7.6 for $N = 10$ switching cycles per cycle of output voltage, and in Figure 7.7 for $N = 20$. Unsurprisingly, as

$$N = \frac{f_{sw}}{f_{o,1}}, \tag{7.6}$$

where f_{sw} and $f_{o,1}$ denote the switching frequency and fundamental output frequency, respectively, the quality of output current increases. Unfortunately, so do switching losses in the inverter switches, and a reasonable tradeoff between the quality and efficiency of operation must be sought. With respect to Eq. (7.6), it must be stressed that the frequency ratio N, not necessarily an integer, has been introduced here to facilitate the comparison of waveforms and spectra. In practice, it is the switching frequency, usually constant, that is the major defining parameter of operation of PWM power converters.

The progression in improvement of operating characteristics of the inverter from the basic square-wave mode, through the optimal square-wave mode, to the PWM mode can best be explained using harmonic spectra of the output voltage and current. All the following spectra pertain to an inverter supplied from a 1 per-unit dc voltage source and supplying an RL load with a per-unit impedance of 1 and the load angle of 30°.

The voltage spectra in the basic and optimal square-wave modes are shown in Figures 7.8a and 7.8b, respectively. Thanks to the half-wave symmetry of the voltage waveforms, only odd harmonics are present (see Appendix B). As pointed out on another occasion, the load inductance acts as a low-pass filter of the output current. Hence, for high quality of that current, low-order harmonics of the output voltage should have possibly low amplitudes. Indeed, the 3rd and 5th harmonics resulting from the optimal distribution of inverter states over the cycle of output voltage are distinctly lower than those in the basic square-wave mode.

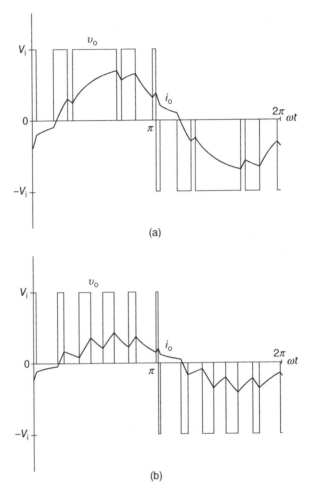

Figure 7.6   Waveforms of output voltage and current in a single-phase VSI in the PWM mode: (a) $m = 1$, (b) $m = 0.5$ ($N = 10$).

Further reduction of the low-order voltage harmonics is obtained in the PWM mode, as illustrated in Figure 7.9 for $m = 1$, especially when the number of switching cycles is high. Notice, for instance, that the lowest-order harmonic whose amplitude is higher than 10% of the fundamental amplitude is the 9th for $N = 10$ (Figure 7.9a) and the 17th for $N = 20$ (Figure 7.9b). However, the maximum available fundamental output voltage has the peak value of 1 per-unit, so that the maximum available rms value, $V_{o,1(max)}$, of this voltage is 0.707 V_i only. This is about 21% lower than the rms voltage in the basic square-wave operation mode and about 15% lower than that in the optimal square-wave mode.

Harmonic spectra of the output currents produced by voltages whose spectra have been shown in Figures 7.8a through 7.9b are depicted in Figures 7.10a through 7.11b, respectively. The highest-amplitude current harmonics in an inverter operating in

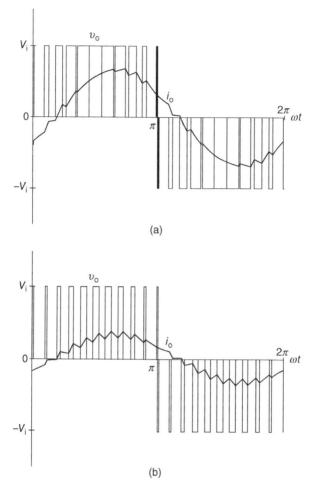

Figure 7.7   Waveforms of output voltage and current in a single-phase VSI in the PWM mode: (a) $m = 1$, (b) $m = 0.5$ ($N = 20$).

either of the square-wave modes are the low-order ones, such as the 3rd, 5th, 7th, so on. In contrast, in a PWM inverter, the highest amplitudes belong to harmonics whose harmonic number is close to the number of switching cycles per cycle of the output voltage. These are the 7th, 11th, and 13th when $N = 10$ (Figure 7.11a), and 17th, 21st, and 23rd when $N = 20$ (Figure 7.11b), all of them with amplitudes more than one order of magnitude lower than that of the fundamental.

The input current, i_i, of the inverter, given by

$$i_i = (a - b)\, i_o \tag{7.7}$$

has a large dc component, $I_{i,dc}$. The product of V_i and $I_{i,dc}$ constitutes the average input power to the inverter. Waveform of the input current in the optimal square-wave mode

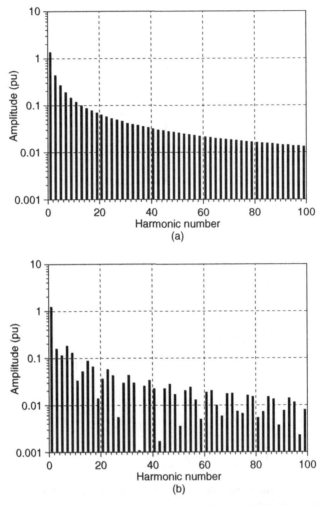

Figure 7.8 Harmonic spectra of output voltage in a single-phase VSI: (a) square-wave mode, (b) optimal square-wave mode.

is shown in Figure 7.12a and that in the PWM mode, with $m = 1$ and $N = 20$, in Figure 7.12b. Harmonic spectra of these currents are shown in Figures 7.13a and 7.13b. The fundamental frequency of the currents is twice the output frequency, and the first harmonic is comparable with the dc component. Other than that, the harmonic spectra in the low-frequency region are similar to those of the respective output currents. Note that, apart from the fundamental, the most prominent harmonics of the input current in a PWM inverter are clustered about the $N/2$ value of the harmonic order. Thus, if the switching frequency is sufficiently high, a small dc link can prevent most of the high-frequency content of the input current from reaching the supply source (power system or battery).

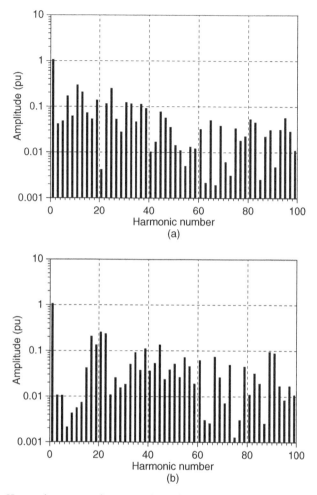

Figure 7.9 Harmonic spectra of output voltage in a single-phase VSI in the PWM mode ($m = 1$): (a) $N = 10$, (b) $N = 20$.

The practically instant changes of the input current in Figure 7.11b confirm the necessity of the dc-link capacitor. Such current could not be drawn directly from the ac grid, via the line-side rectifier, because of all the inductances in the path of that current.

The output frequency in all types of inverters is controlled by the appropriate timing of switching instants. However, control of magnitude of the output voltage in the square-wave mode can only be realized outside the inverter, by adjusting the dc-input voltage. Specifically, a controlled rectifier must be used in place of the diode rectifier shown in Figures 7.1 and 7.2. In contrast, the PWM mode of operation allows the voltage control by adjusting the modulation index, m, which equals the magnitude control ratio, M. Here, the latter parameter represents the ratio of rms fundamental

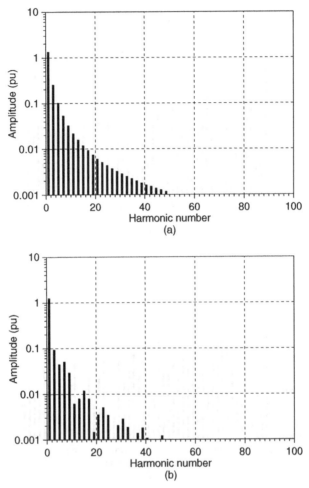

Figure 7.10   Harmonic spectra of output current in a single-phase VSI: (a) basic square-wave mode, (b) optimal square-wave mode.

output voltage, $V_{o,1}$, to the maximum value, $V_{o,1(max)}$, of this voltage available using a given PWM strategy.

7.1.2   Three-Phase VSI

Most of the practical inverters, especially medium- and high-power ones, are of the three-phase type. The power circuit of a three-phase VSI in Figure 7.14 is obtained by adding a third leg to the single-phase inverter. Assuming, as before, that of the two power switches in each leg (phase) of the inverter one and only one is always on, that is, neglecting the short time intervals when both switches are off (dead time), three

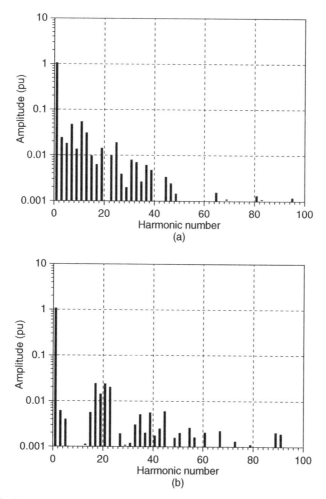

Figure 7.11 Harmonic spectra of output current in a single-phase VSI in the PWM mode ($m = 1$): (a) $N = 10$, (b) $N = 20$.

switching variables, a, b, and c, can be assigned to the inverter. A state of an inverter is designated as abc_2, making for a total of eight states, from state 0, when all output terminals are clamped to the negative dc bus, through state 7, when they are clamped to the positive bus.

It is easy to show that the instantaneous line-to-line output voltages, v_{AB}, v_{BC}, and v_{CA}, are given by

$$\begin{bmatrix} v_{AB} \\ v_{BC} \\ v_{CA} \end{bmatrix} = V_i \begin{bmatrix} 0 & -1 & 0 \\ 0 & 1 & -1 \\ -1 & 0 & 1 \end{bmatrix} \begin{bmatrix} a \\ b \\ c \end{bmatrix}. \tag{7.8}$$

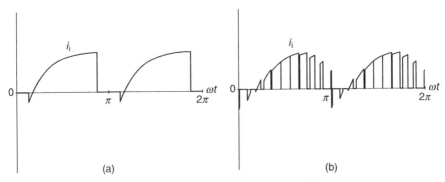

Figure 7.12 Waveforms of input current in a single-phase VSI: (a) optimal square-wave mode, (b) PWM mode ($m = 1, N = 20$).

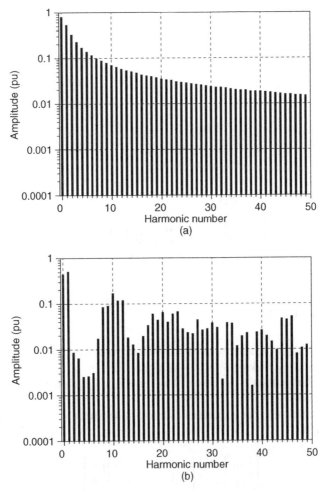

Figure 7.13 Harmonic spectra of input current in a single-phase VSI: (a) optimal square-wave mode, (b) PWM mode ($m = 1, N = 20$).

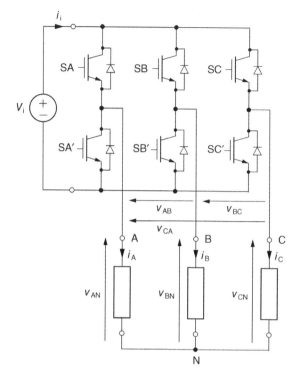

Figure 7.14   Three-phase VSI.

In a balanced three-phase system, the instantaneous line-to-neutral output voltages, v_{AN}, v_{BN}, and v_{CN}, can be expressed as

$$\begin{bmatrix} v_{AN} \\ v_{BN} \\ v_{CN} \end{bmatrix} = \frac{1}{3} \begin{bmatrix} 1 & 0 & -1 \\ -1 & 1 & 0 \\ 0 & -1 & 1 \end{bmatrix} \begin{bmatrix} v_{AB} \\ v_{BC} \\ v_{CA} \end{bmatrix}, \qquad (7.9)$$

which, when combined with Eq. (7.8), yields

$$\begin{bmatrix} v_{AN} \\ v_{BN} \\ v_{CN} \end{bmatrix} = \frac{V_i}{3} \begin{bmatrix} 2 & -1 & -1 \\ -1 & 2 & -1 \\ -1 & -1 & 2 \end{bmatrix} \begin{bmatrix} a \\ b \\ c \end{bmatrix}. \qquad (7.10)$$

Equations (7.8) and (7.10) allow the determination of the line-to-line and line-to-neutral output voltages for all states of an inverter. The results are assembled in Table 7.1. The line-to-line voltages can only assume three values, 0 and $\pm V_i$, while the line-to-neutral voltages can assume five values, 0, $\pm V_i/3$, and $\pm 2V_i/3$.

Table 7.1 States and Voltages of the Three-Phase VSI

State	abc	v_{AB}/V_i	v_{BC}/V_i	v_{CA}/V_i	v_{AN}/V_i	v_{BN}/V_i	v_{CN}/V_i
0	000	0	0	0	0	0	0
1	001	0	−1	1	−1/3	−1/3	2/3
2	010	−1	1	0	−1/3	2/3	−1/3
3	011	−1	0	1	−2/3	1/3	1/3
4	100	1	0	−1	2/3	−1/3	−1/3
5	101	1	−1	0	1/3	−2/3	1/3
6	110	0	1	−1	1/3	1/3	−2/3
7	111	0	0	0	0	0	0

If the $5-4-6-2-3-1-\ldots$ sequence of states is imposed, each state lasting one-sixth of the desired period of the output voltage, the individual line-to-line and line-to-neutral voltages acquire waveforms shown in Figure 7.15. Notice that the voltages, although not sinusoidal, are balanced. This is the square-wave mode of operation, in which each switch of the inverter is turned on and off once within the cycle of output

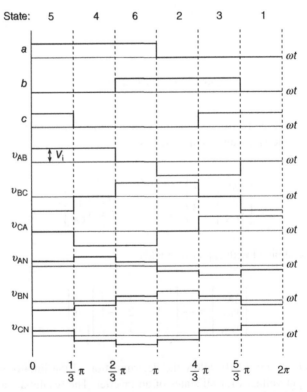

Figure 7.15 Switching variables and waveforms of output voltages in a three-phase VSI in the square-wave mode.

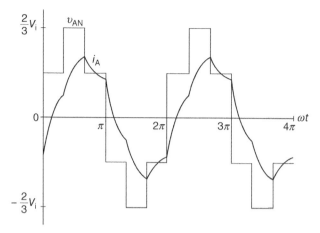

Figure 7.16 Waveforms of output voltage (line-to-neutral) and current in a three-phase VSI in the square-wave mode (RL load).

voltage. No wonder that most of the SCR-based inverters of the past were operated in this mode only. The peak value, $V_{LL,1,p}$, of the fundamental line-to-line output voltage equals approximately 1.1 V_i and that, $V_{LN,1,p}$, of the line-to-neutral voltage, 0.64 V_i. Both voltages have the same total harmonic distortion, THD, of 0.31. The ratio $V_{LL,1,p}/V_i$ represents the voltage gain of the inverter, which in this case is greater than unity. As in the square-wave single-phase inverter, the magnitude control of the output voltage must be realized on the dc supply side.

Waveforms of the output voltage and current for one phase of the inverter are shown in Figure 7.16. They are similar to those in a single-phase inverter in the same mode and rich in low-order odd harmonics, except for the triple ones. The input current, i_i, is related to the output currents, i_A, i_B, and i_C, as

$$i_i = ai_A + bi_B + ci_C, \tag{7.11}$$

and it is illustrated in Figure 7.17. Its fundamental frequency is six times higher than that of the output currents.

Apart from the high fundamental output voltage, the advantage of the square-wave operation consists in the low switching frequency, which equals that of the output voltage. However, output currents are of distinctly lower quality than those in inverters operating in the PWM mode. Therefore, the square-wave mode is primarily reserved for high-power inverters with slow switches, such as GTOs. It is also sporadically employed in PWM inverters when the maximum possible output voltage is needed.

Example pulse trains of the switching variables and waveforms of output voltages in a three-phase PWM VSI are shown in Figure 7.18. To illustrate the impact of the load angle on the output current, waveforms of one of the line-to-neutral voltages and the corresponding output current are shown in Figure 7.19. The load angle is 30° in

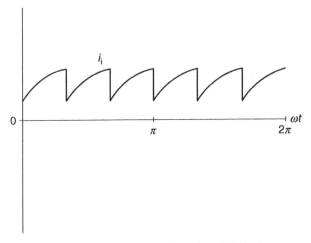

Figure 7.17 Waveform of input current in a three-phase VSI in the square-wave mode.

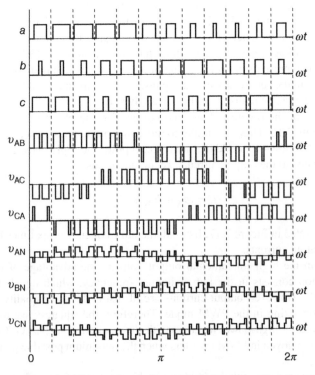

Figure 7.18 Switching variables and waveforms of output voltages in a three-phase VSI in the PWM mode.

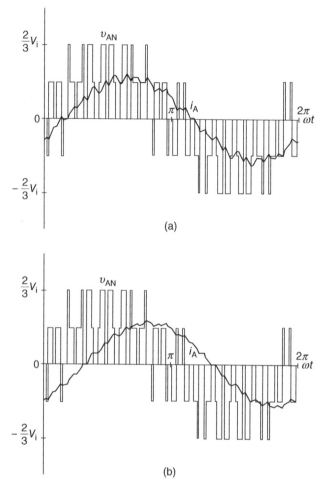

Figure 7.19 Waveforms of output voltage and current in an RL load of a three-phase VSI in the PWM mode: (a) load angle of 30°, (b) load angle of 60°.

Figure 7.19a and 60° in Figure 7.19b. The higher the load inductance is, the smoother the current waveform becomes.

Input current waveforms corresponding to output currents in Figure 7.19 are shown in Figure 7.20. Interestingly, the fundamental frequency of the input current in three-phase PWM inverters is three times higher than the output frequency, while, as already mentioned, in inverters operating in the square-wave, it is six times higher. As already mentioned before, the possibility of a negative input current in VSIs necessitates the capacitor in the dc link, as semiconductor devices in the supply rectifier could not pass such current. Similar to the single-phase inverter analyzed in the preceding section, the input current in the PWM mode of a three-phase inverter contains, apart

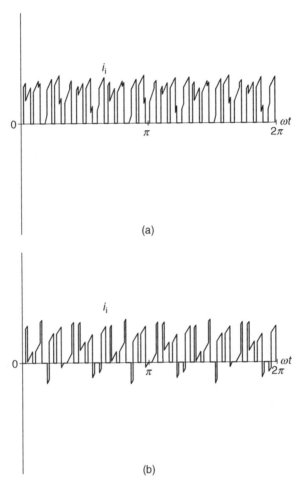

Figure 7.20 Waveforms of input current in a three-phase VSI in the PWM mode: (a) load angle of 30°, (b) load angle of 60°.

from the dc component, mostly the high-order harmonics. Therefore, to attenuate the high-frequency harmonic content of the current drawn from the supply source, the components of the dc link can be significantly smaller than those needed for a comparable square-wave mode inverter.

When supplied from a diode rectifier, a VSI has no means for reversing the direction of power flow. This can be inconvenient in certain applications, such as adjustable speed ac drives, in which the load is capable of power generation. Then, either a braking resistor in the dc link, such as that used in chopper systems (see Figure 6.25), or a controlled rectifier must be employed. Use of the IGBT power module of Figure 2.28 in an inverter-fed ac-drive system capable of four-quadrant operation is described in Section 7.6.

7.1.3 Voltage Control Techniques for PWM Inverters

The most popular voltage control techniques for PWM VSIs are described below. Desired control characteristics include:

(1) Good utilization of the dc supply voltage, that is, a possibly high value of the maximum voltage gain, $K_{V(max)}$, defined here as

$$K_{V(max)} \equiv \frac{V_{LL,1,p(max)}}{V_i}, \tag{7.12}$$

where $V_{LL,1,p(max)}$ denotes the maximum peak value of the fundamental line-to-line output voltage available using the technique under consideration.

(2) Linearity of the voltage control, that is,

$$V_{LL,1,p}(m) = m V_{LL,1,p(max)}, \tag{7.13}$$

where m denotes the modulation index, which in inverters is identical to the magnitude control ratio. As usual, the magnitude control ratio is defined as the ratio of the actual output voltage (line-to-line or line-to-neutral, peak or rms value) to the maximum available value of this voltage.

(3) Low amplitudes of low-order harmonics of the output voltage to minimize the harmonic content of the output current.

(4) Low switching losses in the inverter switches.

(5) Sufficient time allowance for proper operation of the inverter switches and control system.

A good tradeoff between the conflicting requirements (3) and (4) is particularly important. As demonstrated before, the quality of output current increases with the increase in the number of commutations (switchings) per cycle of the output voltage (compare Figures 7.5 and 7.6). On the other hand, each switching is associated with an energy loss in the switch. Thus, the quality of operation of a PWM inverter can be increased at the expense of power efficiency and vice-versa.

Carrier-Comparison PWM Technique. In the 1960s, when first PWM VSIs were built, only analog control systems were available. The simple *carrier-comparison* (*"triangulation"*) technique developed for those inverters has survived today in certain inexpensive modulators. The magnitude and the frequency of the output voltage are the controlled variables. The principle of the carrier-comparison method is illustrated in Figure 7.21. Reference waveforms, r_A, r_B, and r_C, given by

$$\begin{aligned} r_A(\omega t) &= F(m, \omega t) \\ r_B(\omega t) &= F\left(m, \omega t - \frac{2}{3}\pi\right), \\ r_C(\omega t) &= F\left(m, \omega t - \frac{4}{3}\pi\right) \end{aligned} \tag{7.14}$$

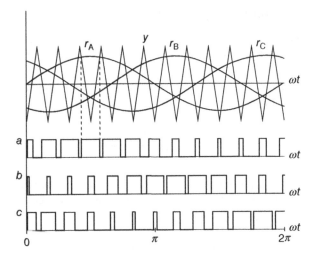

Figure 7.21   Illustration of the carrier-comparison PWM technique ($N = 12$, $m = 0.75$).

where $F(m, \omega t)$ denotes the modulating function employed, are compared with a unity-amplitude triangular waveform y. Values of the switching variables, a, b, and c, change from 0 to 1 and from 1 to 0 at every sequential intersection of the carrier and respective reference waveforms.

The sinusoidal modulating function, $F(m, \omega t) = m \sin(\omega t)$, is simple, but the voltage gain of the inverter can significantly be increased using a non-sinusoidal modulating function, many of which were developed over the years. All those functions consist of a fundamental and triple harmonics, which are not reflected in the

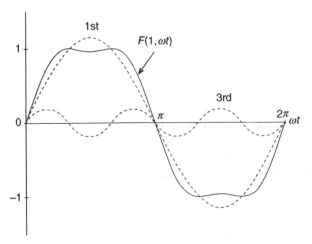

Figure 7.22   Third-harmonic modulating function and its components at $m = 1$.

three-phase output voltages and currents of the inverter. For example, the so-called *third-harmonic modulating function*, shown in Figure 7.22 with $m = 1$, is given by

$$F(m, \omega t) = \frac{2}{\sqrt{3}} m \left[\sin(\omega t) + \frac{1}{6} \sin(3\omega t) \right] \qquad (7.15)$$

having thus only the fundamental and third harmonic. At $m = 1$, the fundamental equals $2/\sqrt{3} \approx 1.15$, which represents a 15% increase in the voltage gain with no changes to the inverter.

Space Vector PWM Techniques. Comparing Figures 4.53 and 7.14, it can be seen that the voltage-type three-phase PWM rectifier and the voltage-source three-phase PWM inverter have the same topology of the power circuit. Consequently, the space vector approach to PWM described in various parts of Section 4.3 can be employed in the VSI control as well. Definitions of switching variables and states of the inverter are identical with those of the voltage-type PWM rectifier (see Section 4.3.2).

The voltage space vector plane with the reference vector $\vec{v}^*$ of line-to-neutral voltages is shown in Figure 7.23a and repeated in Figure 7.23b in the per-unit format, with the maximum available magnitude of that vector as the base. Figure 7.23a is almost identical with Figure 4.56, with V_i in place of V_0. The modulation index, m, constitutes the magnitude of per-unit $\vec{v}^*$. Neglecting the voltage drops in the inverter, the highest available peak value of the output line-to-line voltage equals the dc-input voltage, V_i. Thus

$$m = \frac{V^*_{LL,p}}{V_i}, \qquad (7.16)$$

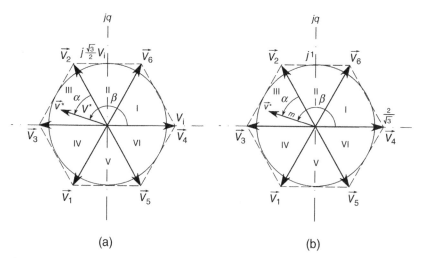

(a)   (b)

Figure 7.23   Voltage space vector plane of a three-phase VSI: (a) in volts, (b) per-unit.

where $V_{LL,p}^*$ denotes the reference peak line-to-line voltage. Equation (7.16) allows determination of the required value of m for specified output voltages of the inverter. It can be seen that with $m = 1$ the maximum available peak value of the fundamental line-to-line output voltage equals the dc-input voltage.

In the steady state, when the fundamental output voltage and current maintain fixed magnitude and frequency, m is constant and $\vec{v}^*$ rotates with a constant speed. However, the space vector PWM technique allows the synthesis of an instantaneous voltage vector, which may change in magnitude and speed from one switching cycle to another. This is required, for instance, in the so-called vector control methods of high-performance ac drives fed from VSIs.

The angular position, β, of the reference vector allows the determination of the sector of the complex plane in which the vector is located within the given sampling cycle of the digital modulator. Specifically,

$$S = \text{int}\left(\frac{3}{\pi}\beta\right) + 1, \tag{7.17}$$

where β is expressed in radians and S is the sector number (I–VI). The in-sector position, α, of $\vec{v}^*$ is then given by

$$\alpha = \beta - \frac{\pi}{3}(S - 1). \tag{7.18}$$

Analogous to the PWM rectifier control, the revolving reference voltage vector is synthesized from stationary active (non-zero) vectors, $\vec{V}_X$ and $\vec{V}_Y$, framing the sector in question, and a zero vector, $\vec{V}_0$ or $\vec{V}_7$. Durations, T_X, T_Y, and T_Z, of states generating those vectors are given by Eqs. (4.77)–(4.79), in the per-unit format with T_{sw} as the base, or, directly, by Eqs. (4.90)–(4.92). For convenience, these equations are repeated below without captions:

$$T_X = mT_{sw}\sin(60° - \alpha), \quad T_Y = mT_{sw}\sin(\alpha), \quad T_Z = T_{sw} - T_X - T_Y.$$

Times T_X, T_Y, and T_Z indicate only how long a given inverter state should last in the switching cycle under consideration, but how the cycle is divided between the employed states must also be specified. The two most commonly used state sequences can be called a *high-quality sequence* and a *high-efficiency sequence*. The high-quality sequence, called so because for a given number of switching cycles per cycle it tends to result in the lowest total harmonic distortion of output currents, is

$$X - Y - Z_1 - Y - X - Z_2 \ldots \tag{7.19}$$

where each state in the sequence lasts half of the allotted time. States Z_1 and Z_2, complementarily 0 and 7, are placed in such an order that a transition from one state to another involves switching in one inverter leg only. In other words, only one switching variable changes its value. For example, in sector II, where $X = 6$ and $Y = 2$ (see Figure 7.22), the high-quality state sequence is $6 - 2 - 0 - 2 - 6 - 7 \ldots$ or, in the binary notation, $110 - 010 - 000 - 010 - 110 - 111$. This sequence results in such a switching pattern that each switching variable has N pulses per cycle of output voltage (see Eq. 7.6). It means that each inverter switch is turned on and off once per switching cycle.

The number of commutations can further be reduced, at the expense of slightly increased distortion of output currents, when the high-efficiency state sequence

$$X - Y - Z - Y - X \ldots \tag{7.20}$$

is employed. Here, sequential states X and Y last $T_X/2$ and $T_Y/2$ seconds respectively, and state Z lasts T_Z seconds. Moreover, $Z = 0$ in the even sectors (II, IV, and VI) and $Z = 7$ in the odd sectors (I, III, and V). If sector II is again considered as an example, the high-efficiency state sequence is $6 - 2 - 0 - 2 - 6\ldots$, or $110 - 010 - 000 - 010 - 110$. Note that in the sector in question variable c is zero throughout all switching cycles. With this state sequence, the average number of pulses of a switching variable per cycle of output voltage is $2N/3 + 1 \approx 2N/3$. As a result, the switching losses decrease by about 30% in comparison with the high-quality state sequence, but the ripple of output current of the inverter slightly increases. Switching patterns generated by the two described variants of space vector PWM technique are illustrated in Figures 7.24 and 7.25. In both cases, the reference vector is assumed to be in sector I.

Programmed PWM Techniques. The best compromise between efficiency and quality of inverter operation is achieved in *programmed*, or *optimal switching pattern*, PWM techniques. To understand their principle, consider the example switching pattern for phase A of an inverter, shown in Figure 7.26. The $a(\omega t)$ waveform has both the half-wave symmetry and quarter-wave symmetry (see Appendix C).

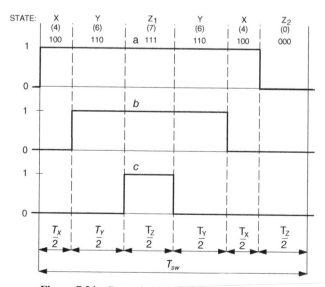

Figure 7.24 Example high-quality state sequence.

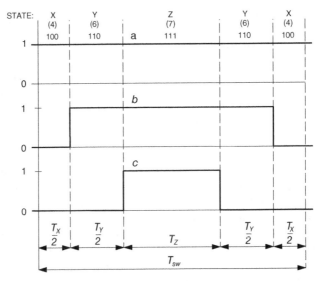

Figure 7.25   Example high-efficiency state sequence.

Consequently, the full-cycle switching pattern is uniquely determined by the switching angles α_1 through α_4 in the first quarter-cycle. These angles, whose number, K, here 4, is arbitrary, will subsequently be called *primary switching angles*. Commutations in the remaining three quarter-cycles and in phases B and C occur at *secondary switching angles*, which are simple linear functions of the primary angles. Switching

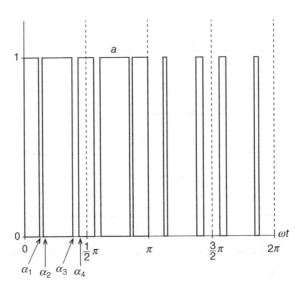

Figure 7.26   Example switching pattern with the half- and quarter-wave symmetries.

patterns $b(\omega t)$ and $c(\omega t)$ differ from $a(\omega t)$ by the appropriate phase shifts only, that is,

$$a\left(\omega t\right) = b\left(\omega t + \frac{2}{3}\pi\right) = c\left(\omega t + \frac{4}{3}\pi\right). \tag{7.21}$$

Condition (7.21) implies that waveforms of all three switching variables have identical harmonics.

According to Eqs. (7.8) and (7.10), output voltages of an inverter depend linearly on the switching variables. Therefore, all these voltages have the same harmonic spectrum as $a(\omega t)$, except for triple harmonics which, thanks to condition (7.21), are absent in the voltage waveforms. The half-wave symmetry of waveforms of switching variables results in the absence of even harmonics as well, while the quarter-wave symmetry allows expressing amplitude of the kth harmonic, $A_{k,\mathrm{p}}$, of $a(\omega t)$ as

$$A_{k,\mathrm{p}} = \frac{4}{k\pi}\left[\sum_{i=1}^{K}(-1)^{i-1}\cos\left(k\alpha_i\right) - \frac{1}{2}\right]. \tag{7.22}$$

It can be seen that the amplitude of each harmonic depends on all the primary switching angles. Therefore, the K angles can be set to such values that the fundamental and $K-1$ higher harmonics have the desired amplitudes (if feasible). This requires solving K equations (7.22) for $\alpha_1, \alpha_2, \ldots, \alpha_K$, with given K values of $A_{k,\mathrm{p}}$, including $k = 1$, and with the constraint $0 < \alpha_1 < \alpha_2 < \ldots < \alpha_K$. The equations are nonlinear and solvable only by numerical computations.

The most common approach to the definition of optimal values of primary switching angles is the *harmonic-elimination* PWM technique. It consists in setting $A_{1,\mathrm{p}}$ to $MA_{1,\mathrm{p(max)}}$, where M is the magnitude control ratio and $A_{1,\mathrm{p(max)}}$ denotes the maximum amplitude of the fundamental available with the given number of primary switching angles. Amplitudes, $A_{5,\mathrm{p}}$, $A_{7,\mathrm{p}}$, $A_{11,\mathrm{p}}$, $\ldots$, of the $K-1$ lowest-order odd and nontriple harmonics are set to zero. For instance, if $K = 5$, the maximum available amplitude of the fundamental, $A_{1,\mathrm{p(max)}}$, of $a(\omega t)$ is 0.583 and the 5th, 7th, 11th, and 13th harmonics can be eliminated, leaving the 17th as the lowest-order one. For comparison, $A_{1,\mathrm{p(max)}}$ in the space vector PWM techniques reaches the value of $1/\sqrt{3} \approx 0.577$, that is, about 1% less.

Optimal primary switching angles for various values of K and M can be found in the technical literature of the subject. The optimal angles in the example case of $K = 5$ are shown in Figure 7.27 as functions of the magnitude control ratio, M. A harmonic spectrum of the output line-to-neutral voltage in Figure 7.28, determined for $M = 1$, demonstrates the desired elimination of low-order, even, and triple harmonics.

The number of pulses of a switching variable in programmed PWM techniques is $2K + 1$ per cycle of the output voltage. With the same number of pulses per cycle, these techniques tend to produce currents of higher quality than do other PWM techniques. This is illustrated in Figure 7.29, which shows three example switching patterns and the corresponding output voltage and current waveforms. These were

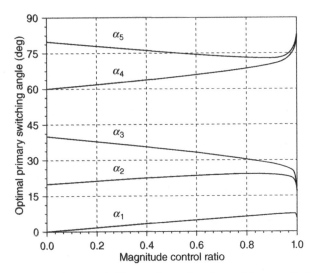

Figure 7.27 Optimal primary switching angles as functions of the magnitude control ratio ($K = 5$).

obtained using: (1) carrier-comparison PWM technique with a sinusoidal reference waveforms, (2) space vector PWM technique with the high-efficiency state sequence, and (3) programmed PWM technique for the harmonic elimination ($K = 5$). In all three cases, an RL load with the load angle of 30° was assumed, and the magnitude control ratio was set to 0.9. The load impedance was adjusted to compensate for unequal fundamental voltages and obtain identical fundamental currents. The increasing from case (1) through (2) to (3) quality of the current can be observed and

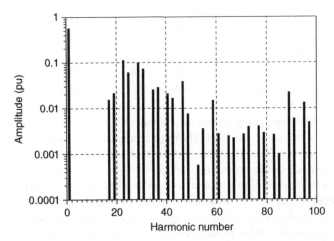

Figure 7.28 Harmonic spectrum of the line-to-neutral voltage with the harmonic-elimination technique ($K = 5$, $M = 1$).

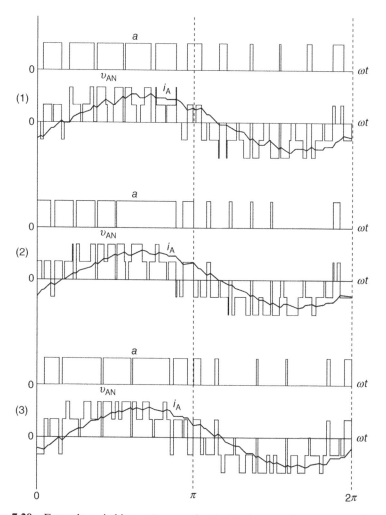

Figure 7.29   Example switching patterns and output voltage and current waveforms: (a) carrier-comparison PWM with sinusoidal reference, (b) space vector PWM with high-efficiency state sequence, (c) programmed PWM with harmonic elimination.

confirmed by the determination of the total harmonic distortion, THD. The respective values of the THD are 11.6%, 9.2%, and 5.2%.

Computation of optimal switching angles is so time consuming that it cannot be done in real time. Therefore, sets of values of switching angles for each value of the magnitude control ratio must be stored in the memory of a digital control system. Besides the need for a large memory, the disadvantage of programmed PWM techniques manifests itself in the disruption of the optimal switching patterns when rapid changes of the reference frequency and/or magnitude of the output voltage are required.

The programmed PWM techniques allow generation of high-quality sinusoidal currents, when the vectors of output voltage and current maintain a fixed magnitude and revolve with a constant speed. However, they are incapable of following a fluctuating reference voltage vector, which the space vector PWM strategies (and, to certain extent, the carrier-comparison method) can easily do. Therefore, practical applications of the programmed PWM techniques are mostly limited to ac power sources, for example, the uninterruptable power supplies, which operate with a constant output frequency and a narrow range of voltage control.

Random PWM Techniques. The subsequent considerations pertain to the space vector PWM techniques, which dominate modern power electronic converters. In practice, the switching cycles are usually made to coincide with the sampling cycles of the digital modulator, so that the switching frequency, f_{sw}, of the inverter equals the fixed sampling frequency, f_{smp}, of the modulator. The constant switching frequency results in clusters of higher harmonics in the frequency spectra of output voltage and current. The harmonic clusters are centered about multiples of f_{sw}.

By randomly varying (dithering) consecutive switching periods, aperiodic switching patterns are generated, and part of the harmonic power is shifted to the continuous spectrum of the output voltages. The random PWM (RPWM) techniques have been found to alleviate undesirable acoustic, vibration, and electromagnetic interference (EMI) effects in electromechanical systems, typically ac drives, fed from PWM inverters. For instance, the typical unpleasant whine of an ac motor supplied from an inverter with fixed switching frequency can be changed to soothing "static" by employing an RPWM method in the inverter.

In the classic RPWM strategy, consecutive switching (and sampling) periods are randomly drawn from a uniform-probability pool of values ranging from $xT_{sw,ave}$ to $(2 - x)T_{sw,ave}$, where $T_{sw,ave}$ denotes the desired average switching period and the coefficient x (usually 0.5) determines the shortest switching period as a fraction of the average period. The results of such period dithering applied to the popular space vector strategy are illustrated in Figure 7.30. Output currents of an inverter controlled using the regular version of this strategy with $N = T_o/T_{sw} = 48$, where T_o denotes the period of output voltage, are shown in Figure 7.30a. Because of the balanced load, all current waveforms are identically rippled. In the random mode, when N is randomly varied in the 24–72 range, the current waveforms are such as in Figure 7.30b. Their shapes change from cycle to cycle and phase to phase. Note that the average number of switching cycles per cycle is the same in both cases, so that the average amount of the current ripple remains unchanged. However, there is a big difference in power spectra of output voltages and currents.

The frequency spectrum of line-to-neutral output voltage shown in Figure 7.31a for the regular PWM is purely discrete. Clusters of higher harmonics appear about multiples of the switching frequency. In contrast, apart from the fundamental harmonic that has not changed, the spectrum of the same voltage of the randomly modulated inverter, taken over 75 cycles of output voltage and shown in Figure 7.31b, displays continuous power density. All higher harmonics have disappeared. If the spectrum is measured over a very high number of cycles, the practically white-noise quality can be observed.

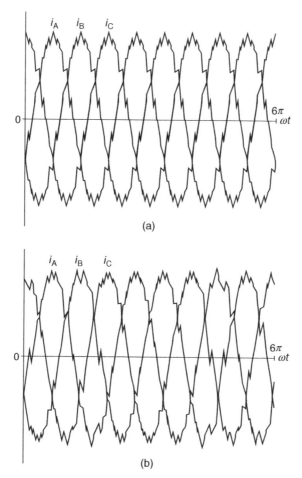

Figure 7.30   Example waveforms of output current in a three-phase VSI: (a) regular PWM, (b) random PWM.

Variations of the sampling frequency in the original RPWM method were disadvantageous. The sampling frequency in digital systems is normally so selected as to represent the best compromise between various operational requirements, especially when the pulse-width modulator is a part of a larger control scheme. Therefore, to maintain a fixed sampling frequency, the so-called variable-delay random pulse width modulation (VD-RPWM) was developed. The switching periods vary, and their average equals the constant sampling period. This is realized by delaying consecutive switching cycles with respect to the corresponding sampling cycles. The delay time, t_{del}, changes from cycle to cycle, being calculated as rT_{smp} where r is a random number from the 0–1 range. It may happen that a given switching period is too short to be feasible. In that case, it is set to the minimum allowable value, $T_{sw,min}$, and, as a result, the lengths of switching cycles vary from $T_{sw,min}$ to $2T_{smp}$. Principles

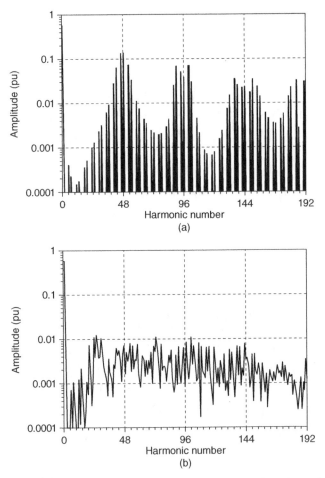

Figure 7.31 Frequency spectra of line-to-neutral output voltage in a three-phase VSI: (a) regular PWM, (b) random PWM.

of the two described random PWM techniques are illustrated in Figure 7.32 along the regular PWM with fixed switching frequency.

7.1.4 Current Control Techniques for VSIs

The PWM techniques described perform open-loop, or *feedforward*, control of output voltages in a VSI. If output currents, which depend not only on the voltages but also on the load, are to be controlled instead, a feedback from current sensors must be provided. A control system compares the actual currents with the reference currents and generates appropriate switching variables, a, b, and c for individual phases of the inverter. As a result, the switching variables and output voltages are pulse width modulated such that the output current waveforms follow the reference waveforms.

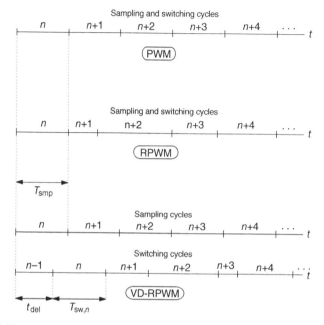

Figure 7.32   Comparison of random PWM techniques with the regular PWM method.

High-performance current control is a challenging task because in most practical cases the inverter load is unknown and varying. A successful current control technique should ensure:

(1) Good utilization of the dc supply voltage, meant here as the feasibility of producing a possibly high current in a given load.
(2) Low static and dynamic control errors.
(3) Low switching losses in the inverter, that is, possibly infrequent switching.
(4) Sufficient time allowance for proper operation of the inverter switches and control system.

It can be seen that requirements (1), (3), and (4) closely match those for the voltage control PWM techniques. As is the case with those techniques, many current control methods have been proposed in the technical literature. The concepts range from very simple to quite complex, the latter involving machine-intelligence systems such as neural networks and fuzzy-logic controllers. Here, four classic approaches to the current control in three-phase inverters are discussed.

Hysteresis Current Control. A VSI with the simplest version of the so-called *hysteresis control* of output currents is shown in Figure 7.33. The output currents, i_A, i_B, and i_C, of an inverter are sensed and compared with the respective reference current waveforms, i_A^*, i_B^*, and i_C^*. Current errors, Δi_A, Δi_B, and Δi_C, are applied to current controllers that produce switching variables, a, b, and c, for the inverter.

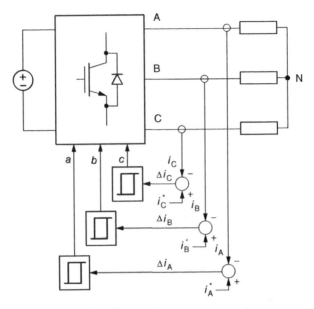

Figure 7.33 Hysteresis current control scheme.

The $a = f(\Delta i_A)$ characteristic of a current controller for phase A of the inverter is shown in Figure 7.34. The characteristic constitutes a hysteresis loop, described as

$$a = \begin{cases} 0 & \text{if } \Delta i_A < -\dfrac{h}{2} \\ 1 & \text{if } \Delta i_A > \dfrac{h}{2} \end{cases}, \tag{7.23}$$

where h denotes the width of the loop. If $-h/2 \leq \Delta i_A \leq h/2$, the value of variable a remains unchanged. Characteristics of controllers for the other two phases are the

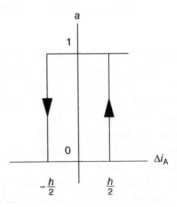

Figure 7.34 Characteristic of the hysteresis current controller.

same and therefore all further considerations will be limited to phase A only. The loop width, h, can be thought of as the width of a tolerance band for the controlled current, i_A, since as long as the current error, Δi_A, remains within this band, no action is taken by the controller. If the error is too high, that is, the actual current is lower from its reference value by more than $h/2$, variable a attains the value of one. According to Eq. (7.10), this makes voltage v_{AN} equal to or greater than zero, which is a necessary condition for current i_A to increase. Analogously, switching a to zero, when the output current is too high, causes v_{AN} to be equal to or less than zero, which is conducive for i_A to decrease.

Interaction between phases of an inverter is a disadvantage of the described scheme. Note that if, for example, variable a is switched from zero to one to increase current i_A, voltage v_{AN} increases as required, but voltages v_{BN} and v_{CN} decrease, which may be detrimental for control of currents i_B or i_C. As a result, in certain loads, such as ac motors, current errors tend to sporadically exceed the tolerance of $\pm h/2$. The value of h affects the average switching frequency, which increases with the decrease in the width of tolerance band. This is illustrated in Figure 7.35, which shows waveforms of output currents in an inverter with an RL load and with h equal to 20% (Figure 7.35a) and 10% (Figure 7.35b) of the peak value of the sinusoidal reference currents.

The hysteresis control is an excellent choice when a fast response to rapid changes of the reference current is required, because the current controllers have negligible inertia and delays. An example situation, when the magnitude, frequency, and phase of reference currents are simultaneously changed, is illustrated in Figure 7.36 for current i_A. It is only the time constant of the load that limits the speed of transition of the current to the new waveform.

In practice, the width, h, of the tolerance band is often made proportional to the magnitude control ratio, M, defined here as

$$M \equiv \frac{I^*_{o,p}}{I^*_{o,p(max)}}, \tag{7.24}$$

where $I^*_{o,p}$ denotes the peak value of the reference output current and $I^*_{o,p(max)}$ is the maximum available peak value of that current. If $h(M) = Mh(1)$, then the ripple factor of the current is maintained at an approximately constant level, independently on the magnitude control ratio. When the peak value of reference currents is increased beyond $M = 1$, an inverter transits from the PWM mode to the square-wave mode of operation.

Instead of using three current controllers, one for each phase of an inverter, two controllers can be employed in the space vector scheme depicted in Figure 7.37. Based on the Park transformation given by Eq. (4.73) and applied to the output currents of an inverter, components i_d and i_q of the space vector, $\vec{i}$, of output currents are computed and compared with respective components, i^*_d and i^*_q, of the reference current vector, $\vec{i}^*$. The current controllers have three-level outputs, z_d and z_q, as illustrated in Figure 7.38 for the d-component controller. Signals z_d and z_q are applied

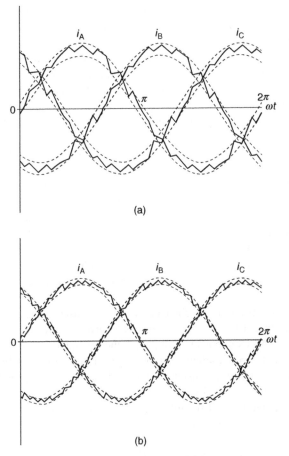

Figure 7.35 Waveforms of output currents in a VSI with hysteresis current control: (a) 20% tolerance, (b) 10% tolerance.

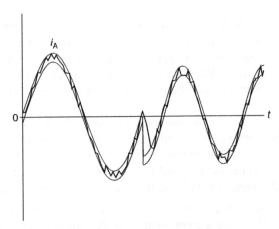

Figure 7.36 Waveform of output current in a VSI with hysteresis current control at a rapid change of the magnitude, frequency, and phase of the reference current.

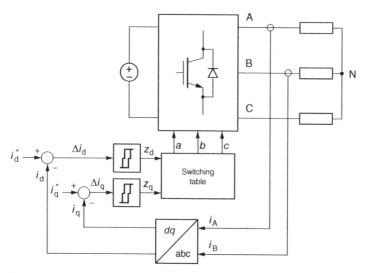

Figure 7.37   Space vector version of the hysteresis current control scheme.

to a switching table, which then selects a specific state of the inverter. The best control effects are obtained when state 0 or 7 is imposed for $(z_d, z_q) = (0, 0)$, state 1 for $(0, 1)$ and $(1, 1)$, state 2 for $(1, -1)$, state 3 for $(1, 0)$, state 4 for $(-1, 0)$, state 5 for $(-1, 1)$, and state 6 for $(-1, -1)$ and $(0, -1)$.

Ramp-Comparison Current Control. The hysteresis control systems are characterized by unnecessarily high switching frequencies, especially at low values of the magnitude control ratio, since the three current controllers act independently from each other. Also, the somewhat chaotic operation of the inverter can be perceived as a disadvantage. To stabilize the switching frequency, the so-called *ramp-comparison control* can be employed. A block diagram of this control scheme is shown in Figure 7.39 for phase A of an inverter.

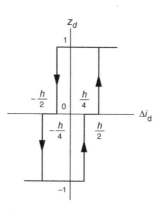

Figure 7.38   Characteristic of the current controller for the space vector version of the hysteresis current control scheme.

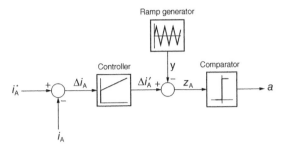

Figure 7.39   Ramp-comparison scheme for current-controlled VSI.

The current error, $\Delta i_A = i_A^* - i_A$, is applied to the input of a linear controller, usually of the proportional-integral (PI) type. The output signal, x_A, of the controller is, in turn, compared with a triangular ramp signal, y, similar to that used in the carrier-comparison PWM technique for voltage-controlled inverters. The difference, z_A, of those two signals activates a comparator, which generates the switching variable a according to the equation

$$a = \begin{cases} 0 & \text{if } z_A \leq 0 \\ 1 & \text{if } z_A > 0 \end{cases}.$$ (7.25)

Identical control loops are used for the other two phases.

It is easy to see that the ramp-comparison control can be thought of as the carrier-comparison PWM technique covered in Section 7.1.3 but with the processed current error, $\Delta i_A'$, as the modulating function. If, for instance, $i_A < i_A^*$ and voltage v_{AN} should be increased to boost i_A, switching variable a is modulated by signal $\Delta i_A'$ in such a way that wide pulses are interspersed with narrow notches. For illustration, go back to Figure 7.21 and imagine that signal r_A is replaced with $\Delta i_A'$.

Example waveforms of the output current of an inverter with the ramp-comparison current control are shown in Figure 7.40 for two values of the ratio of the ramp signal frequency, f_r, to the fundamental output frequency, f_1, of the inverter. The frequency ratio, f_r/f_1, is 10 in Figure 7.40a and 20 in Figure 7.40b, and the same RL load as before is employed. Comparing the waveforms with those in Figure 7.35 for the hysteresis current control, greater regularity of the switching pattern and stability of the switching frequency can be discerned.

Significant advantage of this control scheme lies in the increased number of adjustable parameters. In contrast to the hysteresis control, in which only the width of the hysteresis band of the controllers can be adjusted, the ramp-comparison system allows tuning of the settings of the linear controller, as well as of the amplitude and frequency of the ramp signal. Disadvantages of the ramp-comparison control involve somewhat reduced speed of response to rapid changes of reference currents (this can be improved by using a high-gain proportional controller) and, generally, an increased average switching frequency. The latter characteristic results from the absence of zero states (state 0 or state 7) of the inverter. Indeed, since all the three

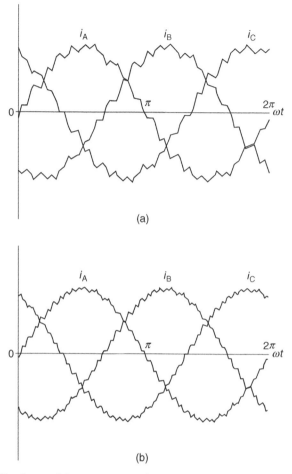

Figure 7.40 Waveforms of the output current in a VSI with ramp-comparison current control: (a) $f_r/f_1 = 10$, (b) $f_r/f_1 = 20$.

current errors cannot simultaneously be of the same polarity, the values of switching variables imposed by the comparators are never all zero or one.

The ramp-comparison technique is usually implemented in an analog system. A related discrete PWM scheme, the so-called *current-regulated delta modulator*, is shown in Figure 7.41. In the digital delta modulator, the ramp signal generator is replaced with a sample-and-hold circuit, which allows fixing the switching frequency of an inverter. Typically no linear controller is used, and the comparator acts as an infinite-gain proportional (P) controller. Again, detrimentally to the efficiency of operation of the modulator, no zero states are imposed.

Predictive Current Control. An optimal switching pattern for a given set of values of the reference currents could be determined if parameters of the load were known. This assumption underlies the principle of the so-called *predictive current controllers*.

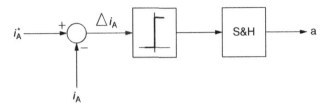

Figure 7.41   Current-regulated delta modulation scheme for a current-controlled VSI.

Two basic types of these controllers minimize either the average switching frequency, f_{sw}, or the total harmonic distortion of the controlled currents at a fixed value of f_{sw}. In each step of operation, based on the response of the load to known voltage changes, a predictive controller performs the estimation of the load parameters to optimize the selection of the next state of the inverter. Microprocessors or digital signal processors are used to handle the computations involved.

Linear Current Control. In the three control schemes described so far, the current feedback directly enforces switching patterns in the inverter. A different, indirect approach, illustrated in Figure 7.42, involves traditional linear controllers of the PI type. Similar to the space vector version of hysteresis control, the reference current is expressed as a space vector $\vec{i}^*$, whose components, i_d^* and i_q^*, serve as reference signals for the respective components, i_d and i_q, of the actual vector, $\vec{i}$, of output currents. Control errors, Δi_d and Δi_q, are converted by the linear controllers into components, v_d^* and v_q^*, of the voltage reference vector, $\vec{v}^*$, (line-to-line or line-to-neutral), to be realized by the inverter using the voltage space vector PWM technique described in

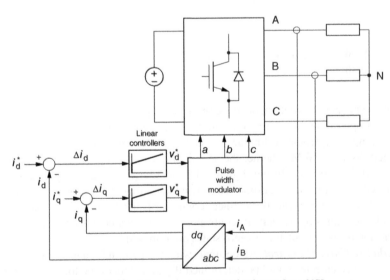

Figure 7.42   Linear current control scheme for a VSI.

Section 7.2.1. Indeed, given the magnitude and phase of $\vec{v}^*$, the corresponding values of m and α required in Eqs. (4.77)–(4.79) can easily be determined.

7.2 CURRENT-SOURCE INVERTERS

The freewheeling diodes, typical for VSIs, become redundant if an inverter is supplied from a dc current source. Then, the current entering any leg of the inverter cannot change its polarity and, therefore, it can only flow through the semiconductor power switches. Use of the current source prevents overcurrents, even in case of a short circuit in the inverter or load. However, in order to avoid hazardous overvoltages across the dc-link inductance, continuity of the current must be preserved during commutation between switches. Therefore, if fully controlled switches are used, switching signals of the outgoing switch and incoming switch overlap a little.

The absence of freewheeling diodes reduces the size and weight of the power circuit and further increases the reliability of the CSI. The practical current source consists of a controlled rectifier, usually SCR based, and an inductive dc link. The output current of the rectifier is maintained at a constant level employing a closed control loop. The dc-link inductor attenuates the current ripple. In the subsequent considerations, an ideal dc-input current, I_i, is assumed.

In practice, CSIs are mostly used for control of ac motors. As such, they are invariably of the three-phase type and, therefore, only those inverters are covered here. If needed, a single-phase CSI can be obtained by removing one leg from the three-phase inverter. Control strategies for the single-phase CSI can be developed by proper adaptation of those for three-phase inverters, described later in this section.

7.2.1 Three-Phase Square-Wave CSI

The block diagram of a three-phase CSI supplied from a three-phase ac line through a controlled rectifier and a dc link is shown in Figure 7.43. The rectifier uses a feedback loop to maintain a constant current in the link. The power circuit of the inverter is depicted in Figure 7.44. The input current cannot be reversed, so a power flow from the load to the source requires reversal of the input voltage. For protection against reverse bias, asymmetrical power switches must be connected in series with blocking diodes. Symmetrical switches, such as nonpunch-through IGBTs, do not require blocking diodes.

As simultaneous conduction of both switches in an inverter leg is allowed, six switching variables, one for each of the inverter switches, must be introduced. In the subsequent considerations, variables a, b, and c represent switches SA, SB, and SC (e.g., $a = 1$ means that SA is in the on-state), while a', b', and c' are assigned to switches SA', SB', and SC'.

The high-power CSIs, typically based on relatively slow GTOs, operate in the square-wave mode only. Within the consecutive one-sixths of the period of output voltage, the conducting switch pairs are SA–SB', SA–SC', SB–SC', SB–SA', SC–SA', and SC–SB'. Note that the same conduction sequence is employed in a

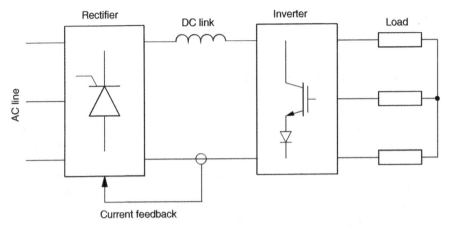

Figure 7.43   CSI supplied from a controlled rectifier.

six-pulse rectifier. Indeed, if the inverter is used to feed an ac motor, which generates a three-phase counter-EMF of rotation, it can be thought of as a six-pulse rectifier operating in the inverter mode. The controlled dc voltage source and dc-link inductor supplying the inverter directly correspond to the commonly used RLE load of the

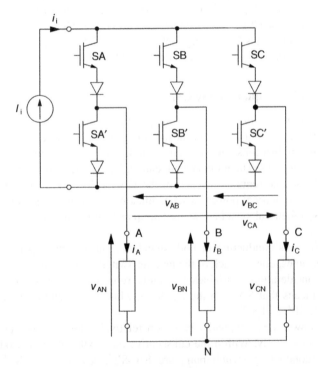

Figure 7.44   Three-phase CSI.

rectifier, while the ac load EMF constitutes a counterpart of the rectifier's supply voltage. Vice-versa, a rectifier supplied from an ac generator and feeding a dc motor represents an inverse of a CSI fed from a dc generator and supplying an ac motor.

It is easy to show that the output currents, i_A, i_B, and i_C, of the inverter are given by

$$\begin{bmatrix} i_A \\ i_B \\ i_C \end{bmatrix} = \begin{bmatrix} a - a' \\ b - b' \\ c - c' \end{bmatrix} I_i \tag{7.26}$$

while currents i_{AB}, i_{BC}, and i_{CA} in a balanced delta-connected load are given by

$$\begin{bmatrix} i_{AB} \\ i_{BC} \\ i_{CA} \end{bmatrix} = \frac{1}{3} \begin{bmatrix} 1 & -1 & 0 \\ 0 & 1 & -1 \\ -1 & 0 & 1 \end{bmatrix} \begin{bmatrix} i_A \\ i_B \\ i_C \end{bmatrix}. \tag{7.27}$$

Waveforms of switching signals for the square-wave operation are shown in Figure 7.45, and those of the corresponding currents in a wye-connected load (i_A, i_B, and i_C) and delta-connected load (i_{AB}, i_{BC}, and i_{CA}) are shown in Figure 7.46. The *current gain*, K_I, defined as the ratio of the peak value, $I_{L,1p}$, of the fundamental output line current to the input current, I_i, is $2\sqrt{3}/\pi \approx 1.1$.

Rapid changes of the output currents during transitions from one state to another would cause high-voltage spikes across inductive loads. Therefore, in practice, the commutation between inverter switches is purposefully prolonged, in order to limit the spikes to an allowable, safe level. Also, if possible, low-reactance loads are selected, such as induction motors with low leakage inductances. The output voltage waveform depends on the load. This is illustrated in Figure 7.47, depicting example

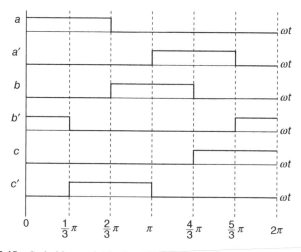

Figure 7.45 Switching variables in a three-phase CSI in the square-wave mode.

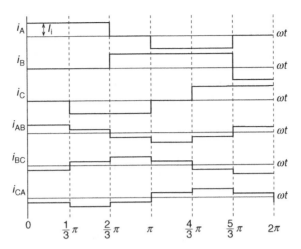

Figure 7.46 Waveforms of output currents in a three-phase CSI in the square-wave mode.

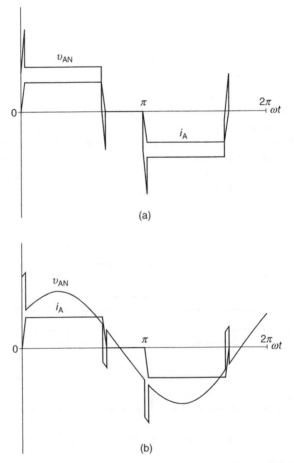

Figure 7.47 Waveforms of output voltage and current in a three-phase CSI in the square-wave mode: (a) RL load, (b) LE load.

waveforms of output voltage and current in one phase of an inverter. Figure 7.47a shows the voltage and the current with an RL load, while waveforms in Figure 7.47b correspond to an LE load, that is, a series connection of an inductance and an ac EMF. The later load is commonly used to model ac motors in adjustable speed drives, which represent an almost exclusive application of CSIs.

The square-wave mode of operation does not provide for magnitude control of the output current within the inverter, hence the current must be controlled in the rectifier supplying the inverter. The rectifier also allows the bidirectional power flow, which is an important advantage of CSIs. As already mentioned, the input current is always positive, so the negative power flow requires a negative average output voltage of the rectifier. Apart from the simplicity and reliability of the power circuit, the excellent dynamics of current control, important in high-performance ac drives, represents a distinct advantage of CSIs. On the other hand, the highly distorted, stepped waveforms of the output current constitute an obvious weakness.

7.2.2 Three-Phase PWM CSI

PWM CSIs are feasible although less common in practice than their voltage-source counterparts. Block diagram of such inverter is shown in Figure 7.48. Comparing it with the square-wave inverter in Figure 7.43 note the addition of capacitors between the output terminals. The capacitors act as low-pass filters for output currents, shunting most of the high-frequency harmonic content of pulsed currents produced by the inverter switches. Clearly, a PWM CSI represents an exact inverse of the current-type PWM rectifier depicted in Figure 4.45.

Control strategies for PWM CSIs differ from those for VSIs. The pulsed output currents i'_A, i'_B, and i'_C are generated using switching patterns such that only two switches, one in the common-cathode group and one in the common-anode group, are simultaneously on. Typically, the two conducting switches belong to different legs of the inverter, so that the currents flow in two phases. However, in certain applications, the shoot-through situations, when both switches in a given phase are on, are included in the control strategy in order to ease the voltage stress on the output capacitors.

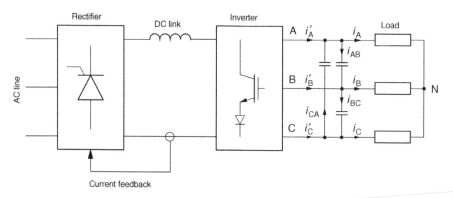

Figure 7.48   Three-phase PWM CSI.

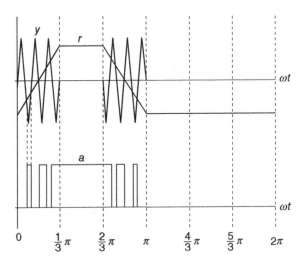

Figure 7.49 Carrier-comparison method for the PWM CSI ($P = 5$).

The inverter can be supplied from a fixed-current or adjustable-current source. Subsequent considerations apply to the latter case. As a result of a given PWM technique a number of current pulses appear in each phase of the inverter within a single cycle of output current. This number equals $2P$, where P denotes the always-odd number of pulses of each switching variable. A fixed switching pattern is generated, while the magnitude control of the output currents is realized in the supplying rectifier. Similarly as in PWM VSIs, the waveforms of output currents, i_A, i_B, and i_C, of a PWM CSI are rippled sinusoids. Both the open-loop control and the closed-loop control of output currents are feasible.

Two PWM techniques have found widespread application in CSIs. The first one resembles the classic carrier-comparison method for VSIs. However, both the carrier and modulating function are different here. Generation of a switching pattern with $P = 5$, which results in 10 pulses of currents i'_A, i'_B, and i'_C per cycle, is illustrated in Figure 7.49. It shows waveforms of the triangular carrier, y, trapezoidal reference signal, r, and switching variable a. Waveforms of other switching variables are identical, but shifted in phase. Specifically, those of b and c are delayed by 120° and 240° with respect to $a(\omega t)$, while waveforms of variables a', b', and c' are shifted with respect to those of a, b, and c, by a half cycle. Note that the switching pattern for a square-wave inverter displays the same property (see Figure 7.45). The best attenuation of low-order harmonics in the output currents is achieved when the ratio of peak values of the reference and carrier signals is 0.82.

The second technique is a programmed, harmonic-elimination PWM method. By proper selection of the primary switching angles, specific harmonics of the output currents can be cancelled. A switching pattern, again with $P = 5$, is shown in Figure 7.50. Two optimal primary switching angles, α_1 and α_2, equal 7.93° and 13.75°, respectively, allow elimination of the 5th and 7th harmonics. If the 11th harmonic was to be

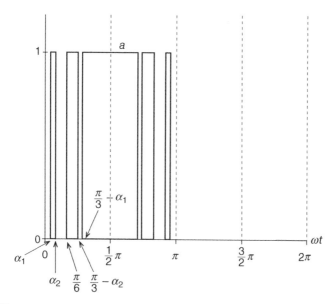

Figure 7.50 Optimal switching pattern for the PWM CSI with two primary switching angles.

cancelled as well, 14-pulse ($P = 7$) currents would have to be produced using three optimal switching angles of 2.24°, 5.60°, and 21.26°. An optimal switching pattern, $a(\omega t)$, must have the quarter-wave and half-wave symmetry and must be symmetrical about 30° and 150°. No switching is permitted in the 60° to 120° interval.

Example waveforms of the output currents, i_A and i'_A, capacitor current, i_{AB}, and output voltage, v_{AN}, in an inverter with a wye-connected RL load and the carrier-comparison PWM with $P = 9$ are shown in Figure 7.51. The ripple of the output current can be reduced further by employing larger output capacitors or increasing the number of switching pulses, P. The output voltage waveform depends on the load. Interestingly, the current gain, K_I, for the pulsed currents, i'_A, i'_B, and i'_C, may exceed unity. For instance, with the programmed method for elimination of the 5th and 7th harmonics, K_I of 1.029 is higher than that, of 0.955, for the square-wave inverter. However, the output capacitors divert a part of fundamentals of the pulsed currents from the output, so that the current gain for the output currents i_A, i_B, and i_C is lower than unity.

If a fixed-current supply source for the rectifier is employed, a space vector PWM technique can be used to control the output currents. The identity of such techniques for control of voltage-type PWM rectifiers and voltage-source PWM inverters has already been pointed out in Section 7.1.3. For the current-source PWM rectifier considered here, the analogy with the current-type PWM rectifier described in Section 4.3.3 is exploited. Specifically, the space vectors of currents shown in Figure 4.46 and the corresponding formulas (4.90)–(4.92) are used.

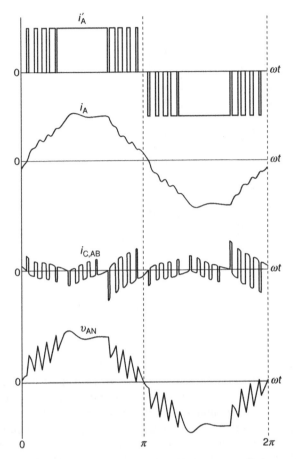

Figure 7.51   Waveforms of the output current, capacitor current, and output voltage in a three-phase PWM CSI (wye-connected RL load, $P = 9$).

7.3   MULTILEVEL INVERTERS

Three-phase VSIs described in Section 7.1 are by far the most common dc-to-ac power converters encountered in practice. As explained later, they can be classified as *two-level* inverters. Although first proposed decades ago, the so-called *multilevel* VSIs have enjoyed a widespread interest only in recent years. The multilevel inverters offer better performance than two-level inverters, and their voltage ratings exceed those of the employed power switches, including power diodes. Specifically, the voltage stress on individual switches equals $V_i/(l-1)$, where V_i denotes the input dc voltage and l is the number of levels. Therefore, multilevel inverters are mostly applied in medium- and high-voltage power electronic systems.

The "two-level" adjective describing the VSIs presented so far has arisen from the fact that the potential of any output terminal can assume two values only. Indeed, the

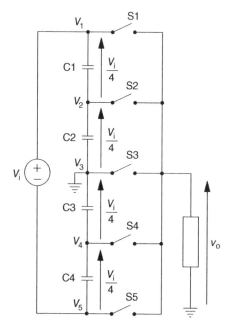

Figure 7.52   Generic five-level inverter.

switches of an inverter connect each terminal to either the positive or the negative dc bus. Consequently, the line-to-line voltage can assume three values, and the line-to-neutral voltage five values. This, in the square-wave mode, which is highly efficient, limits the options for minimizing distortion of output voltage waveforms. If more than two voltage levels were obtainable at the output terminals, the stepped waveforms could be made to better resemble the sinewaves.

The idea of a single-phase multilevel inverter, here a five-level one, is illustrated in Figure 7.52. The four capacitors, C1 through C4, make up a voltage divider, which constitutes a dc link for the inverter. The center node of the link and one terminal of the load are grounded. Five switches, S1 through S5, of which one and only one is assumed to be on at any time, allow applying any of the five fractional voltages, V_1 through V_5 to the non-grounded load terminal. Thus, the output voltage, v_o, can assume any of these five values, from $-V_i/2$ to $V_i/2$, including $v_o = V_3 = 0$.

Note that even if the number of levels of the generic multilevel inverter were limited to two, by elimination of switches S2 through S4 and capacitors C2 and C3 between them, the resultant topology would not be equivalent to that of the single-phase inverter in Figure 7.2. The power circuit of that inverter is of the bridge type, while the inverter in Figure 7.52 has the so-called *half-bridge* topology. A practical, although of marginal use, half-bridge single-phase inverter is shown in Figure 7.53. Clearly, the inverter is capable of a two-level output voltage only, that is, $v_o = \pm V_i/2$, while the full-bridge inverter of Figure 7.2 may have the output voltage of either $-V_i$, 0, or V_i.

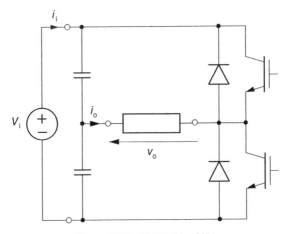

Figure 7.53   Half-bridge VSI.

7.3.1   Diode-Clamped Three-Level Inverter

The unidirectional semiconductor devices in practical multilevel inverters require somewhat more complicated topologies. This and the next section describe two basic topologies of three-phase *three-level inverters*, leaving those with five or more levels to more advanced literature.

The power circuit of the so-called *diode-clamped* (or *neutral-clamped*) inverter is shown in Figure 7.54. Each of the three legs of the inverter is composed of four semiconductor power switches, S1 through S4, four freewheeling diodes, D1 through D4, and two clamping diodes, D5 and D6. Theoretically, the four switches in each inverter leg imply the possibility of 2^4 states of a leg, and 2^{12} (that is, 4,096!) states of the whole inverter. However, only three states of a leg are allowed, which makes for the total of 27 states of the inverter. The limiting condition is that two and only two adjacent switches must be ON at any time. A ternary switching variable can thus be assigned to each inverter phase and, for phase A, defined as

$$a = \begin{cases} 0 & \text{if S3 and S4 are ON} \\ 1 & \text{if S2 and S3 are ON}. \\ 2 & \text{if S1 and S2 are ON} \end{cases} \tag{7.28}$$

Switching variables b and c for the other two phases are defined analogously.

It is easy to see that the potential of a given output terminal of the inverter with respect to the "ground" (inverter's neutral), G, can be expressed in terms of the associated switching variable and input voltage. For instance, the voltage, v_A, at terminal A is

$$v_A = \frac{a-1}{2}V_i. \tag{7.29}$$

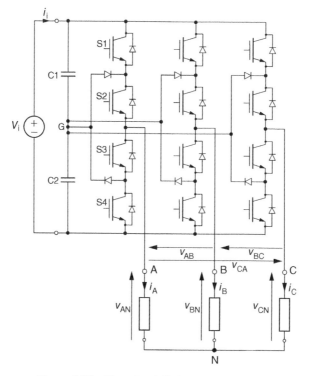

Figure 7.54 Three-level diode-clamped inverter.

Consequently, the output line-to-line voltages are given by

$$\begin{bmatrix} v_{AB} \\ v_{BC} \\ v_{CA} \end{bmatrix} = \frac{V_i}{2} \begin{bmatrix} 1 & -1 & 0 \\ 0 & 1 & -1 \\ -1 & 0 & 1 \end{bmatrix} \begin{bmatrix} a \\ b \\ c \end{bmatrix} \tag{7.30}$$

and, based on Eq. (7.9), the line-to-neutral voltages by

$$\begin{bmatrix} v_{AN} \\ v_{BN} \\ v_{CN} \end{bmatrix} = \frac{V_i}{6} \begin{bmatrix} 2 & -1 & -1 \\ -1 & 2 & -1 \\ -1 & -1 & 2 \end{bmatrix} \begin{bmatrix} a \\ b \\ c \end{bmatrix} \tag{7.31}$$

Listing all possible values of the line-to-line and line-to-neutral voltages it can be seen that they can assume five and nine values, respectively. Generally, these numbers in an l-level inverter are $2l - 1$ and $4l - 3$.

Control methods for the three-level inverter allow changes from only 0 to 1, 1 to 2, and vice-versa for each switching variable, while transitions from 0 to 2 and 2 to 0 are forbidden. This ensures smooth commutation and reduces the chances for a shoot-through, since out of the four switches in a leg only two change their conduction states

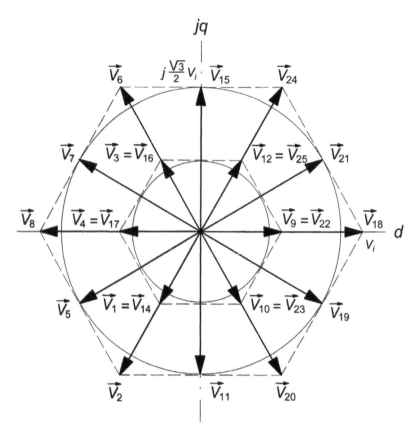

Figure 7.55 Voltage space vectors of a three-level neutral-clamped inverter.

simultaneously. The dc voltage is shared by at least two switches so that, as already mentioned, their voltage ratings can be significantly lower than those of switches in a regular, two-level inverter supplied from the same source.

Analogously to the two-level inverter, a state of the three-phase three-level inverter can be designated as abc_3. For example, an inverter is said to be in state 14 if $a = 1$, $b = 1$, and $c = 2$, since $112_3 = 14$. Out of the 27 states, states 0, 13, and 26 are zero states, yielding zero voltages at all three output terminals. Space vectors of the line-to-neutral voltages across a wye-connected load, corresponding to individual states of the inverter are shown in Figure 7.55. It can be seen the active voltage vectors can be divided into three groups. The high-voltage vectors, such as $\vec{V}_{18}$, have the magnitude of V_i, the medium-voltage vectors, such as $\vec{V}_{21}$, are $\sqrt{3}V_i/2$ strong, and the magnitude of the low-voltage vectors, such as $\vec{V}_9$, is $V_i/2$.

Comparing the vector diagram with that in Figure 7.23 for the two-level inverter, it is easy to guess that the 18–21–24–15–6–7–8–5–2–11–20–19–... state sequence, each state maintained for one-twelfth of the desired cycle of output voltage, represents the square-wave operation mode. The corresponding waveforms of switching

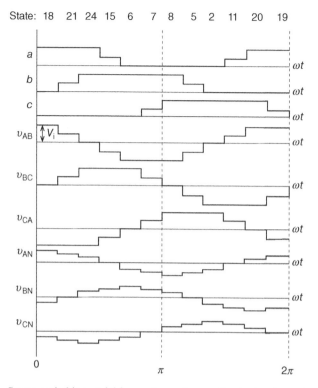

State: 18 21 24 15 6 7 8 5 2 11 20 19

Figure 7.56 States, switching variables, and waveforms of output voltages in a three-level neutral-clamped inverter in the square-wave mode.

variables and output voltages of the inverter are shown in Figure 7.56. Although the voltage gain is 1.065, that is, slightly lower than the 1.1 value for a two-level inverter, higher quality of the output voltage waveforms in the three-level inverter is obvious (see Figure 7.16 for comparison). The low-distortion waveform of output current i_A, shown in Figure 7.57 with that of voltage v_{AN}, confirms this conclusion.

Even higher quality of operation can be achieved by PWM, with significantly lower switching frequencies than those typical for two-level inverters. Several PWM techniques based on the space vector approach have been developed. Note that the number of stationary voltage vectors to be used for synthesis of the revolving reference vector is here much higher than in the classic two-level VSI. This allows more freedom in designing effective space vector PWM strategies.

7.3.2 Flying-Capacitor Three-Level Inverter

The *flying-capacitor inverter* is somewhat similar to the diode-clamped one described in the preceding section. A three-level version of that inverter is shown in Figure 7.58. It can be seen that the clamping diodes in Figure 7.54 are replaced here with a capacitor C3.

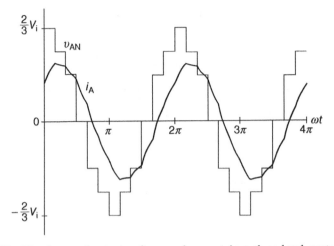

Figure 7.57 Waveforms of output voltage and current in a three-level neutral-clamped inverter in the square-wave mode.

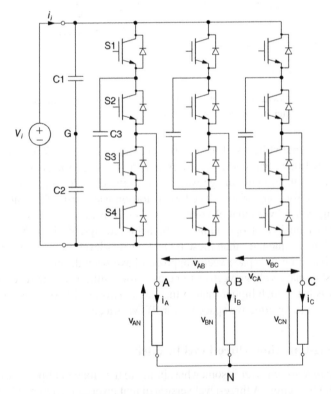

Figure 7.58 Three-level flying-capacitor inverter.

Allowing two and only two (but not necessarily adjacent) switches in a given inverter leg to be ON, the switching variable, a, for phase A of the inverter is given by

$$a = \begin{cases} 0 & \text{if S3 and S4 are ON} \\ 1 & \text{if S1 and S3 or S2 and S4 are ON} \\ 2 & \text{if S1 and S2 are ON} \end{cases} \qquad (7.32)$$

and switching variables, b and c, for the other two phases are defined analogously. Then, for calculation of voltages, Eqs. (7.29) through (7.31) can be employed, and voltage vectors are the same as in Figure 7.55. The clamping capacitor is charged when S1 and S3 are ON and discharged when S2 and S4 are ON.

7.3.3   Cascaded H-Bridge Inverter

The *cascaded H-bridge* inverter, shown in Figure 7.59, is assembled from N single-phase inverters, already described in Section 7.1.1 and depicted in Figure 7.2. Each constituent inverter (H-bridge) is supplied from a separate dc source. The sources have usually the same voltage level, although the so-called asymmetrical inverters, in which individual source voltages differ, are also feasible. Identical dc sources are assumed in the subsequent considerations.

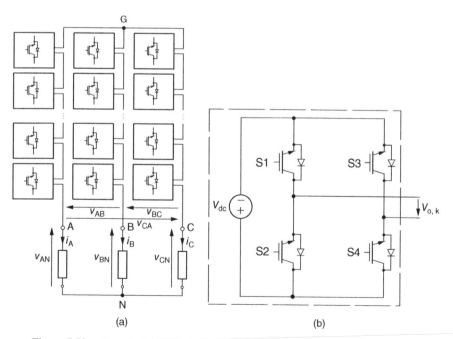

Figure 7.59 Cascaded H-bridge inverter: (a) block diagram, (b) constituent bridge.

An H-bridge can generate three voltage levels between its output terminals, namely $-V_{dc}$, 0, and V_{dc}. The number of H-bridges in an l-level inverter is $(l-1)/2$, thus N bridges form an inverter with $2N + l$ levels. Topologies of diode-clamped or flying-capacitor inverters with a high number of levels are quite complex and require sophisticated control methods, so the cascaded H-bridge inverter has a distinct advantage here.

In the constituent H-bridge two and only switches can be ON at any time. For the kth bridge in a leg of a three-phase inverter, the ternary switching variable, a_k, is defined as follows:

$$a_k = \begin{cases} 0 & \text{if S2 and S3 are ON} \\ 1 & \text{if S1 and S3 or S2 and S4 are ON} \\ 2 & \text{if S1 and S4 are ON} \end{cases} \tag{7.33}$$

The output voltage, $v_{o,k}$, of the bridge is then given by

$$v_{o,k} = (a_k - 1)V_{dc}. \tag{7.34}$$

The H-bridges in a leg of the inverter are connected in series, so the voltage of terminal A, v_A, with respect to the inverter's neutral, G, is a sum of output voltages of all the bridges. Consequently, a switching variable of phase A of the inverter can be defined as

$$a = \sum_{k=1}^{N} a_k \tag{7.35}$$

where, depending on control of the individual bridges, a can assume any integer value from the 0 to $2N$ range. Then,

$$v_A = (a - N)V_{dc}. \tag{7.36}$$

Switching variables b and c are defined analogously, yielding the following equations for the line-to-line and line-to-neutral output voltages of the inverter:

$$\begin{bmatrix} v_{AB} \\ v_{BC} \\ v_{CA} \end{bmatrix} = V_{dc} \begin{bmatrix} 1 & -1 & 0 \\ 0 & 1 & -1 \\ -1 & 0 & 1 \end{bmatrix} \begin{bmatrix} a \\ b \\ c \end{bmatrix} \tag{7.37}$$

and

$$\begin{bmatrix} v_{AN} \\ v_{BN} \\ v_{CN} \end{bmatrix} = \frac{V_{dc}}{3} \begin{bmatrix} 2 & -1 & -1 \\ -1 & 2 & -1 \\ -1 & -1 & 2 \end{bmatrix} \begin{bmatrix} a \\ b \\ c \end{bmatrix}. \tag{7.38}$$

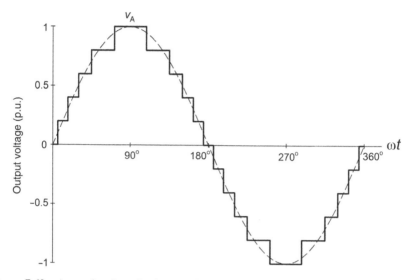

Figure 7.60   Approximation of a sinewave by a stepped waveform in the H-bridge cascaded inverter.

Note the similarity of Eqs. (7.36)–(7.38) and Eqs. (7.29)–(7.31) when $V_i = 2V_{dc}$.

The cascaded H-bridge inverter allows generation of output voltages whose multistep waveforms closely approximate sinewaves. The square-wave operation resulting in those waveforms is highly efficient, as each switch turns on and off only once per cycle of the output voltage. This is illustrated in Figure 7.60, where a reference sinusoid of v_A is approximated with a high degree of accuracy by a stepped waveform produced in a five-bridge inverter.

The square-wave operation mode of individual bridges does not allow for smooth magnitude control of output voltages. If K bridges are involved, the peak value of the generated voltage equals KV_{dc}. However, reducing K increases the THD of the produced waveform. Therefore, the square-wave operation is recommended in systems where magnitude control is not required, or where the dc sources can be adjusted as needed. If the magnitude adjustment is to be carried out in the inverter, a simple PWM technique, already described in Section 7.1.1, can be employed in each bridge with the same modulation index, m. See Eq. (7.5) for phase A of the cascaded H-bridge inverter. The arguments α_n of the modulating functions for phases B and C should be expanded by 120° and 240°, respectively.

The large number of available values of switching variables a, b, and c of the cascaded H-bridge inverters allows numerous SVPWM switching strategies. Endpoints of space vectors of the line-to-neutral voltage corresponding to all combinations of values of switching variables in a two-bridge (five-level) inverter are shown in Figure 7.61. As each variable can assume five values, 0 through 4, $5^3 = 125$ vectors are generated, of which 64 are repeated, leaving 61 distinct ones. This set includes, for instance, the six active (non-zero) vectors of the classic, three-phase two-level inverter, shown in Figure 7.23.

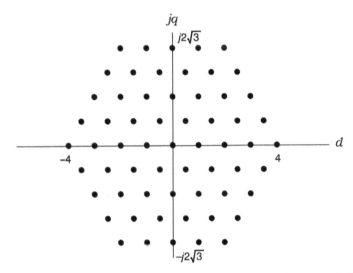

Figure 7.61   Endpoints of line-to-neutral voltage vectors of two-bridge cascaded inverter.

In certain applications, the constituent converters are supplied from ac sources rather than dc ones. Such inverters combine a number of cascaded "cells," which usually constitute the H-bridges augmented with front-end rectifiers. Two such cells, with diode rectifiers, are shown in Figure 7.62. If the capability of bi-directional power flow is required, PWM rectifiers are used instead as shown in Figure 7.63.

Multilevel inverters are particularly suitable for high-power, high-voltage applications. High performance can be attained at significantly lower switching frequencies than those in the classic two-level inverters. Thus, the relatively sluggish GTOs can be employed and still produce high-quality currents. Several hybrid topologies

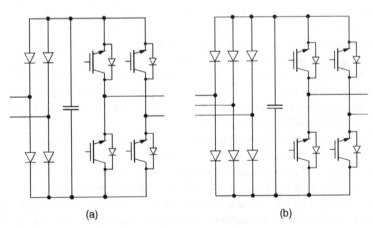

(a)   (b)

Figure 7.62   Cells with diode rectifiers for ac-supplied cascaded inverter: (a) single-phase, (b) three-phase.

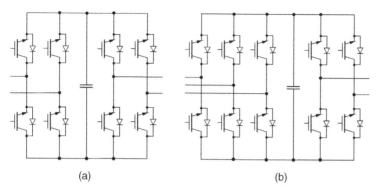

Figure 7.63   Cells with PWM rectifiers for ac-supplied cascaded inverter: (a) single-phase, (b) three-phase.

combining sub-circuits of differing types of the basic multilevel inverters described have been proposed.

7.4   SOFT-SWITCHING INVERTERS

All the inverters and other power converters presented so far are characterized by the so-called *hard switching*. It means that the transition of their switching devices from one conduction state to another occurs in the presence of nonzero voltages and currents. Consider, for instance, phase A of the VSI in Figure 7.14. When switch SA is off and diode DA is not conducting, the full input dc voltage, V_i, appears across them. Conversely, if switch SA is on and diode DA' is not conducting, current i_A flows through the switch. Thus, a switch turns on with a nonzero voltage across it and turns off when carrying a non-zero current. As a result, as already explained in Chapter 2, each switching is associated with certain energy loss, which depends on the values of current and voltage involved.

If the switching frequency were high enough, currents practically free from ripple could be produced in VSIs. Then, however, the switching losses would increase dramatically. Therefore, in practice, the switching frequency is usually limited not as much by the dynamic properties of the switches as by the thermal effects. Hard switching is also detrimental from the point of view of the radiated EMI, as electromagnetic waves are generated due to rapid current changes.

If the voltage across a device transitioning from the off-state to the on-state were zero, or no current flew through the device to be turned off, the resultant *soft switching* would cause minimum losses and EMI. As subsequently described in Chapter 8, these two types of soft switching, the zero-voltage switching (ZVS) and zero-current switching (ZCS), are employed in the so-called *resonant converters* for low-power switching power supplies. The phenomenon of electrical resonance is utilized there to impose lossless, ZVS or ZCS conditions. Significant reduction of switching losses in resonant converters allows for high switching frequencies, leading to minimization

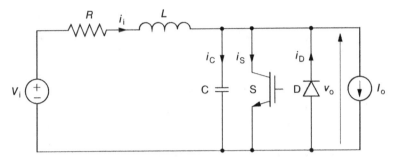

Figure 7.64 Switched network for illustration of the operating principle of resonant dc link.

of filters, magnetic components, and heat sinks, and, consequently, of the whole converters.

A *resonant dc-link inverter* represents an endeavor of implementing the ZVS principle in voltage-source dc-to-ac converters. To explain the basic idea of the resonant dc link, the switched network in Figure 7.64 will first be considered. It consists of a supply source producing an ideal dc-input voltage, V_i, a resonant LC circuit, a semiconductor switch S, and a freewheeling diode D. The resonant circuit is non-ideal, having the resistance of R ohms. The current source at the network's output represents a load, whose inductance is assumed to be much higher than that, L, of the resonant circuit. Consequently, the output current, I_o, can be assumed constant during the short cycle of operation of the network.

The operation cycle can be divided into three subcycles. In the first of them, the switch is closed for a period of t_1, and the inductance, L, is charged with the electromagnetic energy due to the increased flow of input current, i_i. As explained later, the switch current does not start flowing immediately after the switching signal, g, is applied to the switch at $t = 0$, but after a short delay, t_0. The capacitor, C, has been discharged in the preceding cycle, so its current, i_C, is zero, and so are the diode current, i_D, and output voltage, v_o. Since

$$V_i = Ri_i + L\frac{di_i}{dt} \tag{7.39}$$

and

$$i_S = i_i - I_o \tag{7.40}$$

then

$$i_i = \frac{V_i}{R}\left(1 - e^{-\frac{R}{L}t}\right) + I_1 e^{-\frac{R}{L}t} \tag{7.41}$$

and

$$i_S = \frac{V_i}{R}\left(1 - e^{-\frac{R}{L}t}\right) + I_1 e^{-\frac{R}{L}t} - I_o \tag{7.42}$$

where I_1 denotes the initial input current, $i_i(t_0)$. At the end of the considered subcycle, the input current attains the value of I_2, equal

$$I_2 = i_i(t_1) = \frac{V_i}{R}\left(1 - e^{-\frac{R}{L}t_1}\right) + I_1 e^{-\frac{R}{L}t_1}. \tag{7.43}$$

In the second subcycle, the switch opens and the energy stored in the inductance is discharged through the capacitor. Now, both the switch and diode currents are zero, while the capacitor current and output voltage are given by

$$i_C = i_i - I_o \tag{7.44}$$

and

$$v_o = \frac{1}{C}\int_0^t i_C dt. \tag{7.45}$$

The Kirchhoff Voltage Law for the resonant circuit can be written as

$$V_i = Ri_i + L\frac{di_i}{dt} + v_o \tag{7.46}$$

and substituting Eq. (7.44) in Eq. (7.45), and Eq. (7.45) in Eq. (7.46), yields

$$V_i = Ri_i + L\frac{di_i}{dt} + \frac{1}{C}\int_0^t i_i dt - \frac{I_o}{C}t. \tag{7.47}$$

Assuming that the resonant circuit is underdamped, solving Eq. (7.47) for i_i, and finding i_C and v_o from Eqs. (7.44) and (7.45), gives

$$i_C = \left\{\left[\frac{V_i - RI_o}{L\omega_r} - \frac{\alpha}{\omega_r}(I_2 - I_o)\right]\sin(\omega_r t) + (I_2 - I_o)\cos(\omega_r t)\right\}e^{-\alpha t} \tag{7.48}$$

and

$$v_o = V_i - RI_o + \left\{\left[\frac{\alpha}{\omega_r}(V_i - RI_o) + \frac{I_2 - I_o}{C\omega_r}\right]\sin(\omega_r t)\right.$$
$$\left. -(V_i - RI_o)\cos(\omega_r t)\right\}e^{-\alpha t}, \tag{7.49}$$

where

$$\alpha = \frac{R}{2L} \tag{7.50}$$

and

$$\omega_r = \sqrt{\frac{1}{LC} - \alpha^2} \qquad (7.51)$$

denotes the damped resonance frequency.

If quality of the resonant circuit is high, then $RI_o \ll V_i$ and $\alpha \ll \omega_r$, and Eqs. (7.48) and (7.50) can be simplified, to give

$$i_C = Ie^{-\alpha t} \sin(\omega_r t + \varphi_1) \qquad (7.52)$$

and

$$v_o = V_i - Ve^{-\alpha t} \cos(\omega_r t + \varphi_2), \qquad (7.53)$$

where

$$I = \sqrt{\left(\frac{V_i}{L\omega_r}\right)^2 + (I_2 - I_o)^2} \qquad \varphi_1 = \tan^{-1}\left[\frac{L\omega_r(I_2 - I_o)}{V_i}\right] \qquad (7.54)$$

and

$$V = \sqrt{V_i^2 + \left(\frac{I_2 - I_o}{C\omega_r}\right)^2} \qquad \varphi_2 = \tan^{-1}\left(\frac{I_2 - I_o}{C\omega_r V_i}\right). \qquad (7.55)$$

As $L\omega_r \approx 1/(C\omega_r)$, then $\varphi_1 \approx \varphi_2$. At the beginning of the subcycle in question, the output voltage is zero, then it increases to a certain peak value, $V_{o,p}$, and decreases back to zero. If not for the freewheeling diode, the voltage would next become negative, completing the cycle of resonant oscillation. However, once the diode becomes forward biased, it shorts the output of the network and the output voltage remains at the zero level.

The instant at which the output voltage gets back to zero after the half-wave oscillation marks the beginning of the third subcycle of the operating cycle. Denoting the duration of the second subcycle by t_2, the input current at the end of the subcycle assumes the value of I_3, equal

$$I_3 = i_1(t_2) = I_o + i_C(t_2) = I_o + Ie^{-\alpha t_2} \sin(\omega_r t_2 + \varphi_1). \qquad (7.56)$$

In the third subcycle, whose duration is t_3, the switch and capacitor currents are zero, and so is the output voltage, while the diode current, i_D, is given by

$$i_D = I_o - i_i \qquad (7.57)$$

where i_i can be found from the again valid Eq. (7.39). As a result, the input current can be expressed as

$$i_i = \frac{V_i}{R}\left(1 - e^{-\frac{R}{L}t}\right) + I_3 e^{-\frac{R}{L}t} \qquad (7.58)$$

and the diode current as

$$i_D = I_0 - \frac{V_i}{R}(1 - e^{-\alpha t}) - I_3 e^{-\alpha t}. \qquad (7.59)$$

Obviously, Eq. (7.52) is only valid as long as it yields a positive value of i_D. When the left-hand side of that equation reaches zero, the diode ceases to conduct.

The next cycle of operation of the network should begin before the diode current reaches zero, so that the switch is turned on under the zero-voltage condition. On the other hand, the flow of the switch current, i_S, can only begin when the diode has ceased to conduct, since simultaneous conduction of these two anti-parallel devices is not possible. This explains the delay, t_0, of the switch current in the first subcycle. In retrospect, the three subcycles can be called *charging, resonance*, and *dead time* (the latter should not to be confused with that used for avoidance of the shoot-through in hard-switching converters). Waveforms of the voltage and currents considered are illustrated in Figure 7.65. The ZVS conditions for both the switch and diode can easily be observed.

Equation (7.55) indicates strong dependence of the output voltage, v_0, on the I_2–I_0 difference. Note that the output current, I_0, has been assumed constant only within a single cycle of operation of the considered network. Therefore, the control system of a resonant dc-link inverter must monitor both the output current and the input current at the instant t_1 of each operating cycle.

In a practical inverter, the output voltage, v_0, of the resonant dc link is fed to a regular VSI. The three legs of the power circuit will thus alternately play the role of the switch-diode pair in the hypothetical network analyzed before. The shoot-through states of each leg are employed in the charging subcycle. The average value of the pulsed voltage v_0 is only slightly lower than the dc supply voltage, V_i, because of the low power loss in the resistance of the dc link. However, the peak value of v_0, which can be up to three times as high as the dc voltage supplied to the resonant circuit, may require inverter switches and diodes with unacceptably high voltage ratings. Therefore, an additional clamping arrangement is often needed to shave off the peaks of the voltage pulses.

The circuit diagram of a three-phase resonant dc-link inverter with an active clamp is shown in Figure 7.66. The active clamp consists of a clamping capacitor, C_{cl}, precharged to $(k_{cl}-1)V_i$ volts and an anti-parallel connection of a switch, S, and diode, D. The *clamping voltage ratio*, k_{cl}, is usually set to 1.2–1.4. As in the hypothetical switching network considered before, the dc bus is shorted in the first subcycle by switches in one leg of the inverter.

In the second subcycle, the link voltage across the resonant capacitor, C, increases toward its natural resonant peak. However, when the voltage reaches the clamping

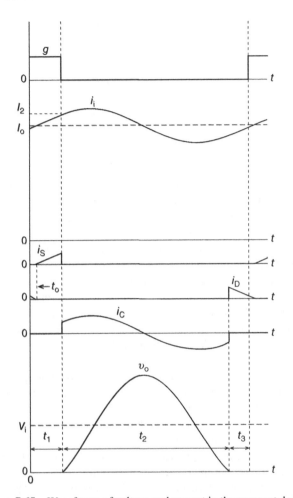

Figure 7.65   Waveforms of voltage and current in the resonant dc link.

value, V_{cl}, equal $k_{cl}V_i$, diode D starts conducting. As a result, the clamping capacitor shunts the resonant inductor, L, and the bus voltage becomes clamped to $V_i + (k_{cl}-1)V_i = V_{cl}$. The diode current charges the clamping capacitor. Next, switch S is turned on under the ZVS condition, the diode current eventually transfers to the switch, and the clamping capacitor loses its extra charge. The switch is turned off as soon as the net charge received by the clamping capacitor reaches zero, so that the clamping circuit is ready for the next cycle of operation. The dc bus voltage decreases to zero, to complete the resonance subcycle and, after the dead time, appropriate switches of the inverter are turned on again, initiating the next operating cycle of the inverter.

The resonance frequency, ω_r, must be at least two orders of magnitude higher than the output frequency of the inverter, so that continuous rectangular pulses of the line-to-line voltages of hard-switching VSIs can be replaced by series of separate

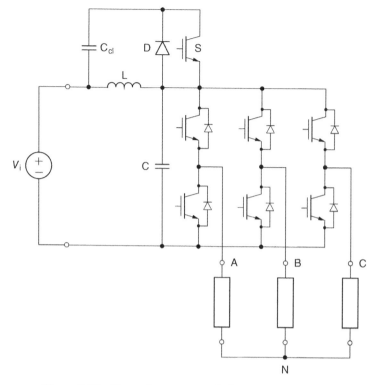

Figure 7.66   Three-phase resonant dc link with active clamp.

resonant pulses. This is called a *discrete pulse modulation*, as an integer number of voltage pulses appear within each switching cycle. The discrete pulse modulation in a resonant dc-link inverter with an active clamp is illustrated in Figure 7.67 that shows example line-to-line output voltages of the inverter. For good resolution, only a fragment of a cycle is depicted, the resonance frequency being here one hundred times higher than the output frequency.

In low-voltage inverters, the unclamped version of the resonant dc link is preferable, because of the higher, approaching unity, voltage gain. Shaving off the peaks of the pulsed voltage produced by the link significantly reduces this factor.

The requirement that all switches must change their states in synchronism with the resonating dc bus constitutes a major disadvantage of the resonant dc-link inverters. The resultant discrete PWM is less precise that that in hard-switching inverters, limiting the quality and control range of output currents. It would be more convenient to independently activate individual switches while maintaining the soft-switching conditions, especially in high-power applications. This type of operation has been achieved in the so-called *auxiliary resonant commutated pole inverter*, where the term "pole" refers to the "totem pole" arrangement of inverter switches (see Figure 3.14).

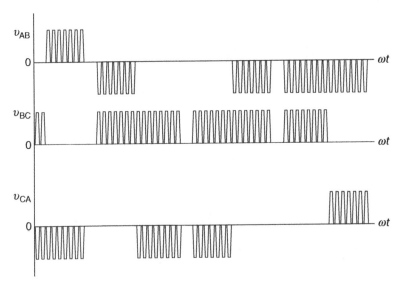

Figure 7.67 Example waveforms of line-to-line output voltages in a resonant dc-link inverter.

Power circuit of one phase of the auxiliary resonant commutated pole inverter is shown in Figure 7.68a and the whole three-phase inverter in Figure 7.68b. The auxiliary circuit, AC, which triggers the resonances, is connected between the midpoint of two dc-link capacitors, C_i, and the center of the pole. Each inverter pole consists of two main switches, S1 and S2, with freewheeling diodes, and two resonant capacitors, C_r. The auxiliary circuit is comprised of a bidirectional switch, S, and a resonant inductor, L_r. The other inductor, L_f, in each phase of the inverter constitutes a part of the output filter and plays a role in shaping the voltage and current transients during the resonance.

The bidirectional switch allows freedom in timing the resonance. The zero-crossing of the voltage of a resonant capacitor allows lossless commutation of the parallel main switch. The zero-crossing of the resonant inductor allows lossless commutation of the auxiliary switch. The efficiency of the inverter is high and the switching frequency is controllable. However, control is difficult because of the complexity of the three possible commutation modes, two of which require different approaches for low and high currents.

In comparison with hard-switching inverters, the soft-switching ones are characterized by lower losses and lower level of parasitic side effects, such as the EMI associated with switching of large currents. On the other hand, the soft-switching inverters suffer from increased complexity of the power circuit and control algorithm, higher voltage stresses on switches, and narrower control ranges. Many topological and control variations have been proposed, but most of the power industry has remained faithful to the mature technology of hard-switching inverters.

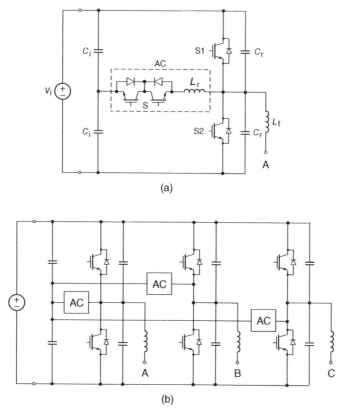

(a)

(b)

Figure 7.68 Auxiliary resonant commutated pole inverter: (a) one phase with the auxiliary circuit, (b) whole inverter.

7.5   DEVICE SELECTION FOR INVERTERS

The rated voltage, V_{rat}, of power switches in the regular, two-level, hard-switching inverters must be at least equal to the peak value, $V_{i,p}$, of the input voltage which, generally, is not ideally constant in time. If an inverter is supplied from a rectifier, it is the peak value of the ac voltage feeding the rectifier that should be taken as $V_{i,p}$. Thus,

$$V_{rat} \geq (1 + s_V)V_{i,p}. \tag{7.60}$$

As already mentioned in Section 7.1.3, the dc-input voltage, V_i, in PWM VSIs, equals the maximum available peak value of the fundamental line-to-line output voltage, $V_{LL,1,p(max)}$. In the square-wave operation mode, $V_i \approx 0.9\, V_{LL,1,p}$. These relations facilitate selection of the dc supply source.

Switches in CSIs are subjected to switching-generated voltage spikes (see Figure 7.47). Denoting the maximum expected instantaneous value of the line-to-line output voltage by $V_{\text{LL(max)}}$, the rated voltage of the switches must satisfy the condition

$$V_{\text{rat}} \geq (1 + s_{\text{V}})V_{\text{LL(max)}}. \tag{7.61}$$

In multilevel inverters, when correctly controlled, the maximum voltage stress across any device does not exceed the value of $V_{\text{i}}/(l - 1)$. However, under faulty operating conditions, some devices may be subjected to a much higher voltage. The safety margin is, therefore, left to the designer's discretion. Often, multilevel inverters are selected precisely for their high voltage capability, and the correct control is assumed.

In soft-switching converters, the maximum voltage stresses depend on the specific topology. For instance, in the resonant dc-link inverters, if the resonant pulses of the input voltage are not clamped, their peak value may be up to 2.5 times as high as the dc voltage at the input to the resonant circuit. With clamping, the typical voltage stresses on the devices are of order of 1.3–1.5 of the dc voltage. The right-hand side of condition (7.60) should be modified accordingly.

The highest current stresses in VSIs occur in the square-wave mode of operation, in which each switch may be forced to conduct the output current for the whole half cycle of output voltage. This applies also to the freewheeling diodes in inverters employed in systems with the bidirectional power flow. If an inverter is to operate with unidirectional power flow only, the average current of the freewheeling diodes depends on the load angle. As illustrated in Figure 7.69a, with a purely resistive load (R load) the diodes do not conduct at all, and each switch must pass the whole half-wave of the current. Thus, similarly as in ac voltage controllers (see Eq. 5.54), the rated current, $I_{\text{S(rat)}}$, of the switches should satisfy the condition

$$I_{\text{S(rat)}} \geq \frac{\sqrt{2}}{\pi}(1 + s_{\text{I}})I_{\text{L(rat)}}, \tag{7.62}$$

where $I_{\text{L(rat)}}$ denotes the rated rms value of the line output current of the designed inverter. For the freewheeling diodes, the worst-case scenario, depicted in Figure 7.69b, is a purely inductive load (L load), when each switch and each diode in a given leg of the inverter conduct a quarter-wave of the output line current. As a result, the required rated current, $I_{\text{D(rat)}}$, of the diodes can be as low as half of $I_{\text{S(rat)}}$.

The pulses of the line output currents in CSIs operating in the square-wave mode are one-third of the cycle long (see Figure 7.46). Thus, the average switch current amounts to one-third of the input current, I_{i}. On the other hand, the rms value, $I_{\text{L,1}}$, of the fundamental line current equals $(\sqrt{6}/\pi)I_{\text{i}}$. Consequently, the necessary condition for the rated current of switches is

$$I_{\text{S(rat)}} \geq \frac{\pi}{3\sqrt{6}}(1 + s_{\text{I}})I_{\text{L(rat)}}. \tag{7.63}$$

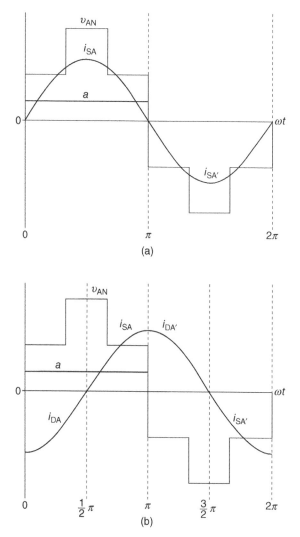

Figure 7.69 Idealized line-to-neutral voltage and line current waveforms in a VSI in the square-wave mode: (a) R-load, (b) L-load.

Condition (7.62) can be employed for determination of current ratings of power switches in multilevel inverters. In diode-clamped inverters, even with a purely inductive load, the diodes, both the freewheeling and clamping ones, conduct average currents of not more than a quarter of the maximum average current that can flow through a switch. It is so because the freewheeling diodes in a given leg of the inverter share the output current with the clamping diodes when the current cannot flow through the switches. Thus, diodes with the rated current, $I_{D(rat)}$, equal $I_{S(rat)}/4$ can safely be selected.

Resonant circuits in soft-switching inverter do not significantly affect currents in the main switches. For instance, the current conducted by the both switches in a leg of a resonant dc-link inverter during the charging subcycle is simply too low in comparison with the output current. Therefore, the same rules as for hard-switching inverters can be employed for the selection of switches and diodes.

To evaluate the shortest on- and off-times of switches in PWM inverters, information about the specific PWM strategy employed is required. Sometimes, as in the case of hysteresis current control, the switching patterns are difficult to predict by theoretical analysis and computer simulations are needed. Therefore, no generalized formulas can be derived. Example calculations of $t_{ON(min)}$ and $t_{OFF(min)}$ are presented in Example 7.3 at the end of this chapter.

7.6 COMMON APPLICATIONS OF INVERTERS

Generally, power inverters are used in either direct dc-to-ac or indirect ac-to-ac power conversion systems. For instance, a battery-powered electric vehicle driven by an ac motor employs a direct dc-to-ac conversion scheme, in which a VSI interfaces the battery with the motor. Another example may involve an off-grid farm supplied (among other sources) from a photovoltaic array, where an inverter provides three-phase voltage for the motor driving an irrigation pump. Such systems usually include a battery that is charged by the array during daylight, and which feeds the inverter.

If, in another case, a photovoltaic array were used to supplement power supply from an ac system, a system shown in Figure 7.70 could be used. In this photovoltaic utility interface, a PWM VSI receives the dc voltage from a photovoltaic array and converts it into high-frequency ac voltage applied to a transformer. The transformer provides electrical isolation and voltage matching between the array and ac power system. Thanks to the high frequency involved, the transformer is much smaller than a comparable 60-Hz one. The secondary voltage of the transformer is converted back to a dc voltage by a diode rectifier, and this voltage is fed to a phase-controlled rectifier, separated from the diode rectifier by an inductive filter. The controlled rectifier, connected to the utility grid, operates in the inverter mode, the filter and the dc voltage representing an RLE load with a negative EMF. Thus, the power flows into the grid.

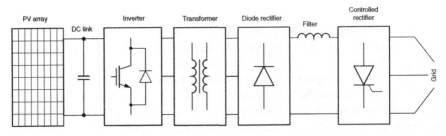

Figure 7.70 Block diagram of photovoltaic utility interface.

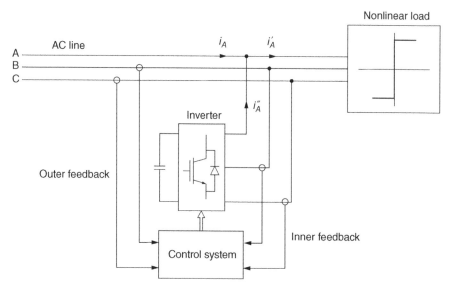

Figure 7.71   Block diagram of active power filter.

Inverters are widely used in various renewable energy systems, and the described photovoltaic utility interface is only one member of a large family of such interfaces. Some of them are presented in Chapter 9, devoted to "green" applications of power electronics.

Active power filters for elimination of harmonic currents in a power system represent another practical example of the direct dc-to-ac conversion scheme. A block diagram of an active power filter is shown in Figure 7.71. The filter maintains sinusoidal currents in a power line that supplies a "nonlinear" load, most often a rectifier. Two feedback loops are employed: an outer loop for control of the line current, and an inner loop for the current-controlled inverter that produces currents compensating the current harmonics in the line. Based on the measured line currents and voltages, a control system establishes sinusoidal reference current signals for individual phases of the line. The reference currents are in phase with the respective phase voltages for a unity power factor. When the reference line current signals are subtracted from actual current signals, reference signals for inverter output currents are obtained. For instance, denoting by i_A, i_A', and i_A'' the line, load, and inverter currents in phase A, respectively, the reference inverter current, $i_A''^*$, is given by

$$i_A''^* = i_A' - i_A^*,$$ (7.64)

where i_A^* denotes the reference line current.

Voltage and current waveforms in an active power filter connected to an ac line that feeds a three-phase controlled rectifier are shown in Figure 7.72. For clarity, the current waveforms are idealized, that is, the current drawn by the rectifier has a purely square wave and the line current is ideally sinusoidal, assuming that the

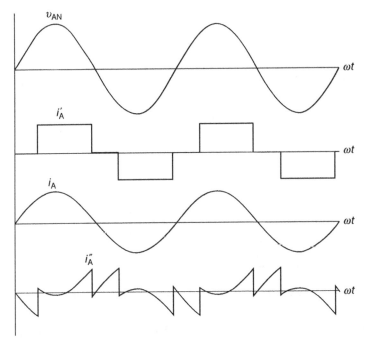

Figure 7.72   Waveforms of voltage and current in an active power filter.

inverter provides exactly the current required. In reality, the instantaneous changes of inverter current shown cannot be realized in a VSI, so that complete elimination of harmonics is not possible. Note the difficult operating conditions of the control system that must compute running waveforms of the reference currents.

Since the load is to be fully powered by the ac line, no real power is required from the inverter. Therefore, only a capacitor is connected to the inverter's input as a dc link, without any supply source, and the losses in the inverter are covered by a low amount of real power drawn from the system. Control of the inverter includes a provision for maintaining a constant voltage across the capacitor. Often, to match the voltage and current ratings of the filter with those of the ac line, a transformer is used as an interface between the inverter and the line. A similar arrangement, the so-called instantaneous VAr controller, can be used for reactive power control in a power system. In that case, the inverter produces sinusoidal currents that lead respective voltage sinewaves in phase, to provide compensation for inductive loads in the system and improve the power factor.

Uninterruptable power supply (UPS) systems are critical in such facilities as hospitals, communication and computer centers, airports, or military installations, which require a constant supply of electric power, even in the case of failure of the grid. These facilities have their own backup battery rooms and diesel-engine or fuel-cell powered ac generators. Two basic types of UPS are standby (off-line) and on-line ones. In the standby UPS system, the load is supplied from the grid, while the battery

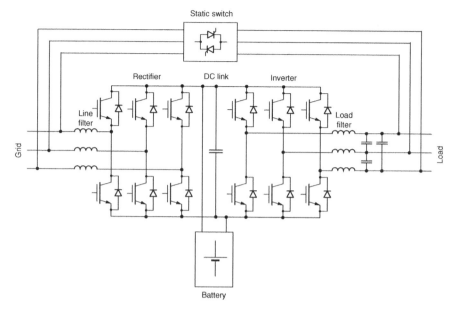

Figure 7.73   UPS system.

is trickle charged to sustain a full charge. When the supply from the power system is interrupted, a VSI is automatically activated to convert the dc voltage from the battery into the regular ac voltage used in the facility. This arrangement is maintained until the normal power supply is restored or a standby generator is brought into operation. Then the inverter returns to its passive status and the battery is charged for fast recovery of full readiness.

The on-line UPS system, apart from providing backup power when needed, isolates a sensitive load from the grid to protect the load from line transients and harmonic pollution. A rectifier-dc-link-inverter cascade separates the grid from the load, while the battery connected in parallel to the dc-link transformer maintains full charge. The ac voltage produced by the inverter is then supplied to the load through a low-pass filter, which mitigates the higher harmonics and makes the load voltage sinusoidal. Thus, power quality fed to the load is high, and independent of possible disturbances in the grid. Diagram of a UPS system is shown in Figure 7.73. If the static power switch is normally on, the system operates as the off-line UPS, while if it is normally off, as the on-line UPS. An isolation transformer (not shown) is often used between the inverter and load.

Adjustable-speed ac drives, based on induction motors and, increasingly, on permanent-magnet synchronous motors, constitute the most common application of inverters. Various types of ac motor drives have been developed over the years for the purpose of control of speed, torque, and position. Depending on the quality of control, the drive systems can be classified as low- or high-performance and, considering the control principles, as scalar- or vector-controlled. Generally, speed of an ac motor depends on the frequency of the stator voltage while the developed torque is related to the stator current.

In scalar-controlled drive systems, only the frequency and magnitude of the fundamental stator voltage or current are adjusted, which precludes high performance of the system under transient operating conditions. Therefore, scalar-controlled, low-performance drives are employed in such machinery as adjustable-speed pumps, fans, blowers, mixers, or grinders, where high control quality would be superfluous. It is worth mentioning that those drives consume most of the electrical energy spent in industry.

High performance is required from motor drives used in electric traction, hybrid and electric cars, lifts and elevators, or the adjustable-torque applications such as winders in paper, plastic, and textile factories, and cold-rollers in steel mills. These drives are vector controlled, which means that those are the instantaneous values of the stator voltage or current that are continuously adjusted. As a result, the transient current waveforms, for example, during a speed reversal of a drive, often do not resemble the sinusoids typical for the steady-state operation. Current-controlled VSIs are usually employed.

In the variable-speed induction motor drives, in order to maintain the maximum available torque at a constant level, the ratio of stator voltage to frequency should be kept constant, which is known as the *Constant Volts per Hertz* (CVH) control. Above the rated speed, adherence to the CVH principle would mean increasing the stator voltage above the rated value, which is not permitted. Therefore, with frequencies higher than rated, the voltage is maintained constant, at the rated level. This area of operation is called *field weakening*, because the intensity of the magnetic field decreases when a frequency increase is not accompanied by a voltage increase. The maximum frequency allowed is that which causes the motor to rotate with the maximum permitted speed. At very low frequencies, the stator voltage must be somewhat higher than that indicated by the CVH rule to compensate for the voltage drop across the stator resistance.

A block diagram of a scalar speed control system with an induction motor is shown in Figure 7.74. The angular velocity, ω_M, of the motor is measured and compared with the reference velocity, ω^*. The speed error signal, $\Delta\omega_M$, is applied to a slip controller whose output variable, ω_{sl}^*, constitutes the reference slip velocity of the motor, that is, the desired difference between the angular velocity of the revolving magnetic field of the stator and the rotor speed. The slip velocity of a motor must be limited for stability purposes and for protecting the motor from too high currents in the stator and rotor windings. For that purpose the slip controller's input–output characteristic exhibits saturation. When the signal ω_{sl}^* is added to the motor speed signal, ω_M, the reference signal for synchronous speed, ω_{syn}^*, of the stator field is obtained. The synchronous speed signal is now multiplied by the number, p_p, of pole-pairs of the stator to result in the reference radian frequency, ω^*, for the stator voltage to be produced by the inverter. The reference magnitude signal, V^*, for that voltage is generated in a voltage controller in dependence on the level of ω^*. The CVH principle is obeyed below the rated frequency and the voltage is made constant in the field weakening area.

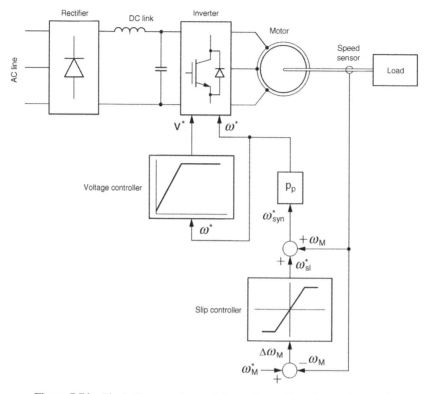

Figure 7.74   Block diagram of an ac drive system with scalar speed control.

In certain applications, such as lift drives or electric ac traction, a drive system is required to have the bidirectional power flow capability to absorb the potential or kinetic energy of the mechanical part of the system during braking. In low-power drives, a braking resistor can be used to dissipate the energy transferred by the inverter from the load. Such an ac drive system, based on the modular frequency changer with IGBTs in Figure 2.28, is shown in Figure 7.75a. The diode in series with the seventh switch serves as a freewheeling diode for the parasitic inductance of the braking resistor circuit. As shown in Figure 7.75b, the same switch and diode, and an external inductor, can serve as a step-up chopper to boost the input voltage to the inverter. The inverter supply capacitor is charged by the chopper to a voltage higher than that across the dc-link capacitor.

In order to recover the braking energy and return it to the supply line, a controlled rectifier is required. SCR-based rectifiers are commonly used in practice, but they require large line filters to improve waveforms of the currents drawn from the line. More elegant solutions, shown in Figure 7.76 and allowing four-quadrant operation of the motor, involve a PWM rectifier whose power circuit is a copy of that of the inverter. The current-type rectifier in Figure 7.76a, identical with the CSI in Figure 7.44, has already been covered in Section 4.3.3. The voltage-type rectifier in

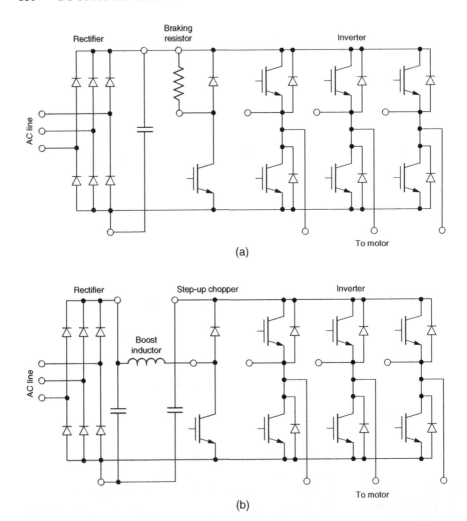

Figure 7.75   Use of the modular frequency changer of Figure 2.28 in an ac drive: (a) system with braking resistor, (b) system with step-up chopper.

Figure 7.76b can be thought of as a VSI operating with a reversed power flow and an inductive-EMF load (LE load). The cascade of a PWM rectifier, dc link, and a PWM inverter forms a high-performance frequency changer whose applications can be extended beyond the field of adjustable-speed drives (see Figure 7.73).

Single-phase VSIs and CSIs are used for induction heating in industry, providing high-frequency currents for the heating coil. Often, a capacitor is connected in series or in parallel with the coil to create a resonant circuit at the inverter output. The output frequencies vary from less than a hundred hertz to a few hundred kilohertz, depending on the application. Inverters are also used in electric arc welding equipment, which

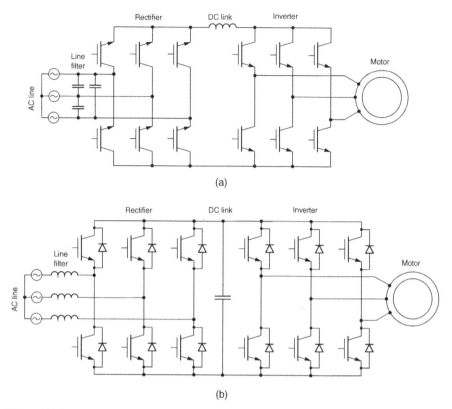

Figure 7.76   PWM rectifier–inverter cascades for bidirectional power flow in ac motor drives: (a) current-type rectifier, inductive dc link, and CSI, (b) voltage-type rectifier, capacitive dc link, and VSI.

requires an isolation transformer between the welding electrodes and utility supply line. The supply voltage is rectified, converted into a high-frequency ac voltage in the inverter, transformed, rectified again, and applied to the electrodes.

In recent years, low-power sources of electricity utilizing the hydro and wind power have been gaining popularity in the USA, and many small independent operators sell the power to local utilities. The power plant may consist of a synchronous generator, driven by a water wheel or a wind turbine as the prime mover, and a utility interface. Usually the speed of the prime mover varies, which causes fluctuations of the frequency of generated voltage. Also, the magnitude of the voltage may be affected by the speed changes. Therefore, the ac voltage produced by the generator is rectified and fed to a VSI that stabilizes the frequency and magnitude of the output voltage, as long as the prime energy does not fall below a specific minimum level. An isolation transformer is often used between the inverter and the power system. More details about wind power systems can be found in Chapter 9.

SUMMARY

The dc-to-ac power conversion is performed by inverters. Depending on the characteristics of the dc source employed, VSIs and CSIs can be distinguished. Apart from the dissimilar dc links, these two types of dc-to-ac converters differ by the absence of freewheeling diodes in the CSIs. PWM CSIs require output capacitors to smooth the output current waveforms. Inverters can be built with any number of phases, three-phase VSIs being most common in practice.

Inverters can be made to operate in the square-wave or PWM mode. The square-wave mode of operation is simple and characterized by a low number of commutations per cycle of the output voltage. However, higher quality of the output current is obtained in PWM inverters, and a number of PWM techniques have been developed. Feedforward (open-loop) voltage control and feedback (closed-loop) current control are employed in VSIs, while feedforward current control is typical for CSIs.

Multilevel VSIs provide high-quality output currents in the square-wave mode and are particularly useful in high-voltage applications. Switching losses are minimized in soft-switching inverters, which are also characterized by reduced EMI effects. On the other hand, these converters are more complex than the hard-switching ones and they have lower resolution of control of output voltage and current.

Inverters have found many applications in the direct dc-to-ac and indirect ac-to-ac power conversion schemes. They form the crucial part of adjustable-speed ac drives, improve the quality of currents in power systems, provide uninterruptable ac power supply, and interface photovoltaic arrays, and wind- and water-driven ac generators with utility lines. Inverters are also used in such industrial processes as induction heating and electric arc welding.

EXAMPLES

Example 7.1 A VSI is supplied from a 620-V dc source and feeds a balanced wye-connected load. At a certain instant, the inverter is in state 3 and the output currents in phases A and B are −72 and 67 A, respectively. Neglect the voltage drops in the inverter, and determine all the output voltages (line-to-neutral and line-to-line) and the input current.

Solution: In state 3, the switching variables, a, b, and c, of the inverter are 0, 1, and 1, respectively, since $011_2 = 3$. The line-to-line output voltages can now be calculated from Eq. (7.8) as

$$\begin{bmatrix} v_{AB} \\ v_{BC} \\ v_{CA} \end{bmatrix} = 620 \begin{bmatrix} 1 & -1 & 0 \\ 0 & 1 & -1 \\ -1 & 0 & 1 \end{bmatrix} \begin{bmatrix} 0 \\ 1 \\ 1 \end{bmatrix} = \begin{bmatrix} -620 \\ 0 \\ 620 \end{bmatrix} V$$

and, from Eq. (7.10), the line-to-neutral voltages as

$$
\begin{bmatrix} v_{AN} \\ v_{BN} \\ v_{CN} \end{bmatrix} = \frac{620}{3} \begin{bmatrix} 2 & -1 & -1 \\ -1 & 2 & -1 \\ -1 & -1 & 2 \end{bmatrix} \begin{bmatrix} 0 \\ 1 \\ 1 \end{bmatrix} = \begin{bmatrix} -414 \\ 207 \\ 207 \end{bmatrix} V.
$$

The individual line output currents are $i_A = -72$ A, $i_B = 67$ A, and $i_C = -i_A-i_B$ = 5A. Thus, according to Eq. (7.11),

$$
i_i = 0 \times (-72) + 1 \times 67 + 1 \times 5 = 72 \text{ A}.
$$

Example 7.2 A PWM inverter, supplied from a 310-V dc source, is controlled using the voltage space vector technique with the switching frequency of 4 kHz and the high-efficiency state sequence. In certain switching cycle, the per-unit reference voltage vector is $0.75\angle 280°$. Determine the switching pattern of the inverter switches in this cycle. Assuming that the modulation index does not change, what is the rms fundamental line-to-line output voltage of the inverter?

Solution: The switching period is

$$
T_{sw} = \frac{1}{f_{sw}} = \frac{1}{4 \times 10^3} = 2.5 \times 10^{-4} \text{s} = 250 \text{ μs}
$$

and the reference voltage vector is in sector V, which extends from 240° to 300° and is framed by stationary vectors $\vec{V}_1$ and $\vec{V}_5$ (see Figure 7.23). Thus, the in-sector angle, α, of the vector is $280° - 240° = 40°$, and the durations of the involved inverter states, 1, 5, and 7, are:

$$
T_X = T_1 = 0.75 \times 250 \times \sin(60° - 40°) = 64 \text{ μs}
$$
$$
T_Y = T_5 = 0.75 \times 250 \times \sin(40°) = 121 \text{ μs}
$$

Thus, the required sequence of states is: state 1 for 32 μs, state 5 for 60.5 μs, state 7 for 65 μs, state 5 for 60.5 μs, and state 1 for 32 μs.

The switching pattern described is illustrated in Figure 7.77. It can be seen that no switching occurs in phase C. Also, as long as the reference voltage vector remains in the same sector, the next switching cycle begins with the same state 1 that concluded the previous cycle. Consequently, no switching takes place in the inverter during the cycle to cycle transition. As a result, the switching losses are reduced by a third in comparison with the high-quality state sequence.

As the maximum available peak fundamental line-to-line output voltage, $V_{LL,1p(max)}$, equals, ideally, the supply dc voltage, here 310 V, then

$$
V_{LL,1p} = 0.75 \times 310 = 232.5 \text{ V}
$$

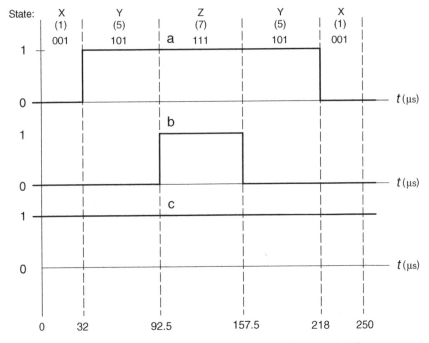

Figure 7.77   Switching pattern of the inverter in Example 7.2.

and

$$V_{LL,1} = \frac{232.5}{\sqrt{2}} = 164 \text{ V}.$$

Example 7.3   Certain space vector PWM technique for the three-level neutral-clamped inverter requires that for the modulation index less than 0.5, the low-voltage vectors, such as $\vec{V}_3$ or $\vec{V}_4$, be used for synthesis of the reference voltage vector, $\vec{v}^*$. Find the three vectors and duty ratios of the corresponding states of the inverter, which will produce a reference vector whose magnitude corresponds to the modulation index, m, of 0.3, and whose phase angle, β, is 45°.

Solution: Figure 7.78 shows a fragment of interest of the diagram of per-unit voltage vectors of the inverter with the input voltage, V_i, taken as the base (see Figure 7.55). The reference voltage vector, $\vec{v}^*$, lies between vectors $\vec{V}_9$ (or $\vec{V}_{22}$) and $\vec{V}_{12}$ (or $\vec{V}_{25}$). In state 9, $abc = 100$ (as $9 = 1 \times 3^2 + 0 \times 3^1 + 0 \times 3^0$), and in state 12, $abc = 110$, while $abc = 211$ in state 22 and 221 in state 25. Both pairs of stationary voltage vectors can be employed, as the transition between states 9 and 12 and between states 22 and 25 changes only one switching variable, b. Let us assume that vectors $\vec{V}_9$ and $\vec{V}_{12}$ will be used along the zero vector $\vec{V}_{13}$, generated when $abc = 111$.

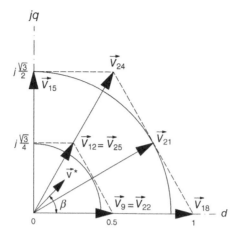

Figure 7.78 Per-unit voltage vectors of the three-level inverter in Example 7.3.

The modulation index $m = 0.3$ implies that the absolute magnitude of the reference voltage vector equals 0.3 of the maximum possible magnitude, which is $(\sqrt{3}/2)V_i$. Thus, in the per-unit format,

$$\vec{v}^* = 0.3 \times \frac{\sqrt{3}}{2} \cos(45°) + j\,0.3 \times \frac{\sqrt{3}}{2} \sin(45°) = 0.184 + j\,0.184$$

while

$$\vec{V}_9 = 0.5 + j\,0$$

and

$$\vec{V}_{12} = 0.25 + j\frac{\sqrt{3}}{4} = 0.25 + j\,0.433.$$

Denoting by d_9, d_{12}, and d_{13} duty ratios of states 9, 12, and 13, the following three equations must be satisfied:

$$d_9 \times 0.5 + d_{12} \times 0.25 = 0.184$$
$$j(d_9 \times 0 + d_{12} \times 0.433) = j\,0.184$$
$$d_9 + d_{12} + d_{13} = 1$$

which, when solved, yield $d_9 = 0.155, d_{12} = 0.424$, and $d_{13} = 0.421$.

The same results can be obtained using Eqs. (4.77)–(4.79) employed in the control of two-level PWM rectifiers and inverters. However, because the low-voltage vectors are used, $2m$ should be substituted for m. Then,

$$d_{12} = d_Y = 2 \times 0.3 \times \sin(45°) = 0.424$$
$$d_{13} = 1 - 0.155 - 0.424 = 0.421.$$

The example illustrates the feasibility of application of space vector PWM techniques to multi-level inverters. Only minor modifications must be made: for example, should the medium-voltage vectors be used for synthesis of the reference voltage vector, $(2/\sqrt{3})m$ would have to be substituted for m in Eqs. (4.77)–(4.79).

Example 7.4   A resonant dc-link inverter is supplied from a 200-V dc source and operates with the switching frequency of 10 kHz. The inductance, L, and capacitance, C, of the resonant circuit are 0.24 mH and 1 µF, respectively, and the circuit resistance is negligible. The current, I_0, drawn by the inverter is constant at 200 A and the input current, I_2, at the beginning of the operation cycle of the dc link, is 209 A. Find:

(1) the average output voltage, V_0, of the resonant dc link when the pulses of the voltage are not clipped and

(2) the same voltage when an active clamp shaves off the peaks of the voltage pulses to the level of 1.3 of the dc-input voltage, V_i, to the resonant dc link.

Solution:

(a) With the assumed absence of damping in the resonant circuit, the resonance frequency, ω_r, can be calculated as

$$\omega_r = \frac{1}{\sqrt{LC}} = \frac{1}{\sqrt{0.24 \times 10^{-3} \times 10^{-6}}} = 64550 \text{ rad/s}$$

and the term $e^{-\alpha t}$ can be removed from expression (7.53) for the output voltage, v_0, of the resonant dc link. Parameters V and φ_2 in that expression, as given by Eq. (7.55), are

$$V = \sqrt{200^2 + \left(\frac{209 - 200}{10^{-6} \times 64550}\right)^2} = 243.8 \text{ V}$$

and

$$\varphi_2 = \tan^{-1}\left(\frac{209 - 200}{10^{-6} \times 64550 \times 200}\right) = 0.609 \text{ rad.}$$

Thus, the output voltage of the link applied to the inverter is

$$v_0 = 200 - 243.8 \cos(64550t + 0.609) \text{ V}.$$

Based on the equation above, the length, t_2 (see Figure 7.65), of the resonance subcycle of operation of the link can be determined as

$$t_2 = \frac{2\pi - \cos^{-1}\left(\frac{200}{243.8}\right) - 0.609}{64550} = 7.85 \times 10^{-5}\text{s} = 78.5\ \mu\text{s}$$

and the average value, V_o, of v_o can be calculated as

$$V_o = \frac{1}{T_{sw}} \int_0^{t_2} v_o dt,$$

where T_{sw} is the switching period, equal $1/10$ kHz $= 10^{-4}$ s $= 100\ \mu$s. Hence, $V_o = \frac{1}{10^{-4}}\{200 \times 7.85 \times 10^{-5} - \frac{243.8}{64550}[\sin(64550 \times 7.85 \times 10^{-5} + 0.609) - \sin(0.609)]\} = 200$ V $= V_i$.

The result is not surprising in view of the purported lossless operation of the link.

(b) When the peak of the voltage pulse is shaved off to the level of $1.3V_i$, that is, to 260 V, the resultant voltage reduction, ΔV_o, is given by

$$\Delta V_o = \frac{1}{T_{sw}} \int_{t_A}^{t_B} (v_o - 1.3V_i)$$

$$dt = \frac{1}{10^{-4}} \int_{t_A}^{t_B} [200 - 243.8\cos(64550t + 0.609) - 260]dt,$$

where t_A and t_B are instants at which the voltage clipping begins and ends, respectively. They can be determined as

$$t_A = \frac{\cos^{-1}\left(\frac{-60}{243.8}\right) - 0.609}{64550} = 1.88 \times 10^{-5}\text{s} = 18.8\ \mu\text{s}$$

and

$$t_B = \frac{2\pi - \cos^{-1}\left(\frac{-60}{243.8}\right) - 0.609}{64550} = 5.97 \times 10^{-5}\text{s} = 59.7\ \mu\text{s}$$

yielding

$$\Delta V_o = \frac{1}{10^{-4}}\{-60 \times (5.97 - 1.88) \times 10^{-5} - \frac{243.8}{64,550}[\sin(64,550 \times 5.97$$
$$\times 10^{-5} + 0.609) - \sin(64,550 \times 1.88 \times 10^{-5} + 0.609)]\} = 48.6\ \text{V}.$$

As a result of the clamping,

$$V_o - \Delta V_o = 200 - 48.6 = 151.4 \text{ V},$$

that is, the voltage gain of the whole inverter is reduced by about 24%.

Example 7.5   A 150-kVA PWM VSI designed for unidirectional power flow is supplied from a six-pulse diode rectifier fed from a 460-V line. Assuming no voltage drops in the system and using safety margins of 0.25 for the rated current and 0.4 for the rated voltage, find the minimum required ratings of switches and diodes for the inverter.

Solution: According to Eq. (4.4), the average dc voltage provided by the rectifier as an input voltage, V_i, of the inverter is

$$V_i = \frac{3}{\pi} \times \sqrt{2} \times 460 = 621 \text{ V}.$$

Assuming a unity voltage gain of the inverter, the maximum available peak value, $V_{LL,1p}$, of the fundamental line-to-line output voltage is also 621 V, and the rated rms value, $V_{LL,1(rat)}$, of this voltage is $621/\sqrt{2} = 439$ V. Thus, the rated rms fundamental line output current, $I_{L,1(rat)}$, can be calculated from the rated power the inverter as

$$I_{L,1(rat)} = \frac{150 \times 10^3}{\sqrt{3} \times 439} = 197 \text{ A}.$$

In agreement with Eq. (7.61), the rated voltage, V_{rat}, of the semiconductor power switches and diodes must satisfy the condition

$$V_{rat} \geq (1 + 0.4) \times \sqrt{2} \times 460 = 911 \text{ V}$$

while condition (7.62) for the rated current, $I_{S(rat)}$, of the switches yields

$$I_{S(rat)} \geq \frac{\sqrt{2}}{\pi} (1 + 0.25) \times 197 = 111 \text{ A}.$$

The rated current, $I_{D(rat)}$, of the freewheeling diodes can be taken as 50% of $I_{S(rat)}$, that is,

$$I_{D(rat)} \geq 0.5 \times 111 = 56 \text{ A}.$$

PROBLEMS

P7.1   A single-phase VSI is supplied from a 300-V dc source. Find the fundamental output voltage when the inverter operates in the basic and optimal square-wave modes.

P7.2   Using the input dc voltage as a base, calculate the per-unit peak values of the first through seventh harmonics of the output voltage in a single-phase VSI in the basic and optimal square-wave modes.

P7.3   A three-phase VSI is supplied from a 600-V dc source and operates in the square-wave mode. Find the rms values of fundamental line-to-line and line-to-neutral output voltages of the inverter.

P7.4   Starting with state 1, determine the state sequence of a three-phase VSI that would result in a negative phase sequence of output voltages.

P7.5   A three-phase VSI operates with a switching frequency of 5 kHz, using the voltage space vector PWM technique. The inverter is supplied from a 600-V dc source and it is to produce the line-to-line output voltage of 400 V with a frequency of 90 Hz. Within certain switching cycle, the reference voltage vector has the phase angle of 140°. Determine the high-quality sequence and durations of inverter states in that interval.

P7.6   Repeat Problem 7.5 for the high-efficiency state sequence.

P7.7   Which switches are not switched in the cycle considered in Problem 7.6?

P7.8   When the programmed, harmonic elimination PWM strategy is used with the magnitude control ratio of 1, the primary switching angles are 9.48°, 14.80°, 87.93°, and 89.07°. Determine and sketch the full-cycle switching pattern for phase A.

P7.9   A three-phase CSI, supplied from a 400-A dc current source, operates in the square-wave mode. The wye-connected load of the inverter represents a 1-Ω/phase resistance in series with a 2.5-mH/phase inductance. Find rms values of the fundamental output current and fundamental line-to-line and line-to-neutral output voltages.

P7.10   A three-phase PWM CSI, fed from an adjustable-current source, is controlled using the harmonic-elimination PWM technique with three primary switching angles. For a single cycle of operation, find all switching angles for both switches in phase A.

P7.11   A three-level neutral-clamped inverter is supplied from a 3.3-kV ac line via a six-pulse diode rectifier. Assume that the input dc voltage of the inverter equals the average output voltage of the rectifier and find rms values of the line-to-line and line-to-neutral output voltages of the inverter in the square-wave operation mode.

P7.12 Which switches in a three-level neutral-clamped inverter are on and which are off when the inverter is in state 19? Also, taking the input voltage as unity, determine values of all the line-to-line and line-to-neutral output voltages of the inverter in that state.

P7.13 Repeat Example 7.3 for $m = 0.85$ and $\beta = 220°$. High-voltage vectors are to be used for synthesis of the reference voltage vector.

P7.14 Switching variables for individual bridges of a seven-level cascaded H-bridge inverter are: $a_1 = 1, a_2 = 2, a_3 = 0, b_1 = 2, b_2 = b_3 = 1, c_1 = 0, c_2 = c_3 = 2$.
Find the output line-to-line voltage vector for this state and the three line-to-neutral output voltages, all in the per-unit format.

P7.15 Consider the resonant dc-link inverter in Example 7.4 and determine the maximum voltage stress on semiconductor devices of the inverter if:

  (a) the inverter operates with the input dc voltage of 200 A and no clamp in the resonant dc-link circuit and

  (b) the inverter operates with the clamp as specified in the example, but the input dc voltage has been increased to compensate for the clipping of the link voltage waveform.

P7.16 A 10-kVA three-phase VSI is supplied from a 230-V ac line through a six-pulse diode rectifier. The inverter is designed for a unidirectional power transfer, from dc to ac. Assume the voltage safety margin of 40%, the current safety margin of 25%, and determine the minimum required voltage and current ratings of switches and diodes in the inverter.

P7.17 A 500-kVA three-level neutral-clamped inverter is supplied from a 2.4-kV ac line through a six-pulse diode rectifier. Assuming the same operating conditions and safety margins as in Problem 7.16, determine the minimum required voltage and current ratings of switches and diodes in the inverter.

COMPUTER ASSIGNMENTS

CA7.1* Run PSpice programs *Sqr_Wv_VSI_1ph.cir* and *Opt_Sqr_Wv_VSI_1ph .cir* for a single-phase VSI in the basic and optimal square-wave modes. For each mode and for both the output voltage and current, find:

  (a) rms value,

  (b) rms value of the fundamental, and

  (c) total harmonic distortion.

  Observe oscillograms of the input current.

CA7.2* Run PSpice program *Sqr_Wv_VSI_3ph.cir* for a three-phase VSI in the square-wave mode. For the output voltages (line-to-line and line-to-neutral) and current, find:

(a) rms value,

(b) rms value of the fundamental, and

(c) total harmonic distortion.

Observe the oscillogram of the input current.

CA7.3   Develop a computer program for calculation of switching angles in a PWM VSI employing the voltage space vector PWM technique. For given values of the switching frequency (for convenience, it can be a multiple of the output frequency) and magnitude control ratio, the program should determine pulse trains of all three switching variables of the inverter. Provide an option for selection of either the high-quality state sequence or the high-efficiency sequence.

CA7.4*   Run PSpice programs *PWM_VSI_9.cir* and *PWM_VSI_18.cir* for a three-phase voltage source with 9 and 18 switching cycles per cycle, respectively. In both cases, determine for the line output current:

(a) the most prominent higher harmonic,

(b) rms value,

(c) rms value of the fundamental, and

(d) total harmonic distortion.

Observe oscillograms of the input current.

CA7.5*   Run PSpice program *Progr_PWM.cir* for a three-phase VSI with programmed PWM (harmonic-elimination), with four primary switching angles. Measure the harmonic spectrum of the line-to-neutral output voltage and find which low-order harmonics have been eliminated. For the line output current, find:

(a) rms value,

(b) rms value of the fundamental, and

(c) total harmonic distortion.

Note that the number of pulses of each switching variable is nine per cycle and, for comparison, repeat the measurements for the respective inverter in Assignment 7.6.

CA7.6*   Run PSpice program *Hyster_Curr_Contr.cir* for a three-phase VSI with hysteresis current control. Set the width of the tolerance band to 5% of the peak reference current and determine:

(a) number of pulses of a switching variable per cycle,

(b) rms value of the line output current,

(c) rms value of the fundamental line output current, and

(d) total harmonic distortion of the line output current.

Repeat the measurements for the width of the tolerance band of 10% of the peak reference current.

CA7.7*   Run PSpice program *Sqr_Wv_CSI.cir* for a three-phase CSI in the square-wave mode. For the line output current, find:

(a) rms value,

(b) rms value of the fundamental, and

(c) total harmonic distortion.

Observe oscillograms of the output voltage and load EMF.

CA7.8*   Run PSpice program *PWM_CSI.cir* for a three-phase PWM CSI. For the line output current, find:

(a) rms value,

(b) rms value of the fundamental, and

(c) total harmonic distortion.

Observe oscillograms of the voltages and currents in the inverter.

CA7.9*   Run PSpice program *Half_Brdg_Inv.cir* for a half-bridge VSI with hysteresis current control. For the output current, determine:

(a) rms value,

(b) rms value of the fundamental, and

(c) total harmonic distortion.

Observe power spectra of the output voltage and current. Notice that although the current ripple is sharply defined, the spectrum is not, because of the randomness of individual switching instants.

CA7.10*   Run PSpice program *Three_Lev_Inv.cir* for a three-level neutral-clamped inverter. For the line-to-neutral output voltage and line output current, determine:

(a) rms value,

(b) rms value of the fundamental, and

(c) total harmonic distortion.

Compare the results with those for the two-level inverter in C7.2*.

CA7.11*   Run PSpice program *Reson_DC_Lnk.cir* for a resonant dc-link network. Observe oscillograms of the voltages and currents. Check if the pulsed voltage waveform conforms to Eq. (7.46).

FURTHER READING

[1] Bellar, M. D., Wu, T.-S., Tchamdjou, A., Mahdavi, J., and Ehsani, M., A review of soft-switched dc-ac converters, *IEEE Transactions on Industry Applications*, vol. 34, no. 4, pp. 847–860, 1998.

[2] Chang, J. and Hu, J., Modular design of soft-switching circuits for two-level and three-level inverters, *IEEE Transactions on Power Electronics*, vol. 21, no. 1, pp. 131–139, 2006.

[3] Holmes, D. G. and Lipo, T. A., *Pulse Width Modulation for Power Converters: Principles and Practice*, IEEE Press–Wiley, Hoboken, NJ, 2003.

[4] Kouro, S., Malinowski, M., Gopakumar, K., Pou, J., Franquelo, L. G., Wu, B., Rodriguez, J., Perez, M. A., and Leon, J. I., Recent advances and industrial applications of multilevel converters, *IEEE Transactions on Industrial Electronics*, vol. 57, no. 8, pp. 2553–2579, 2010.

[5] Loh, P. C., Boost–buck thyristor-based PWM current-source inverter, *IEE Proceedings—Electric Power Applications*, vol. 153, no. 5, pp. 664–672, 2006.

[6] Malinowski, M., Gopakumar, K., Rodriguez, J., and Perez, M. A., A survey on cascaded multilevel inverters, *IEEE Transactions on Industrial Electronics*, vol. 57, no. 7, pp. 2197–2206, 2010.

[7] Peng, F. Z., Z-source inverter, *IEEE Transactions on Industry Applications*, vol. 39, no. 2, pp. 504–510, 2003.

[8] Trzynadlowski, A. M., Borisov, K., Li, Y., and Qin, L., A novel random PWM technique with low computational overhead and constant sampling frequency for high-volume, low-cost applications, *IEEE Transactions on Power Electronics*, vol. 20, no. 1, pp. 116–122, 2005.

8 Switching Power Supplies

In this chapter, switching dc-to-dc power supplies are presented and the distinction between switched-mode and resonant dc-to-dc converters is explained; basic types of nonisolated switched-mode converters are analyzed in detail, and their isolated counterparts and extensions are shown; the concept of resonant switches is introduced and use of these switches in quasi-resonant dc-to-dc converters is illustrated; and series-loaded, parallel-loaded, and series-parallel resonant converters are described.

8.1 BASIC TYPES OF SWITCHING POWER SUPPLIES

Proliferation of personal computers, small communication devices, and automotive electronics has spurred intensive research and development activity in the area of switching power supplies (precisely, switching *dc* power supplies). The shrinking size of the electronic equipment demands ever increasing *power density* of the supply systems. Power density, meant as the ratio of available power to volume or weight of the power supply, can only be high in highly efficient systems. Otherwise, excessive power losses would cause unacceptable heat stresses on system components, whose compactness severely limits their thermal capacity.

The input voltage for switching power supplies is obtained from batteries, photovoltaic or fuel cells or, via a transformer and rectifier, from the utility grid, and it is likely to fluctuate. The current drawn by the load can vary too, affecting the output voltage. In certain applications, such as laboratory power supplies, wide-range adjustability of the output voltage or current is required. Therefore, a switching power supply is usually equipped with a closed-loop system for control of the output quantities. As explained in Section 1.4, switching converters, based on the principle of pulse width modulation (PWM), are better suited for efficient power control than the once popular linear voltage regulators operated, in essence, as controlled resistors.

Typical switching power supplies are low-power electronic systems and it could be argued whether they truly belong to the mainstream *power* electronics. However, because of the similar approach to the conversion of electric power, the switching power supplies are related to such medium- and high-power electronic converters as those covered in Chapters 4 through 7, particularly to choppers. Actually, as shown later, the distinction between choppers and certain types of low-power dc-to-dc

Introduction to Modern Power Electronics, Third Edition. Andrzej M. Trzynadlowski.
© 2016 John Wiley & Sons, Inc. Published 2016 by John Wiley & Sons, Inc.
Companion website: www.wiley.com/go/modernpowerelectronics3e

converters used in switching power supplies is blurred. Technical journals and conferences devoted to power electronics routinely include papers on the switching power supplies. Therefore, a brief review of these systems deserves a place in this textbook.

Two basic structures of switching power supplies reflect the fact that for safety reasons a transformer can be required to isolate the input from the output. The transformer also provides the desired ratio of the output to input voltage. Therefore, switching power supplies can be divided into two classes: *nonisolated* and *isolated*, depending on whether a transformer is incorporated into the power conversion scheme. Bipolar junction transistors and power MOSFETs are most commonly used as semiconductor power switches.

The dc-to-dc converters used in switching power supplies can be classified as *switched-mode* or *resonant* ones. The switched-mode converters are characterized by hard switching (see Section 7.4), while the phenomenon of electric resonance is utilized in resonant converters to provide conditions for zero-voltage switching (ZVS) or zero-current switching (ZCS). Soft switching makes the resonant converters more efficient, but their control ranges are usually narrower than those of switched-mode converters.

8.2 NONISOLATED SWITCHED-MODE DC-TO-DC CONVERTERS

Basic nonisolated switched-mode dc-to-dc converters are single-switch networks which, besides the switch, contain a diode, one or two inductors, and one or two capacitors, one of them connected across the output terminals. Similar to choppers, switched-mode dc-to-dc converters operate in the steady state with constant duty ratios of their switches. The output capacitor smoothes the output voltage. Therefore, in a properly designed converter, the voltage ripple is negligible and the voltage can be considered to have the ideal dc quality.

Increasing the switching frequency allows reduction in size of the output capacitor. This observation also holds true for other elements, such as inductors and, in isolated converters, transformers. Again, as in the case of other PWM power converters, the switching frequency employed represents a tradeoff between the quality and efficiency of converter performance. However, the switching frequencies are typically two orders of magnitude higher than those in PWM power converters covered in Chapters 4–7, reaching the megahertz range. Also, the switched-mode dc-to-dc converters cannot be thought of as simple networks of switches because of the vital role played by the energy storage elements.

A quality converter should be simple, reliable, efficient, compact, and characterized by low ripple of the output voltage and, possibly, of the input current. In certain applications, such as controlled dc-current sources, a fast response to the magnitude control command is an important consideration as well. The most common topologies of nonisolated switched-mode dc-to-dc converters are *buck*, *boost*, *buck–boost*, Ćuk, *SEPIC*, and *Zeta*. Similar to choppers, dc-to-dc converters are supplied from a diode rectifier, or a battery, via a capacitive dc link.

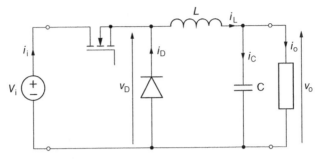

Figure 8.1 Buck converter.

8.2.1 Buck Converter

Circuit diagram of the buck (step-down) converter, shown in Figure 8.1, is identical to that of the step-down, first-quadrant chopper in Figure 6.7 with an addition of a low-pass LC filter at the load terminals. In all the subsequent considerations, difference equations are employed to describe the switched-mode converters (see Section 1.6.1). Additional assumptions include ideal circuit components and negligible ripple of the output voltage.

The difference equation for the inductor can be written as

$$\Delta i_L = \frac{v_D - V_o}{L} \Delta t, \tag{8.1}$$

where Δi_L denotes the change in inductor current, i_L, within the time interval Δt, v_D is the voltage across the diode, V_o is the constant output voltage, and L is the coefficient of inductance of the inductor. During the time when the switch is on, v_D equals the input voltage, V_i, and the inductor current increases by

$$\Delta I_{L(ON)} = \frac{V_i - V_o}{L} dT_{sw}, \tag{8.2}$$

where d denotes the duty ratio of the switch (not the differentiation operator!) and T_{sw} is the switching period, a reciprocal of the switching frequency, f_{sw}. When the switch is off during the $\Delta t = (1 - d)T_{sw}$ interval, the diode is forced to carry the inductor current, so $v_D = 0$. The inductor current decreases, and by the end of the switching cycle it has changed by

$$\Delta I_{L(OFF)} = -\frac{V_o}{L}(1 - d)T_{sw}. \tag{8.3}$$

In the steady state, $\Delta I_{L(ON)} = -\Delta I_{L(OFF)} = \Delta I_L$, where ΔI_L denotes the peak-to-peak amplitude of the inductor current ripple, and the resultant equation

$$\frac{V_i - V_o}{L} dT_{sw} = \frac{V_o}{L}(1 - d)T_{sw} \tag{8.4}$$

can be solved for V_o to give

$$V_o = dV_i. \tag{8.5}$$

The result is not surprising, as it has already been obtained for the first-quadrant chopper (see Eq. (6.6)).

In the subsequent considerations, a *continuous conduction mode* of the converter is assumed. It must be pointed out that the average voltage across an inductor and the average current through a capacitor are zero when averaged over an integer number of cycles. If so, then the average inductor current, I_L, in the buck converter equals the output current, I_o, which can easily be determined if the load is known. This allows finding all the other quantities in the converter from circuit equations. For instance, Eqs. (8.2) and (8.3) and relation $I_L = I_o$ determine the $i_L(t)$ waveform. With $i_L(t)$ known, the capacitor current, $i_C(t)$, can be found as $i_L(t) - I_o$. Finally, the input current, $i_i(t)$, and diode current, $i_D(t)$, equal $xi_L(t)$ and $(1 - x)i_L(t)$, respectively, where x denotes the switching variable associated with the switch. Waveforms of individual currents when the switch is off can be found in a similar way. Selected waveforms, specifically those of the inductor current, i_L, capacitor current, i_C, diode current, i_D, input current, i_i, and voltage, v_D, across the diode, are shown in Figure 8.2.

An ideal dc-output voltage assumed so far would require an infinitely high capacitance, C, of the capacitor or an infinitely high switching frequency, f_{sw}. Therefore, in practice, the output voltage, v_o, is somewhat rippled. Waveforms of the capacitor current and output voltage, the latter given by

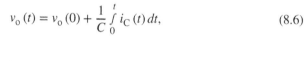

$$v_o(t) = v_o(0) + \frac{1}{C} \int_0^t i_C(t)\, dt, \tag{8.6}$$

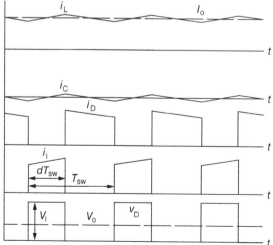

Figure 8.2   Voltage and current waveforms in a buck converter.

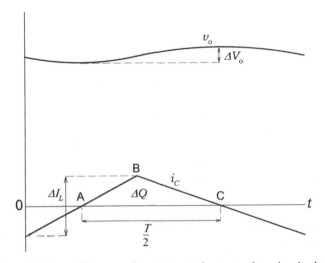

Figure 8.3   Waveforms of the capacitor current and output voltage in a buck converter.

are shown in Figure 8.3. The peak-to-peak amplitude, ΔV_o, of the voltage ripple can be calculated as

$$\Delta V_o = \frac{\Delta Q}{C}, \tag{8.7}$$

where ΔQ is the charge increment in the capacitor, represented in Figure 8.3 by the triangle ABC. Thus,

$$\Delta Q = \frac{1}{2}\frac{T_{sw}}{2}\frac{\Delta I_L}{2} = \frac{T_{sw}}{8}\frac{V_i - V_o}{L}dT_{sw} = \frac{(1-d)V_i}{8L}dT_{sw}^2 = \frac{(1-d)V_o}{8Lf_{sw}^2} \tag{8.8}$$

and, from Eq. (8.7),

$$\frac{\Delta V_o}{V_o} = \frac{1-d}{8LCf_{sw}^2}. \tag{8.9}$$

Equation (8.9) quantifies the impact of parameters L and C and the switching frequency on ripple of the output voltage.

Certain control requirements may result in such low values of the duty ratio, d, that the inductor current becomes discontinuous. As seen in Figure 8.2, this would happen if one-half of the peak-to-peak ripple, ΔI_L, of the inductor current exceeded the average value of this current, I_L. Since $\Delta I_L = |\Delta I_{L(OFF)}|$ and, as already mentioned, $I_L = I_o$, then, based on Eq. (8.3), the condition for the continuous conduction mode can be written as

$$\frac{V_o}{2L}(1-d)T_{sw} < I_o \tag{8.10}$$

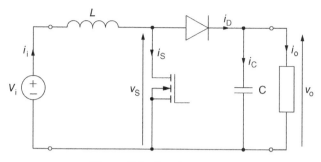

Figure 8.4   Boost converter.

or

$$Lf_{sw} > \frac{V_o}{2I_o}(1-d).\tag{8.11}$$

When designing a buck converter, Eqs. (8.9) and (8.11) facilitate proper selection of the inductance, capacitance, and switching frequency.

8.2.2   Boost Converter

The boost (step-up) converter, shown in Figure 8.4, has the same topology as the step-up chopper (see Figure 6.22) with an output capacitor. The output voltage, as shown in Section 6.3 (see Eq. (6.40)), is given by

$$V_o = \frac{V_i}{1-d}\tag{8.12}$$

that is, always higher than V_i. According to Eq. (6.35), the peak-to-peak amplitude, ΔI_i, of ripple of the input current, i_i, can be expressed as

$$\Delta I_i = \frac{V_i}{L}dT_{sw},\tag{8.13}$$

while the average input current, I_i, can be calculated from the power balance

$$V_iI_i = V_oI_o.\tag{8.14}$$

Taking Eq. (8.12) into account, I_i is found to be

$$I_i = \frac{I_o}{1-d},\tag{8.15}$$

which, with Eq. (8.13), allows the determination of the input current waveform, $i_i(t)$. Other currents and voltages can easily be found as being related to the input and output

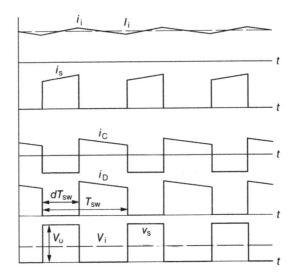

Figure 8.5   Voltage and current waveforms in a boost converter.

quantities. Waveforms of the switch current, i_S, capacitor current, i_C, diode current, i_D, input current, i_i, and voltage, v_S, across the switch are shown in Figure 8.5.

Figure 8.6 illustrates the impact of capacitor current on the ripple of the output voltage. Equation (8.7) can be again be used, in which ΔQ, represented by rectangle ABC0, is given by

$$\Delta Q = dT_{sw}I_o. \tag{8.16}$$

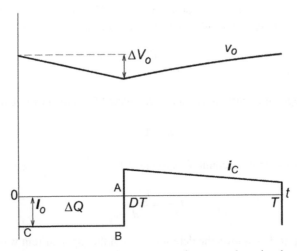

Figure 8.6   Waveform of the capacitor current and output voltage in a boost converter.

Hence,

$$\Delta V_o = \frac{dI_o}{Cf_{sw}} \tag{8.17}$$

or, for a passive load (no load EMF),

$$\frac{\Delta V_o}{V_o} = \frac{d}{RCf_{sw}}, \tag{8.18}$$

where R denotes the load resistance.

For the continuous conduction mode, the average inductor current, I_i, must be greater than $\Delta I_i/2$ (see Figure 8.5). Based on Eqs. (8.12), (8.13), and (8.15), the continuous conduction condition can thus be expressed as

$$\frac{I_o}{1-d} > \frac{(1-d)V_o}{2L}dT_{sw} \tag{8.19}$$

or, after rearrangement,

$$Lf_{sw} > \frac{d(1-d)^2 V_o}{2I_o}. \tag{8.20}$$

As already mentioned in Section 6.3, when d approaches unity, the magnitude of the output voltage saturates at a finite level, depending on the resistances of the converter elements, particularly that of the inductor. For instance, for a boost converter with a passive load, the output voltage can be shown to be

$$V_o = \frac{V_i}{1-d+\frac{r_L}{1-d}}, \tag{8.21}$$

where r_L denotes the ratio of resistance of the inductor to the load resistance. The voltage gain, $K_V \equiv V_o/V_i$, as a function of d for various values of r_L is shown in Figure 8.7.

8.2.3 Buck–Boost Converter

The output voltage of the buck converter cannot be higher than the input voltage and, vice-versa, the output voltage of the boost converter cannot be lower than the input voltage. In contrast, in the buck–boost converter, shown in Figure 8.8, the output voltage can be made less than, equal to, or greater than the input voltage.

As in the boost converter, turning the switch on causes the inductor current, i_L, to increase by

$$\Delta I_{L(ON)} = \frac{V_i}{L}dT_{sw} \tag{8.22}$$

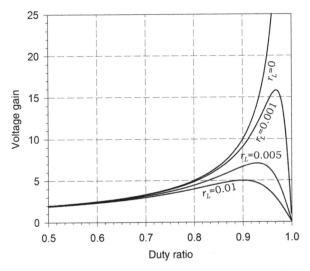

Figure 8.7   Impact of the inductor resistance on the voltage gain of a boost converter.

(see Eq. 8.13). When the switch turns off, the diode is forced to conduct the inductor current, and the output voltage, V_o, appears across the inductor. Now, by the end of the switching cycle, the inductor current decreases by

$$\Delta I_{L(OFF)} = \frac{V_o}{L}(1-d)T_{sw}. \tag{8.23}$$

In the steady state, $\Delta I_{L(ON)} = -\Delta I_{L(OFF)} = \Delta I_L$. Thus,

$$\frac{V_i}{L}dT_{sw} = -\frac{V_o}{L}(1-d)T_{sw}, \tag{8.24}$$

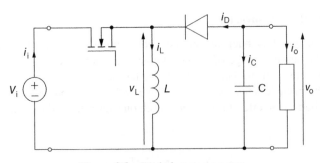

Figure 8.8   Buck–boost converter.

which yields

$$V_o = -\frac{d}{1-d}V_i.$$ (8.25)

As seen from Eq. (8.25), the output voltage is negative, that is, inverted with respect to the input voltage. When the duty ratio, d, of the switch is adjusted from 0 to 0.5, the magnitude of the output voltage changes from zero to the input voltage. Increasing d from 0.5 up causes the output voltage to grow further. As in the boost converter, the maximum voltage gain is limited by the resistances of components (see Figure 8.7).

The average inductor current, I_L, determined from equations

$$I_i = dI_L$$ (8.26)

$$V_i I_i = V_o I_o$$ (8.27)

and Eq. (8.25), is given by

$$I_L = -\frac{I_o}{1-d}.$$ (8.28)

Waveforms of the inductor current, i_L, capacitor current, i_C, diode current, i_D, input current, i_i, and voltage, v_L, across the inductor are shown in Figure 8.9. Notice the apparent similarity of Figures 8.5 and 8.9. In particular, the capacitor current has the same waveform as that in the boost converter (see Figure 8.5). Consequently, the output voltage ripple in the buck–boost converter can be described by Eqs. (8.17) and (8.18) derived for the boost converter.

In the continuous conduction mode,

$$I_L > \frac{\Delta I_L}{2},$$ (8.29)

which, as $\Delta I_L = \Delta I_{L(ON)}$, is tantamount to

$$\frac{|I_o|}{1-d} > \frac{V_i}{2L}dT_{sw},$$ (8.30)

because $I_o < 0$. Taking into account that

$$V_i = \frac{1-d}{d}|V_o|,$$ (8.31)

condition (8.30) can be rearranged to

$$Lf_{sw} > \frac{(1-d)^2 V_o}{2I_o}.$$ (8.32)

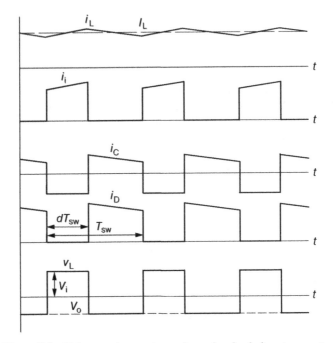

Figure 8.9   Voltage and current waveforms in a buck–boost converter.

8.2.4   Ĉuk Converter

The buck–boost converter allows wide-range control of the output voltage, but the input current is discontinuous and its ripple is high, which in many applications is undesirable. For instance, when the converter is supplied from a rectifier fed from the utility grid, the high-frequency ac component of the drawn current can cause serious electromagnetic interference in the vicinity of transmission lines. To prevent it, a low-pass filter must be installed between the rectifier and the converter or between the grid and the rectifier.

In contrast to the buck–boost converter, instead of the inductor, the Ĉuk converter (so called after its inventor) uses an extra capacitor for storage and transfer of energy. As seen in Figure 8.10, two inductors are employed to smooth the input current, i_i, and the current, i_{L2}, supplying the output stage of the converter. This allows the significant reduction of the output capacitance, C_2, and of the possible line filter. The two inductors are wound on the same core, and the converter is remarkably compact.

The Ĉuk converter is sometimes said to have an optimum topology as voltages and currents at the both ends of the converter are true dc quantities and, as shown later, the voltage gain can theoretically be controlled from zero to infinity. The converter has also other advantages, such as the very high efficiency or the grounded cathode of the switch.

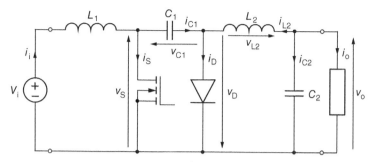

Figure 8.10 Ĉuk converter.

As the initial conditions, constant currents in the inductors and constant voltages across the capacitors are assumed. Specifically, $i_{L1}(t) = I_i$, $i_{L2}(t) = I_{L2}$, $v_{C1}(t) = V_{C1}$, and $v_{C2}(t) = V_o$. With the switch on, the diode is off and

$$i_{C1(ON)} = -I_{L2}. \tag{8.33}$$

When the switch turns off, the diode is forced to conduct the inductor currents, i_{L1} and i_{L2}, and

$$i_{C1(OFF)} = I_i. \tag{8.34}$$

In the steady state, the average charge received by capacitor C_1 over a switching cycle is zero, that is,

$$dT_{sw}i_{C1(ON)} + (1-d)T_{sw}i_{C1(OFF)} = 0 \tag{8.35}$$

and, after rearrangement,

$$\frac{i_{C1(ON)}}{i_{C1(OFF)}} = \frac{d}{1-d}. \tag{8.36}$$

Equations (8.33), (8.34), and (8.36) yield

$$-\frac{I_i}{I_{L2}} = \frac{d}{1-d}. \tag{8.37}$$

The average input power,

$$P_i = V_i I_i \tag{8.38}$$

equals the average power, P_o, delivered to the output stage of the converter and given by

$$P_o = V_o I_{L2}, \tag{8.39}$$

Comparing Eqs. (8.38) and (8.39), yields

$$\frac{V_o}{V_i} = \frac{I_i}{I_{L2}} \tag{8.40}$$

and, using Eq. (8.37),

$$V_o = -\frac{d}{1-d}V_i. \tag{8.41}$$

Relation (8.41) is identical with that for the buck–boost converter (see Eq. 8.25).

When the switch is on, the input voltage, V_i, is impressed across inductor L_1 for the dT_{sw} period of time, causing current i_i to increase by

$$\Delta I_i = \frac{V_i}{L_1}dT_{sw}. \tag{8.42}$$

At the same time, the voltage across inductor L_2 is

$$v_{L2} = V_o + v_{C1}, \tag{8.43}$$

where the voltage across capacitor C_1 equals

$$v_{C1} = V_i - V_o, \tag{8.44}$$

as average voltages across the inductors in the steady state of the converter are zero. Consequently, $v_{L2} = V_i$ and current i_{L2} increases by

$$\Delta I_{L2} = \frac{V_i}{L_2}dT_{sw}. \tag{8.45}$$

Note that when the switch is off, the voltage, v_s, across the switch equals v_{C1}, given by Eq. (8.45). Thus,

$$v_s = V_i - V_o = \frac{V_i}{1-d}. \tag{8.46}$$

The same voltage appears across the diode when the switch is on. Current and voltage waveforms in the Ĉuk converter are shown in Figure 8.11.

The output part of the Ĉuk converter, composed Ĉuk of inductor L_2, capacitor C_2, and the load, is identical with that of the buck converter. Also, current i_{L2} in that

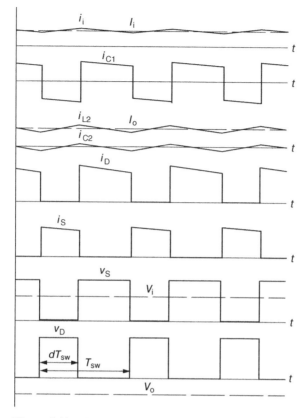

Figure 8.11   Voltage and current waveforms in a converter.

inductor has a continuous waveform with saw-tooth ripple, which is similar to that of inductor current, i_L, in the buck converter. Therefore, Eq. (8.9) for the output voltage ripple in the buck converter is also valid for the Ĉuk converter. In a sense, the Ĉuk converter can be considered a single-switch cascade of the boost and buck converters.

As in the other converters, for continuous conduction, the average current in the inductors must be greater than half of the respective ripple amplitude, ΔI_i and ΔI_{L2}. Based on Eqs. (8.37), (8.41), (8.42), and (8.45), conditions

$$I_i > \frac{\Delta I_i}{2} \tag{8.47}$$

and

$$I_{L2} = I_o > \frac{\Delta I_{L2}}{2} \tag{8.48}$$

can be rearranged to

$$L_1 f_{sw} > \frac{(1-d)^2 V_o}{2dI_o} \tag{8.49}$$

and

$$L_2 f_{sw} > \frac{(1-d) V_o}{2I_o}. \tag{8.50}$$

8.2.5 SEPIC and Zeta Converters

The SEPIC (single-ended primary inductor converter) and Zeta converters, shown in Figure 8.12, represent attempts at improvement of the Ĉuk converter. It can be seen that all the three converters have two inductors, L_1 and L_2, and a capacitor, C_1, separating the output from the input and providing protection from a shorted load. The SEPIC converter employs an N-channel MOSFET and the Zeta converter a P-channel one.

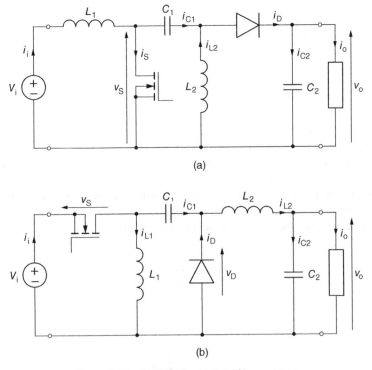

(a)

(b)

Figure 8.12   SEPIC (a) and Zeta (b) converters.

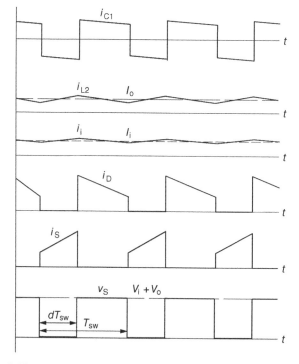

Figure 8.13   Voltage and current waveforms in SEPIC and Zeta converters.

Both the SEPIC and Zeta converters perform a noninverting buck–boost operation, that is,

$$V_o = \frac{d}{1-d} V_i. \tag{8.51}$$

However, Eq. (8.51) is only valid if a zero-voltage drop across the diode is assumed (this comment applies to all dc-to-dc converters with a diode in series with the load). In reality, the output voltage is reduced by the amount of that voltage drop. This is an important consideration if a dc-to-dc converter is designed for application in a low-voltage system, such as a cell phone. Selected voltage and current waveforms in the SEPIC and Zeta converters, which despite different topologies of these converters are quite similar, are shown in Figure 8.13.

8.2.6   Comparison of Nonisolated Switched-Mode DC-to-DC Converters

The dc-to-dc converters can be thought of as dc transformers with an adjustable voltage gain. Indeed, in presence of only minor losses in a converter, its output power approximately equals the input power while the output voltage can be adjusted, often in a wide range. The concept of the dc transformer applies especially well to the Ćuk

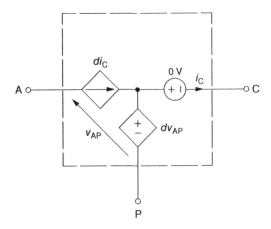

Figure 8.14   Vorperian's switch model.

converter, in which the voltages and currents at the input and output are continuous, with only a minor ripple.

In many instances, a dc-to-dc converter forms a part of a larger system, usually with a closed-loop voltage or current control. To analyze or simulate such systems, it is convenient to use *averaged models* of converters, based on the so-called *Vorperian's switch model*. The averaged model of a converter produces ripple-free output current and voltage waveforms representing values of these quantities averaged over consecutive switching cycles.

The Vorperian's switch model, depicted in Figure 8.14, describes the dc-transformer properties of the converters described in this chapter. Letters "a," "p," and "c" denote the "active," "passive," and "common" terminals. As shown in Figure 8.15, the Vorperian's model is incorporated into converter structure as an analog functional equivalent of the actual switching device. To illustrate the concept of averaged converter model, waveforms of the output current and voltage in a freshly started buck converter are compared in Figure 8.16 with those in the equivalent averaged converter.

All the converters presented were analyzed assuming continuous currents in their inductors, which is the preferred mode of operation and which results in the voltage gain independent of the load. However, under extreme operating conditions, such as I_0 approaching zero, a continuous conduction mode cannot be maintained. The analysis of the discontinuous conduction mode is somewhat more involved, and interested readers are referred to specialized literature.

When choosing a converter for a given application, consideration must be given to various features, such as the required voltage gain, switch utilization, weight and volume of the converter, component cost, or transient response. Generally, if the rated output voltage is lower than the input voltage, the buck converter should be selected and, vice-versa, in the case of the output voltage always higher than the input voltage, the boost converter constitutes the best choice. For a wide range of control of the

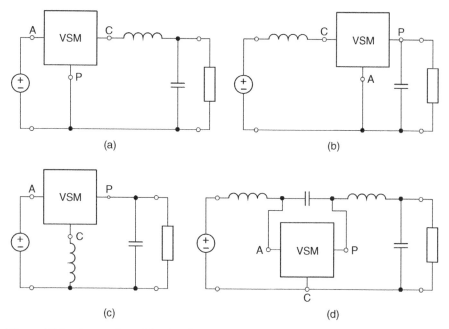

Figure 8.15 Averaged models of switched-mode dc-to-dc converters with the Vorperian's switch model: (a) buck, (b) boost, (c) buck–boost, (d) Ĉuk.

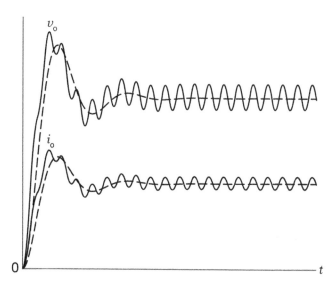

Figure 8.16 Waveforms of the output voltage and current in a buck converter: solid lines—actual converter, broken-lines—averaged model.

output voltage, the Ĉuk converter is preferred over the buck–boost converter. Yet, the Ĉuk converter is not free from certain disadvantages, such as the high component count, high switch current, and a large required capacitor C_1 to handle high-ripple current.

Switch utilization, defined as the ratio of minimum required switch power rating (product of the rated voltage and current) to rated power of the converter, depends on the duty ratio, but its average value for the buck and boost converters is much higher than that for the buck–boost and Ĉuk converters. Therefore, if sufficient for an application, the buck and boost converters offer the highest power density. On the other hand, the dynamic response to voltage control commands of the boost and buck–boost converters is slower than that of comparable buck and Ĉuk converters. Properties of the SEPIC and Zeta converters are similar to those of the Ĉuk converter, although the Zeta converter produces the lowest ripple of load current thanks to inductor L_2 in the path of this current. Rated powers of the commercially available nonisolated switched-mode converters vary widely, but they are usually on the order of a fraction of kilowatt. Typical rated voltages are low, on the order of tens of volts or less.

8.3   ISOLATED SWITCHED-MODE DC-TO-DC CONVERTERS

Most practical switching power supplies are based on switched-mode dc-to-dc converters employing a transformer that provides isolation between the input and output of the converter. For safety and reliability reasons, it is necessary in converters fed from the power grid. As already mentioned, the transformer may also be used for matching the input and output voltages. For instance, if the output voltage is to be much higher than the available input voltage, the voltage gain of the nonisolated boost, buck–boost, or Ĉuk converters may turn out to be insufficient (see Figure 8.7).

The isolated dc-to-dc converters can be classified as single-switch and multiple-switch ones. The single-switch converters use their magnetic components in a unipolar, single-quadrant mode, and are generally restricted to low-power applications. In contrast, transformers in multiple-switch converters are fully utilized magnetically which, at higher power levels (above 100 W), results in the size of a converter reduced by almost half of that of an equivalent single-switch converter.

Two equivalent circuits of a transformer are shown in Figure 8.17. The ideal transformer in Figure 8.17a, also used as a general transformer symbol in circuit diagrams, is represented as a pair of coupled coils and described by the relation

$$\frac{v_1}{v_2} = \frac{i_2}{i_1} = \frac{N_1}{N_2}, \tag{8.52}$$

where N_1 and N_2 denote numbers of turns in the primary and secondary winding, respectively. With the *polarity marks* placed as in the diagrams in Figure 8.17, the input and output voltages, v_1 and v_2, and currents, i_1 and i_2, are in phase. Generally, if the voltage at the dotted end of one winding is positive, the voltage at the dotted end of another winding is positive as well.

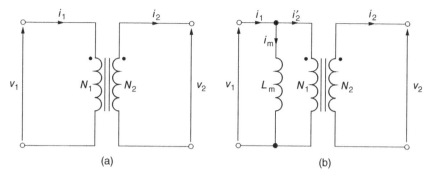

Figure 8.17   Equivalent circuits of a transformer: (a) ideal transformer, (b) transformer with magnetizing inductance included.

For simulation and analysis of isolated switched-mode dc-to-dc converters, the equivalent circuit in Figure 8.17b is more practical as it includes the magnetizing inductance, L_m. In this model of a transformer,

$$\frac{v_1}{v_2} = \frac{i_2}{i'_2} = \frac{N_1}{N_2}, \qquad (8.53)$$

where $i'_2 = i_1 - i_m$.

8.3.1   Single-Switch-Isolated DC-to-DC Converters

Single-switch converters are derived from the buck, buck–boost, and Ĉuk converters. They are the *forward converter, flyback converter,* and *isolated Ĉuk converter,* respectively. A two-switch, so-called *step-up flyback converter,* related to the boost converter, is feasible but of little use, and will not be described.

Forward Converter. The forward converter, which is an isolated version of the buck converter in Figure 8.1, is shown in Figure 8.18. When the switch is on, diode D1 is conducting and electrical energy is stored in the inductor L. Diode D2 provides a current path for the release of that energy when the switch turns off. To remove the energy stored in the magnetizing inductance of the transformer, diode D3 connected to an extra winding is used.

The average output voltage of the forward converter in the continuous conduction mode is given by

$$V_o = k_N d V_i, \qquad (8.54)$$

where

$$k_N \equiv \frac{N_2}{N_1} \qquad (8.55)$$

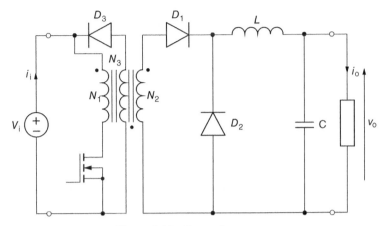

Figure 8.18   Forward converter.

denotes the turn ratio of the transformer. Since the output stage of the forward converter is the same as that of the buck converter, the ripple of the output voltage in both converters is expressed by the same Eq. (8.9). The same applies to the condition (8.11) for continuous conduction. However, the maximum allowable value, d_{max}, of the duty ratio, d, must be less than $N_1/(N_1 + N_3)$ to ensure complete demagnetizing of the transformer core within each switching cycle. Otherwise, the core would saturate after several cycles, and the transformer would cease to work. In practice, usually $N_1 = N_3$ and $d_{max} = 0.5$.

Flyback Converter. A derivative of the buck–boost converter, the flyback converter, is shown in Figure 8.19. The magnetizing inductance, L_m, of the transformer plays the role of inductor L in the buck–boost converter in Figure 8.8. Keep in mind that the magnetizing inductance is a part of the equivalent circuit of a transformer, and not a discrete component. Notice that the placement of the polarity marks indicates that the transformer performs voltage inversion. Therefore, the diode is also inverted with respect to that in Figure 8.8. When the switch is turned on, energy is stored in

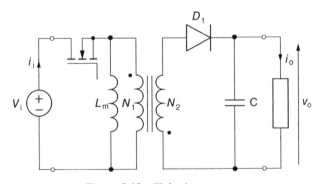

Figure 8.19   Flyback converter.

the core of the transformer and, when the switch turns off, released into the capacitor. An air gap is cut in the core to prevent magnetic saturation.

The output voltage of the flyback converter is given by

$$V_o = k_N \frac{d}{1-d} V_i \tag{8.56}$$

and the ripple of this voltage is given by the same Eqs. (8.17) and (8.18) as for the boost and buck–boost converters. However, condition (8.32) for the continuous conduction mode must be modified to

$$L_m f_{sw} > \frac{(1-d)^2 V_o}{2k_N^2 I_o} \tag{8.57}$$

to account for referring the magnetizing inductance to the primary side of the transformer. The minus sign in Eq. (8.25) for the boost–buck converter is absent in Eq. (8.56) thanks to the mentioned voltage inversion in the transformer.

Isolated Ĉuk Converter. The isolated version of the Ĉuk converter is shown in Figure 8.20. Capacitor C_1 in Figure 8.10, which depicts the nonisolated converter, is split here into two capacitors, C_{11} and C_{12}, large enough to stabilize the voltages across them. As in the flyback converter, the transformer inverts the secondary voltage, which results in the inverted diode and output voltage of the converter. Eqs. (8.56) and (8.9), the latter with $L = L_2$ and $C = C_2$, describe the output voltage and its ripple, respectively. The condition for continuous conduction of the current in inductor L_1 is

$$L_1 f_{sw} > \frac{(1-d)^2 V_o}{2k_N^2 d I_o}, \tag{8.58}$$

while Eq. (8.50) expresses the respective condition for inductor L_2. In practice, usually $N_1 = N_2$, that is, $k_N = 1$.

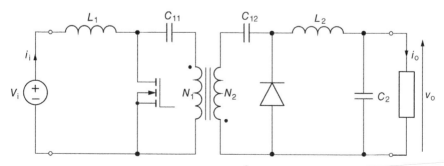

Figure 8.20   Isolated Ĉuk converter.

8.3.2 Multiple-Switch-Isolated DC-to-DC Converters

The three most common multiple-switch-isolated dc-to-dc converters, the *push–pull converter, half-bridge converter*, and *full-bridge converter*, described in this section are related to the buck converter, and their output voltage is always lower than the secondary voltage of the transformer. Each of these converters can be considered a cascade of an inverter, transformer, rectifier, and low-pass filter. A single-phase inverter converts the input dc voltage into a high-frequency square-wave ac voltage, which is next transformed, rectified in a diode rectifier, and smoothed in the filter.

In contrast to the two-pulse rectifier described in Section 1.6.3, rectifiers employed in the multiple-switch dc-to-dc converters are based on two, not four, diodes. This so-called *midpoint rectifier* is depicted in Figure 8.21. It needs a three-wire supply, which is provided by a transformer with a center-tapped secondary winding. If the voltage across a half of the secondary winding is considered as the input voltage, v_i, the output voltage of the rectifier is given by Eq. (4.5), that is, $V_{o,dc} = 2V_{i,p}/\pi \approx 0.637V_{i,p}$, where $V_{i,p}$ denotes the peak value of v_i. It is the same as in the four-diode bridge rectifier in Figure 1.34, but the minimum required rated voltage of the diodes must be twice as high as that of diodes in the bridge arrangement. In the typically low-voltage switching power supplies, it is not a serious problem, though.

Apart from rectifying diodes, circuits of the multiple-switch dc-to-dc converters include freewheeling diodes connected in antiparallel with the semiconductor switches. As explained in Section 7.1.1, these diodes are necessary for conducting reactive load currents in inverters. Here, they would not be required if the inverter fed an ideal transformer, since the current in the output-filter inductor is freewheeled by the rectifying diodes. However, the phenomenon of flux leakage in a practical transformer, irrelevant for the dc-to-dc power conversion, makes the transformer to appear as a slightly inductive load for the inverter. By providing a path for the resultant reactive currents, the freewheeling diodes protect the switches from overvoltages resulting from interrupting such currents. In practice, the freewheeling diodes do not have to

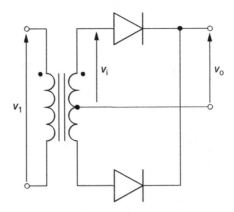

Figure 8.21 Midpoint rectifier.

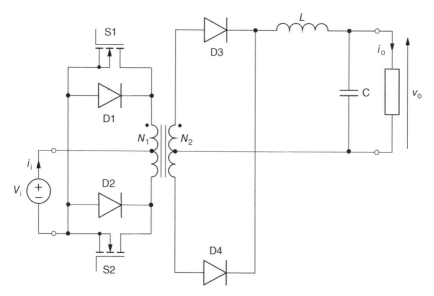

Figure 8.22   Push–pull converter.

be separate devices, but switches with internal diodes, such as power MOSFETs or switch modules, can be used.

Push–Pull Converter. Circuit diagram of the push–pull converter is shown in Figure 8.22. Both switches, S1 and S2, equipped with freewheeling diodes, D1 and D2, operate with the duty ratio, d, and their switching patterns differ by one-half of the switching cycle. The duty ratio is adjustable from 0 to 0.5, so that the switches are never simultaneously on. Diodes D3 and D4 rectify the secondary current of the transformer, supplying dc power to the output stage of the converter. When both switches are off and the secondary voltage of the transformer is zero, the rectifying diodes freewheel the inductor current. It can be shown that with a continuous current in inductor L, the average output voltage is given by

$$V_o = 2k_N dV_i \tag{8.59}$$

and the output voltage ripple by

$$\frac{\Delta V_o}{V_o} = \frac{1 - 2d}{32LCf_{sw}^2}. \tag{8.60}$$

Half-Bridge Converter. The "half-bridge" adjective in the name of the converter comes from the half-bridge inverter on the primary side of the transformer. The half-bridge inverter has already been mentioned in Section 7.3 and depicted in Figure 7.53. Figure 8.23 shows a circuit diagram of the half-bridge dc-to-dc converter. Capacitors C_1 and C_2 form the input voltage divider. Switches S1 and S2 operate in the same

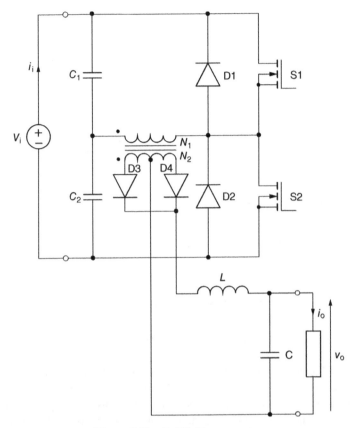

Figure 8.23   Half-bridge converter.

way as those in the push–pull converter. The maximum allowable duty ratio of 0.5
prevents shorting the source by both switches being simultaneously on. Diodes D1
and D2 freewheel reactive currents due to the flux leakage in the transformer, and
diodes D3 and D4 make up the rectifier on the secondary side. The output voltage is
given by

$$V_o = k_N d V_i \qquad (8.61)$$

and, since the circuit on the secondary side of the transformer is the same as in the
push–pull converter, the ripple voltage complies with Eq. (8.60).

Full-Bridge Converter. Analogously to the half-bridge converter, the full-bridge
converter in Figure 8.24 is named after its constituent single-phase bridge inverter,
which was described in Section 7.1.1 and shown in Figure 7.2. Pairs S1–S4 and S2–
S3 are switched alternately with the duty ratio not exceeding 0.5. In comparison with
the half-bridge converter, the full-bridge topology results in a doubled output voltage,
which is given by Eq. (8.59). The ripple of this voltage is expressed by Eq. (8.60).

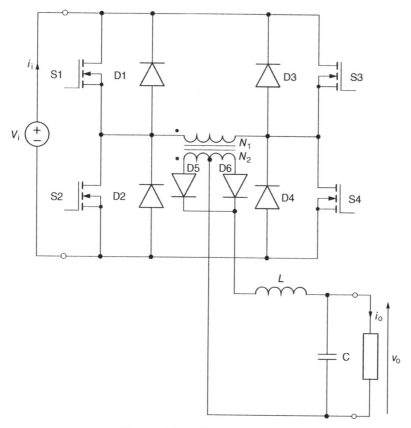

Figure 8.24   Full-bridge converter.

8.3.3   Comparison of Isolated Switched-Mode DC-to-DC Converters

None of the presented isolated switched-mode dc-to-dc converters can be considered definitely superior or inferior to the others, and it is the specific application that dictates selection of the most suitable type. Generally, the single-switch converters are employed at lower power levels than are the multiple-switch converters. The flyback converter, with the lowest component count, is simple and very popular, but its transformer is relatively large and the voltage stress on the switch is high, at twice the input voltage, V_i. The extra inductor in the forward converter allows for a smaller transformer. The Ĉuk, SEPIC, and Zeta converters are highly efficient and their continuous currents result in reduced input and output filter requirements, but the component count is high. Typical rated powers do not exceed 1 kW.

The push–pull converter, suitable for medium-power applications, in the typical range of up to 1 kW, is simple and inexpensive, but the switches and freewheeling diodes are subjected to high-voltage stresses of $2V_i$. In addition, a dc imbalance caused by even slightly differing switching patterns of switches is likely to lead to the saturation of the transformer core. The half-bridge converter, with the voltage stresses of V_i, is the most common switched-mode dc-to-dc converter in the medium power range, up to 2 kW, while application of the full-bridge converters is usually reserved for powers approaching 5 kW. Rated voltages vary in a wide range, typically 5–1000 V.

Regarding other design considerations, the feasibility of grounded, as opposed to floating, drivers for switches is an important advantage, since it simplifies the overall layout of the converter. Selection of the switching frequency depends mostly on the employed semiconductor switches, and the frequencies in most switched-mode dc-to-dc converters range from tens to hundreds of kilohertz. As already stressed, high frequencies allow reduction in size of the electromagnetic components, but they also produce high switching losses. The presence of a transformer in the isolated converters allows for *multiple-output converters*, as several output stages can be supplied from multiple secondary windings.

8.4   RESONANT DC-TO-DC CONVERTERS

Analysis of resonant dc-to-dc converters, many types of which have been developed, takes considerable time and effort as their power circuits change their topology several times during a single switching cycle. Therefore, only the most common types and general operating principles of those converters are subsequently described. Specialized literature is recommended for detailed analytical considerations, while the enclosed Spice circuit files allow close inspection of voltage and current waveforms in the selected converters.

The interest in resonant converters arose mainly from a quest for very high switching frequencies. These, in the switched-mode dc-to-dc converters, have been limited by the tradeoffs between the optimal size, weight, efficiency, reliability, and cost of the converter. In certain applications, such as miniature electronic equipment or space technology, switching frequencies in the megahertz range are desired. MOSFETs can be switched with frequencies of tens of megahertz. The major obstacles are switching losses and switching stresses.

Parasitic inductances in the converters, such as the leakage inductances in transformers and the junction capacitance in the switches, cause the switches to operate with the *inductive turn-off* and *capacitive turn-on*. When a switch turns off with an inductive load, high-voltage spikes are generated by the high di/dt values. Conversely, when turning on, the energy stored in the junction capacitance is trapped and dissipated in the device. Also, the electromagnetic noise generated may interfere with proper operation of the converter or proximate sensitive systems. At high switching frequency levels, the resonant, soft-switching converters offer an attractive alternative to the hard-switching switched-mode converters.

Although many solutions have been proposed, two classes stand out in the field of resonant low-power conversion, namely the *quasi-resonant* or *resonant-switch* converters and *load-resonant* converters. These are presented in the subsequent sections.

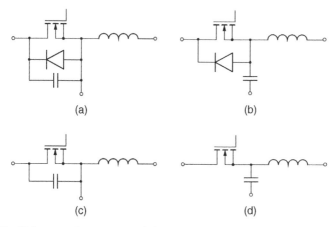

Figure 8.25 Voltage-mode resonant switches: (a) L-type, half-wave, (b) M-type, half-wave, (c) L-type, full-wave, (d) M-type, full-wave.

8.4.1   Quasi-Resonant Converters

A resonant switch is a semiconductor switch, typically a MOSFET or a BJT, with an LC resonant circuit (*LC tank*) and, sometimes, a diode interconnected with the switch. The resultant network makes either the voltage across the switch or current through it to acquire a sinusoidal waveform, instead of the square-wave one typical for switched-mode converters. There are voltage-mode and current-mode resonant switches, each class subdivided into the L-type or M-type switches. In turn, each L-type and M-type switch can be of the half-wave or full-wave variety. The resultant eight topologies are shown in Figures 8.25 and 8.26.

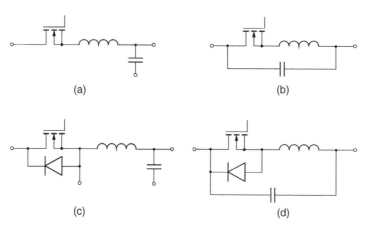

Figure 8.26 Current-mode resonant switches: (a) L-type, half-wave, (b) M-type, half-wave, (c) L-type, full-wave, (d) M-type, full-wave.

As known from the circuit theory, application of a dc voltage, V, to an undamped (lossless) series LC circuit at $t = t_0$ produces a sinusoidal current in the inductor, given by

$$i_L(t) = \frac{V - v_C(t_0)}{Z_0} \sin[\omega_0(t - t_0)] + i_L(t_0) \cos[\omega_0(t - t_0)], \qquad (8.62)$$

where $v_C(t_0)$ and $i_L(t_0)$ denote the initial voltage across the capacitor and initial inductor current, respectively. The *characteristic impedance*, Z_0, and *resonance frequency*, ω_0, are given by

$$Z_0 = \sqrt{\frac{L_r}{C_r}} \qquad (8.63)$$

and

$$\omega_0 = \frac{1}{\sqrt{L_r C_r}}, \qquad (8.64)$$

where L_r and C_r denote the respective values of inductance and capacitance of the resonant circuit. The capacitor voltage is also sinusoidal and it leads the current waveform by 90°, as

$$v_C(t) = V - [V - v_C(t_0)] \cos[\omega_0(t - t_0)] + Z_0 i_L(t_0) \sin[\omega_0(t - t_0)]. \qquad (8.65)$$

Example $v_C(t)$ and $i_L(t)$ waveforms, with $i_L(t_0) > 0$ and $v_C(t_0) < 0$, are shown in Figure 8.27.

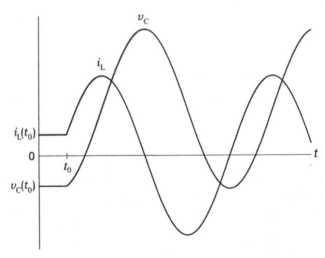

Figure 8.27 Waveforms of the inductor current and capacitor voltage in an undamped resonant circuit.

The voltage-mode resonant switches in Figure 8.25 have the resonating capacitor connected in parallel with the switching device. In the half-wave switches, the clamping diode prevents the voltage, v_C, across the capacitor from becoming negative, and the resonance ends in the midcycle. However, in the full-wave switches, the clamping diode is absent and the voltage can change its polarity.

Turning a switch off initiates the resonance process, and the voltage across the switch acquires a sinusoidal waveform. When, after a half cycle, the voltage reaches zero, the switch can be turned on again under the ZVS conditions. It is particularly advantageous in converters operating with very high switching frequencies, as the capacitive turn-on stresses on the devices are greatly alleviated.

In the current-mode resonant switches in Figure 8.26, an inductor is connected in series with the switch in order to shape the switched current. The inductor and the capacitor form a resonant circuit, whose resonance is triggered by the switch turning on.

In the half-wave switches, the unidirectional conduction ability of the semiconductor power switch prevents the current from changing its polarity. Therefore, only a half cycle of the resonance can be executed, with the current, i_L, flowing in the direction of the load. In the full-wave switches, an antiparallel diode is connected across the power switch so that the current can flow to the source.

When the semiconductor power switch turns on at $t = t_0$, the conducted resonant current increases gradually from zero, following the sinusoidal wave shape. After a half cycle, the current reaches zero and the switch is naturally turned off (commutated). Consequently, both the turn-on and turn-off occur under the ZCS conditions, and the switching losses and stresses are reduced significantly.

An additional advantage of the resonant switches in high-frequency converters consists in the possibility of utilization of parasitic inductances and capacitances. Indeed, since the resonance frequency must be higher than the switching frequency, the required values of the resonant inductance and capacitance become very low. For instance, the resonance frequency of 10 MHz can be obtained using a 15.9-nH inductance and a 15.9-nF capacitance.

All the basic switched-mode dc-to-dc converters described in Sections 8.2 and 8.3 can be converted into quasi-resonant by replacing the regular switches with resonant ones. An example ZVS buck converter with the L-type half-wave switch is shown in Figure 8.28, and a ZCS quasi-resonant boost converter with the M-type, full-wave resonant switch is illustrated in Figure 8.29.

Because of the fixed resonance frequency, control of the duty ratio of a resonant switch is performed by changing the switching frequency. The control range is narrower than that in switched-mode converters, and the EMI, although considerably reduced, is difficult to contain. In general, at high switching frequencies, ZVS is preferable to ZCS. It must also be pointed out that the resonant switches have to carry higher peak currents than their counterparts in the switched-mode dc-to-dc converters. This may not only require higher current ratings of the semiconductor devices but also cause increased conduction losses. As with any engineering system, selection of a given converter type must be based on the best tradeoff between technical and economic advantages and disadvantages.

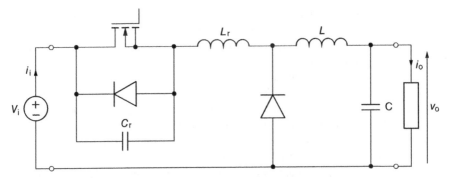

Figure 8.28 Quasi-resonant ZVS buck converter with the L-type half-wave switch.

Distinguishing properties of the ZVS and ZCS quasi-resonant converters are listed in Table 8.1. Generally, the full-wave switches make the converter operation load insensitive. However, the voltage-mode full-wave switches have the disadvantage of trapping energy in the junction capacitance of the semiconductor device. The resultant capacitive turn-on makes those switches inappropriate for converters with very high switching frequencies. The voltage gain, $K_V \equiv V_o/V_i$, of a converter with a full-wave switch is a function of the *frequency ratio*, k_f, defined as

$$k_f \equiv \frac{f_{sw}}{f_0},\tag{8.66}$$

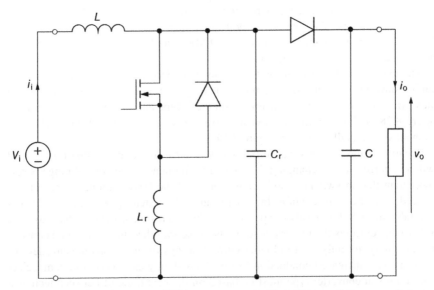

Figure 8.29 Quasi-resonant ZCS boost converter with the M-type full-wave switch.

Table 8.1 Comparison of the ZVS and ZCS Resonant-Switch Converters

Property	ZCS Converters	ZVS Converters
Voltage gain		
Buck converter	k_f	$1 - k_f$
Boost converter	$1/(1 - k_f)$	$1/k_f$
Buck–boost converter	$k_f/(1 - k_f)$	$1/k_f - 1$
Control	Constant t_{ON}	Constant t_{OFF}
Variable t_{ON}	variable t_{OFF}	
Waveform of voltage across the switch	Square	Sinusoidal
Waveform of current through the switch	Sinusoidal	Square
Load range	$1 < r < \infty$	$0 < r < 1$

where f_0 denotes the resonance frequency (in Hz). In converters with half-wave switches, K_V also depends (nonlinearly) on the *normalized load resistance*, r, defined as

$$r \equiv \frac{R}{Z_0}. \tag{8.67}$$

In the circuit theory, r represents a reciprocal of the so-called quality factor, Q, of a series resonant circuit. In all converters with half-wave switches, the voltage gain increases with r, and equals that of the full-wave switch converter only when $r = 1$. When inspecting Table 8.1, note the allowable ranges of r for ZVS and ZCS converters.

8.4.2 Load-Resonant Converters

In load-resonant converters, an LC resonant circuit causes the load voltage and current to oscillate, creating opportunities for ZVS and/or ZCS. Three most common topologies are the *series-loaded* converter, *parallel-loaded* converter, and *series–parallel* converter. All are based on an inverter–rectifier cascade appended with a resonant circuit between these two sub-converters.

Series-Loaded Converter The circuit diagram of a half-bridge series-loaded converter is shown in Figure 8.30. A half-bridge inverter, based on two semiconductor switches, S1 and S2, two freewheeling diodes, D1 and D2, and a capacitive voltage divider, C1 and C2, converts the input dc voltage, V_i, into a square-wave voltage. The inverter is connected via a series resonant circuit, L_r and C_r, to a two-pulse rectifier, composed of four diodes, D3–D6, and an output capacitor, C. The output capacitor is sufficiently large to treat the rectifier-load circuit as a source of fixed voltage, V_o.

Three basic modes of the multiple-stage operation of the converter can be distinguished. If the switching frequency, f_{sw}, is less than half of the resonance frequency, f_0, the converter operates in a *discontinuous mode*, with a discontinuous current in the resonant inductor. The inverter switches turn on with zero current and turn off with both zero current and voltage, making it possible to use SCRs in certain high-power,

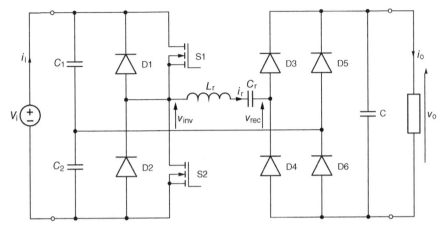

Figure 8.30 Series-loaded resonant converter.

low-frequency applications. If $f_0/2 < f_{sw} < f_0$, the converter operates in the *below-resonant continuous mode*, with the switches turning on at nonzero voltage and current, thus causing turn-on switching losses. However, the turn-off occurs under the zero-voltage and zero-current conditions, so that SCRs can again be used. Finally, in the *above-resonant continuous mode* of operation, with $f_{sw} > f_0$, the switches turn on with zero voltage and current, but they are turned off in the presence of the voltage and current, producing turn-off switching losses.

To derive an approximate relation for the voltage gain, K_V, a simple ac model of the converter can be used. The actual output voltage, $v_{inv}(t)$, of the inverter is an ac square wave with the peak value of $V_i/2$. Analogously, the input voltage, $v_{rec}(t)$, of the rectifier is an ac quasi-square wave with the peak value of V_o. However, current in the resonant circuit, $i_r(t)$, is practically sinusoidal. All three waveforms have the fundamental frequency of f_{sw}. Replacing $v_{inv}(t)$ and $v_{rec}(t)$ with their fundamentals, $v_{inv,1}(t)$ and $v_{rec,1}(t)$, an ac equivalent circuit of the converter, shown in Figure 8.31, is obtained. The peak value, $V_{inv,1,p}$, of $v_{inv,1}(t)$ is given by

$$V_{inv,1,p} = \frac{4}{\pi} \times \frac{V_i}{2} = \frac{2}{\pi} V_i \qquad (8.68)$$

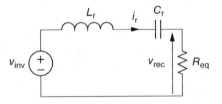

Figure 8.31 AC equivalent circuit of the series-loaded resonant converter.

and that, $V_{rec,1,p}$, of $v_{rec,1}(t)$, by

$$V_{rec,1,p} = \frac{4}{\pi}V_o \qquad (8.69)$$

(see Eq. (1.28)).

Thanks to the practically ideal dc-output voltage, the output current can also be assumed to be of pure dc quality, that is, $i_o(t) = I_o$. It is obtained by the rectification of $i_r(t)$, so that

$$I_o = \frac{2}{\pi}I_{r,p}, \qquad (8.70)$$

where $I_{r,p}$ denotes peak value of $i_r(t)$ (see Eq. 1.15). Dividing $V_{rec,1,p}$ by $I_{r,p}$ yields the equivalent resistance, R_{eq}, of the load,

$$R_{eq} = \frac{V_{rec,1,p}}{I_{r,p}} = \frac{\frac{4}{\pi}V_o}{\frac{\pi}{2}I_o} = \frac{8}{\pi^2}R = \frac{8}{\pi^2}rZ_0 \qquad (8.71)$$

as seen from the input terminals of the rectifier.

From the equivalent circuit in Figure 8.31,

$$\frac{V_{rec,1,p}}{V_{inv,1,p}} = \frac{R_{eq}}{\left|R_{eq} + j(X_L - X_C)\right|}, \qquad (8.72)$$

where

$$X_L = 2\pi f_{sw}L_r = k_f Z_0 \qquad (8.73)$$

and

$$X_C = \frac{1}{2\pi f_{sw}C_r} = \frac{Z_0}{k_f}. \qquad (8.74)$$

Based on Eqs. (8.68) and (8.69), Eq. (8.72) can be rearranged to

$$K_V = \frac{V_o}{V_i} = \frac{1}{2\sqrt{1 + \left(\frac{X_L - X_C}{R_{eq}}\right)^2}} \qquad (8.75)$$

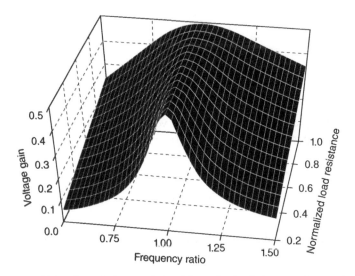

Figure 8.32 Control characteristic of the series-loaded resonant converter.

and, substituting Eqs. (8.71), (8.73), and (8.74), to

$$K_V = \frac{1}{2\sqrt{1 + \left[\frac{\pi^2}{8r}\left(k_f - \frac{1}{k_f}\right)\right]^2}}. \tag{8.76}$$

A three-dimensional graph of the control characteristic (8.76) is shown in Figure 8.32. The highest voltage gain of 0.5 is obtained when $k_f = 1$, that is, $f_{sw} = f_0$. The discontinuous mode is most advantageous because no switching losses occur in the converter.

Alternative topologies of the series-loaded resonant dc-to-dc converter include converters with a full-bridge inverter and converters with a transformer between the resonant circuit and the rectifier. A *load-commutated inverter* is obtained by eliminating the rectifier and connecting the inverter to the load directly, via a resonant circuit. Such inverter can be based on SCRs, which are being naturally commutated, while the resonant circuit results in quasi-sinusoidal waveforms of the output voltage and current.

Parallel-Loaded Converter. The parallel-loaded resonant dc-to-dc converter, whose circuit diagram is shown in Figure 8.33, differs from the series-loaded converter by the connection of the resonant capacitor across the input terminals of the rectifier. Another difference consists in an output inductor, L, that smoothes and stabilizes the current supplied to the final stage of the converter. Consequently, the rectifier can be considered a source of a constant current, I_o.

Like the series-loaded resonant converter, the parallel-loaded converter operates in the discontinuous mode when $f_{sw} < f_0/2$, in the below-resonant continuous mode when $f_0/2 < f_{sw} < f_0$, and in the above-resonant continuous mode when $f_{sw} > f_0$.

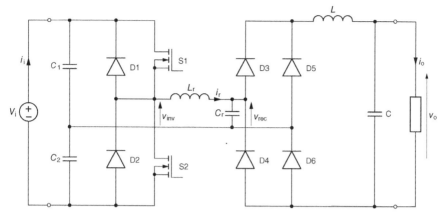

Figure 8.33 Parallel-loaded resonant converter.

Again, the discontinuous mode is characterized by no switching losses, the below-resonant continuous mode by no turn-off losses, and the above-resonant mode by no turn-on losses. The ac equivalent circuit of the converter is shown in Figure 8.34. In the actual converter, $v_{rec}(t)$ and $i_{rec}(t)$ have ac square waveforms, while the waveform of $v_{inv}(t)$ is sinusoidal. Consequently, Eq. (8.68) is valid again and, since V_o is the average value of rectified $v_{rec}(t)$, then

$$V_o = \frac{2}{\pi} V_{rec,1,p}. \tag{8.77}$$

The peak value of $i_{rec}(t)$ equals the output current, I_o, thus the peak value, $I_{rec,1,p}$, of the fundamental of $i_{rec}(t)$ is given by

$$I_{rec,1,p} = \frac{4}{\pi} I_o. \tag{8.78}$$

Based on Eqs. (8.77) and (8.78), the equivalent load resistance, R_{eq}, of the parallel-loaded converter can be expressed as

$$R_{eq} = \frac{V_{rec,1,p}}{I_{rec,1,p}} = \frac{\frac{\pi}{2} V_o}{\frac{4}{\pi} I_o} = \frac{\pi^2}{8} R = \frac{\pi^2}{8} r Z_0. \tag{8.79}$$

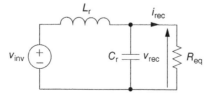

Figure 8.34 AC equivalent circuit of the parallel-loaded resonant converter.

Finally, from the ac equivalent circuit of Figure 8.34,

$$\frac{V_{\text{rec},1,\text{p}}}{V_{\text{inv},1,\text{p}}} = \frac{1}{\left| 1 - \frac{X_{\text{L}}}{X_{\text{C}}} + j\frac{X_{\text{L}}}{R_{\text{eq}}} \right|} \tag{8.80}$$

or

$$K_{\text{V}} = \frac{V_{\text{o}}}{V_{\text{i}}} = \frac{4}{\pi^2\sqrt{\left(1 - \frac{X_{\text{L}}}{X_{\text{C}}}\right)^2 + \left(\frac{X_{\text{L}}}{R_{\text{eq}}}\right)^2}}, \tag{8.81}$$

which, after substituting Eqs. (8.73), (8.74), and (8.79), yields

$$K_{\text{V}} = \frac{1}{\frac{\pi^2}{4}\sqrt{\left(1 - k_{\text{f}}\right)^2 + \left(\frac{8k_{\text{f}}}{\pi^2 r}\right)}}. \tag{8.82}$$

Control characteristic (8.82) of the parallel-loaded resonant converter is illustrated in Figure 8.35. As in the series-loaded converter, the half-bridge inverter can be replaced with a full-bridge one and a transformer can be added to provide electrical isolation between the input and the output.

Series–Parallel Converter. A clone of the parallel-loaded resonant converter, called a *series–parallel resonant converter*, is shown in Figure 8.36. Addition of the series resonant capacitor increases the range of control of the output voltage and improves the efficiency of the converter with light loads. If $C_{\text{r1}} = C_{\text{r2}} = 1/(\omega$

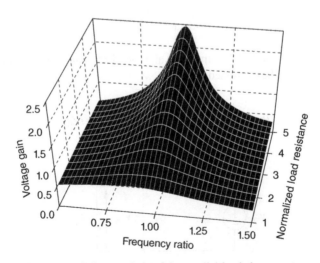

Figure 8.35 Control characteristic of the parallel-loaded resonant converter.

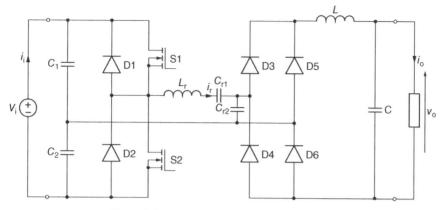

Figure 8.36   Series–parallel resonant converter.

L_r), then the approximate control characteristic of the series–parallel converter is given by

$$K_V = \frac{1}{\frac{\pi^2}{4}\sqrt{\left(2 - k_f^2\right)^2 + \left[\frac{8}{\pi^2}\left(k_f - \frac{1}{k_f}\right)\right]^2}} \qquad (8.83)$$

and illustrated in Figure 8.37.

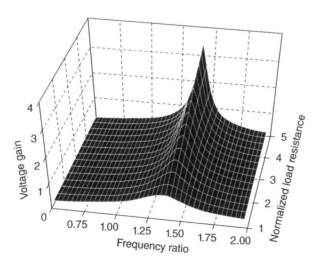

Figure 8.37   Control characteristic of the series–parallel resonant converter.

8.4.3   Comparison of Resonant DC-to-DC Converters

The analogy between the single-switch switched-mode dc-to-dc converters and quasi-resonant converters is obvious, as the latter ones are obtained by replacing regular switches by the ZVS or ZCS resonant switches. There is also an analogy between the multiple-switch switched-mode converters and load-resonant converters, as both classes are based on a single-phase inverter feeding, indirectly, a two-pulse rectifier. Consequently, the analogous converters tend to share similar applications. Generally, single-switch dc-to-dc converters are used at low-power levels, while the multiple-switch ones are mostly used for medium-power loads, typically on the order of 1 kW and more.

Comparing the individual types of load-resonant converters, it must be reminded that Eqs. (8.76), (8.82), and (8.83), based on the representation of square-wave quantities by their fundamentals, are not exact. Nevertheless, important functional differences between series-loaded and parallel-loaded dc-to-dc converters can be discerned from the control characteristics in Figures 8.30 and 8.33. First, the voltage gain of the series-loaded converter is limited to the zero to 0.5 range, while that of the parallel-loaded converter can exceed unity, depending on the load. Second, to allow control of the output voltage, the load resistance, R, should be less than the characteristic impedance, Z_0, in the series-loaded converter and higher than the characteristic impedance in the parallel-loaded converter.

The load resistance affects the voltage gain of the converters in different ways: although the maximum voltage gain of 0.5 in the series-loaded converter can be realized with any load, the control range of the output voltage increases when the load decreases; in contrast, in the parallel-loaded converter, the control range and maximum voltage gain increase with the load. In the continuous operating modes, the resonant converters perform "firm" switching, that is, the switching losses are reduced when compared with the hard-switching converters, but not completely eliminated.

By an exact analysis of operation, it can be shown that the series-loaded converter in the preferred discontinuous mode is impervious to short circuits in the load. However, the converter loses its control capability with light loads, including the no-load situation. On the other hand, the parallel-loaded converter in the transformer version is better suited for multiple outputs, but with light loads its efficiency is poor. The series–parallel converter offers certain improvements over the functionally similar parallel-loaded converter.

SUMMARY

Switching power supplies are employed as sources of stable or adjustable dc voltage in numerous applications, mainly those involving electronic circuits. The input dc voltage is obtained from batteries, photovoltaic or fuel cells, or rectifiers, and controlled in switching dc-to-dc converters. These can be divided into hard-switching switched-mode converters and soft-switching (or "firm-switching") resonant converters.

The switched-mode dc-to-dc converters are built in the nonisolated and isolated versions. A transformer is used to provide electrical isolation and, if required, voltage matching between the input and output terminals. Also, a number of output stages can be supplied from multiple secondary windings of the transformer. The basic topologies of nonisolated converters, such as the buck, boost, buck–boost, and Ĉuk converters, employ a single semiconductor switch, while certain isolated converters, such as the push–pull, half-bridge, and full-bridge converters, are based on two or more switches. The multiple-switch converters are composed of a single-phase inverter coupled with a single-phase rectifier through a transformer.

High power density is a very important consideration for switching power supplies, such as those for the portable communication equipment or spacecraft power systems. High switching frequencies allow the reduction of the electromagnetic and electrostatic components, but they cause high switching losses. Miniature converters cannot dissipate much heat, so the size reduction must be accompanied by the reduction of losses. This is the main rationale for resonant converters, in which the phenomenon of electric resonance is harnessed to provide ZCS and ZVS conditions. The resonant converters can be classified as quasi-resonant (resonant-switch) converters and load-resonant converters. Resonant switches, constructed by augmentation of regular semiconductor switches with an LC tank and, in certain types, a diode, come in various configurations. Employing a resonant switch in a switched-mode topology results in a ZVS or ZCS resonant converter.

Similar to the multiple-switch switched-mode converters, the load-resonant converters are based on the inverter–rectifier cascade, with a resonant LC circuit inserted between these two stages. An isolation transformer can be used at the input to the rectifier.

Because of the variety of types of switching power supplies, it is difficult to formulate general rules for selection of semiconductor devices. As usual, the rated voltage should exceed the highest voltage expected anywhere in the converter. Certain types of converters employ the chopper and inverter topologies covered in the previous chapters, so that current ratings of the switches can be selected similarly. In resonant converters, the semiconductor devices are often forced to carry low-average but high-peak currents. Specialized design and simulation software are recommended for the design and analysis of converters.

It must be stressed that to cover the rich field of switching power supplies in some depth, a whole book, not just a book chapter, is required. Indeed, for many practicing engineers, power electronics means just that field and not much else. Numerous topics, such as the power-factor correction pre-regulators or the Luo converters, have been omitted here for the lack of space and the general focus of the book on medium- and high-power switching converters. For interested readers, several good sources are listed in the Further Reading section.

EXAMPLES

Example 8.1   A buck converter supplied from a 12-V dc source operates with the switching frequency of 10 kHz. The maximum load resistance is 6 Ω, and the output capacitor has the capacitance of 20 μF. Find:

  (a) the minimum inductance of the output inductor required for continuous conduction,

(b) the required duty ratio of the switch to produce the output voltage of 9 V, and

(c) the peak-to-peak amplitude of ripple of the output voltage.

Solution:

(a) If condition (8.11) is satisfied for $d = 0$, it is satisfied for any other values of the duty ratio. Substituting the maximum load resistance, $R_{max} = 6\,\Omega$, for the V_o/I_o ratio, yields

$$L > \frac{R_{max}}{2f_{sw}} = \frac{6}{2 \times 10 \times 10^3} = 3 \times 10^{-3} \text{H} = 3\,\text{mH}.$$

(b) From Eq. (8.5),

$$d = \frac{V_o}{V_i} = \frac{9}{12} = 0.75.$$

(c) From Eq. (8.9),

$$\Delta V_o = \frac{1-d}{8LCf_{sw}^2} V_o = \frac{1 - 0.75}{8 \times 3 \times 10^{-3} \times 20 \times 10^{-6} \times (10 \times 10^3)^2} \times 9$$
$$= 0.047\text{V} = 47\,\text{mV}.$$

At about 0.5% of the output voltage, the ripple is practically negligible.

Example 8.2   A boost converter supplied from a 6-V dc source is to produce the output voltage of 15 V. Assuming an ideal, lossless inductor, find the required duty ratio of the switch. What is the output voltage if the resistance of the inductor is 2% of that of the load?

Solution: From relation (8.12) for an ideal converter,

$$d = 1 - \frac{V_i}{V_o} = 1 - \frac{6}{15} = 0.6.$$

If the inductor resistance is taken into account, Eq. (8.21) should be used instead, giving

$$V_o = \frac{6}{(1 - 0.6)\left[1 + \frac{0.02}{(1-0.6)^2}\right]} = 13.3\,\text{V}.$$

Thus, the inductor resistance causes reduction of the output voltage by about 11% from the theoretical value.

Example 8.3   A half-bridge converter is fed from a 120-V ac line via a two-pulse rectifier with a capacitive output filter, such that the average input voltage to the converter is 165 V. The output voltage of the converter is 24 V, and the turn ratio, N_2/N_1, of the isolation transformer is 1:3. Values of the inductance and capacitance of the output filter are 1 mH and 25 μF, respectively, and the converter operates with the switching frequency of 20 kHz. Find the required duty ratio of converter switches and the peak-to-peak amplitude of ripple of the output voltage.

Solution: From Eq. (8.61),

$$d = \frac{V_o}{k_N V_i} = \frac{24}{\frac{1}{3} \times 165} = 0.436,$$

which is a feasible value, as the duty ratio must be less than 0.5. The amplitude of the voltage ripple can be found from Eq. (8.59) as

$$\Delta V_o = \frac{1 - 2 \times 0.436}{32 \times 1 \times 10^{-3} \times 25 \times 10^{-6} \times \left(20 \times 10^3\right)^2} \times 24 = 0.0096 \, \text{V} = 9.6 \, \text{mV}$$

which amounts to only 0.04% of the magnitude of output voltage.

Example 8.4   Draw a circuit diagram of a ZCS quasi-resonant buck converter with an L-type, full-wave switch.

Solution: To obtain the converter in question, the regular semiconductor switch in the switched-mode buck converter in Figure 8.1 is replaced with the current-mode resonant switch in Figure 8.26c. The resultant circuit is shown in Figure 8.38.

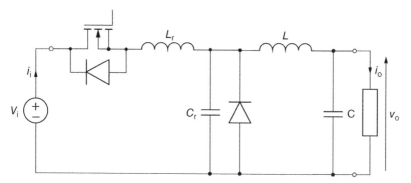

Figure 8.38   Quasi-resonant ZCS buck converter with the L-type, full-wave switch.

Example 8.5   A parallel-loaded resonant converter supplied from a 100-V dc source operates with the switching frequency of 5 kHz. The resonant circuit of the converter is composed of a 0.1-mH inductor and a 4-μF capacitor, and the load resistance is 10 Ω. Find the operating mode of the converter and estimate the average output voltage.

Solution: The characteristic impedance and resonance frequency of the resonant circuit of the converter, calculated from Eqs. (8.63) and (8.64), are

$$Z_0 = \sqrt{\frac{0.1 \times 10^{-3}}{4 \times 10^{-6}}} = 5\,\Omega$$

and

$$f_0 = \frac{1}{2\pi\sqrt{0.1 \times 10^{-3} \times 4 \times 10^{-6}}} = 7958\,\text{Hz} \approx 8\,\text{kHz},$$

respectively. The switching frequency of 5 kHz is less than the resonance frequency but higher than half of it, thus the converter operates in the continuous below-resonant mode. According to definitions (8.66) and (8.67), the frequency ratio, k_f, and normalized load resistance, r, are

$$k_f = \frac{5}{8} = 0.625$$

and

$$r = \frac{10}{5} = 2.$$

Now, the average output voltage can be estimated from Eq. (8.82) as

$$V_o = \frac{100}{\frac{\pi^2}{4}\sqrt{\left(1 - 0.625^2\right)^2 + \left(\frac{8 \times 0.625}{\pi^2 \times 2}\right)^2}} = 61.4\,\text{V}.$$

PROBLEMS

P8.1   A buck converter produces the average output voltage of 6 V. The on-time of the switch is 30 μs and the off-time is 20 μs. Find the input voltage and switching frequency of the converter.

P8.2   A buck converter supplied from a 60-V dc source operates with the switching frequency of 25 kHz, a duty ratio of 0.7, and a 5-Ω load. A 100-μF output capacitor is employed, and the inductor has the inductance twice as high as the minimum required for the continuous conduction mode. Find the average output voltage and the amplitude of its ripple.

P8.3 Repeat Problem 8.1 for a boost converter.

P8.4 Repeat Problem 8.1 for a boost–buck converter.

P8.5 Design a buck converter (select the inductance, capacitance, and switching frequency values) which will convert a 15-V input voltage into a 6-V output voltage with the ripple amplitude less than 0.5% of the output voltage (many solutions are possible).

P8.6 Repeat Problem 8.5 for a boost converter.

P8.7 Repeat Problem 8.5 for a buck–boost converter.

P8.8 A forward converter supplied from a 25-V dc source operates with the switching frequency of 50 kHz and duty ratio of 0.5. The turn ratio of the transformer is 1:1, the output capacitance is 150 μF, and the load draws a current of 2.5 A. Select the minimum inductance for continuous conduction mode and determine the average value and ripple amplitude of the output voltage.

P8.9 A flyback converter is supplied from a 75-V dc source and produces the output voltage of 110 V across a 22-Ω load. The transformer turn ratio is 1:2, and the magnetizing inductance and output capacitance are 0.13 mH and 0.2 mF, respectively. What is the minimum switching frequency ensuring the continuous conduction mode, and what is the corresponding amplitude of ripple of the output voltage?

P8.10 An isolated Ĉuk converter supplied from a 200-V dc source operates with the switching frequency of 50 kHz and duty ratio of 0.45. The transformer turn ratio is 1:1 and the load resistance is 25 Ω. Assuming inductances of the converter at twice the minimum values for continuous conduction, find the average value and ripple of the output voltage if the output capacitance is 30 μF.

P8.11 A push–pull converter supplied from a 100-V dc source operates with the switching frequency of 25 kHz. The transformer turn ratio is 2:1, and the inductance and capacitance of the output filter are 0.1 mH and 20 μF, respectively. Determine the duty ratio of converter switches required to obtain the average output voltage of 150 V. What is the ripple amplitude of that voltage?

P8.12 A half-bridge converter is to produce a 120-V average output voltage. The turn ratio of the isolation transformer is 2:1. What is the minimum required input dc voltage?

P8.13 Repeat Problem 8.12 for a full-bridge converter.

P8.14 Draw the circuit diagram of a quasi-resonant ZCS buck–boost converter with a current-mode L-type half-wave resonant switch.

P8.15 Draw the circuit diagram of a quasi-resonant ZCS flyback converter with a current-mode M-type full-wave resonant switch.

P8.16 Draw the circuit diagram of a quasi-resonant ZVS Ĉuk converter with a voltage-mode L-type full-wave resonant switch.

P8.17 The resonant inductance and capacitance in certain resonant dc-to-dc converter are 0.2 mH and 0.5 μF, respectively, the switching frequency is 45 kHz, and the load resistance is 10 Ω. Determine the resonance frequency, characteristic impedance, frequency ratio, and normalized load resistance of the converter.

P8.18 A series-loaded resonant converter supplied from a 12-V dc source operates with the switching frequency of 25 kHz and resistive load of 2 Ω. The resonant inductance and capacitance are 60 μH and 120 nF, respectively. Estimate the output voltage and current of the converter.

P8.19 A 100-V, 600-VA parallel-loaded resonant converter is designed to operate with the switching frequency of 40 kHz. The resonant inductance and capacitance are 12 μH and 3.3 μF, respectively. Estimate the required input voltage (round it up to the nearest 10 V).

P8.20 A series-parallel resonant converter, with two identical resonant capacitors of 1.5 μF each, is to operate with the switching frequency of 20 kHz. Assuming the normalized load resistance of 4.5, select the resonant inductance that should result in the highest voltage gain of the converter (this problem may require use of a computer).

COMPUTER ASSIGNMENTS

CA8.1* Run PSpice program *Buck_Conv.cir* for a buck converter. Observe oscillograms of voltages and currents, and find the peak-to-peak amplitude of ripple of the output voltage for two different values of the switching frequency.

CA8.2* Run PSpice program *Boost_Conv.cir* for a boost converter. Observe oscillograms of voltages and currents, and determine the average output voltage for two different values of duty ratio of the switch.

CA8.3* Run PSpice program *Buck-Boost_Conv.cir* for a buck–boost converter with two values of the switch duty ratio such that the average output voltage is less and greater than the input voltage. In both cases, find the average output voltage and peak-to-peak amplitude of its ripple.

CA8.4* Run PSpice program *Cuk_Conv.cir* for a Ĉuk converter. Determine the peak-to-peak amplitudes of ripple of the input current and output voltage. Observe other voltage and current waveforms.

CA8.5*   Run PSpice program *Buck_Conv.cir* for a buck converter and program *Buck_Aver_Model.cir* for the average model of the same converter using the Vorperian's switch model. Compare the output voltage waveforms.

CA8.6   Use file *Boost_Conv.cir* to develop a PSpice circuit file for the average model of the boost converter employing the Vorperian's switch model (see *Buck_Aver_Model.cir* for comparison).

CA8.7*   Run PSpice program *Forward_Conv.cir* for a forward converter. Find the average output voltage and peak-to-peak amplitude of its ripple. Observe other voltage and current waveforms.

CA8.8*   Run PSpice program *Flyback_Conv.cir* for a flyback converter. Find the average value and the peak-to-peak amplitude of ripple of output voltage. Observe other voltage and current waveforms.

CA8.9*   Run PSpice programs *Push_Pull_Conv.cir*, *Half_Brdg_Conv.cir*, and *Full_Brdg_Conv.cir* for the push–pull, half-bridge, and full-bridge converters. Notice that all these converters are supplied with the same voltage, have the same output filter parameters, and operate with the same switching frequency and duty ratio of switches. Compare the output voltage waveforms by determining the average values and ripple amplitudes. Observe other voltage and current waveforms.

CA8.10*   Run PSpice program *Reson_Buck_Conv.cir* for a quasi-resonant ZVS buck converter. Find the average value and the peak-to-peak amplitude of ripple of output voltage. Observe, simultaneously, oscillograms of the voltage, V(A,C), across the switch and the gate signal, V(G,C), that well illustrate the ZVS conditions.

CA8.11*   Run PSpice program *Reson_Boost_Conv.cir* for a quasi-resonant ZCS boost converter. Find the average value and the peak-to-peak amplitude of ripple of output voltage. Observe, simultaneously, oscillograms of the current, I(Vsense), in the switch and the gate signal, V(G,C), that well illustrate the ZCS conditions.

CA8.12*   Run PSpice program *Series_Load_Conv.cir* for a series-loaded resonant converter for all three conduction modes (comment out the unneeded .PARAM statements). Observe oscillograms of the switch current, I(Vsense), and gate signal, V(G1,C) that well illustrate the ZCS conditions.

CA8.13*   type="star"Repeat Assignment 8.12* for a parallel-loaded resonant converter in PSpice file *Parallel_Load_Conv.cir.*

CA8.14   Based on the PSpice *Parallel_Load_Conv.cir* file, develop a similar file for the series-parallel resonant converter.

FURTHER READING

[1] Kazimierczuk, M. K. and Czarkowski, D., *Resonant Power Converters*, 2nd ed., John Wiley & Sons, Inc., Hoboken, NJ, 2011.

[2] Maniktala, S., *Switching Power Supply Design and Optimization*, 2nd ed., McGraw-Hill, New York, 2014.

[3] Pressman, A. I., Billings, K., and Taylor, M., *Switching Power Supply Design*, 3rd ed., McGraw-Hill, New York, 2009.

9 Power Electronics and Clean Energy

In this chapter: an overview of "green" applications of power electronics is presented; use of power converters in renewable energy sources and distributed generation systems is outlined; electric and hybrid cars are described; and role of power electronics in energy conservation is explained.

9.1 WHY IS POWER ELECTRONICS INDISPENSABLE IN CLEAN ENERGY SYSTEMS?

The rapidly growing interest in various aspects of clean energy is a reaction to the major energy-related challenges facing humanity. The grim reality of climate change, environmental pollution, and limited resources of fossil fuels have spawned massive research and development efforts devoted to "clean," or "green," energy. They are mostly focused on renewable energy sources, distributed generation systems, and conservation of energy in a variety of systems, such as industrial drives or electric and hybrid cars.

Renewable energy sources can be classified as sustained and intermittent. Geothermal and hydroelectric plants or biofuel distilleries represent typical examples of sustained energy sources. The energy is generated, or concentrated, in a similar fashion as in the traditional power plants or oil refineries, and the role of power electronics is limited. The intermittent sources include the increasingly common wind and photovoltaic (PV) systems, whose randomly varying output makes power electronic converters a vital part. To efficiently interface a renewable energy source with a load or power grid, the converters perform power conditioning, voltage boosting, and control of the flow of power.

As an aside, it is worth to mention that the intermittency of sunlight-dependent sources of renewable energy is significantly alleviated in the so-called concentrated solar power plants, in which parabolic mirrors focus the sunlight to heat up liquid medium, such as fluoride salts, to high temperatures (over 700°C). This allows storing

Introduction to Modern Power Electronics, Third Edition. Andrzej M. Trzynadlowski.
© 2016 John Wiley & Sons, Inc. Published 2016 by John Wiley & Sons, Inc.
Companion website: www.wiley.com/go/modernpowerelectronics3e

the thermal energy, which can be used around the clock to produce steam for electric turbogenerators. Such plants are somewhat similar to binary-cycle geothermal plants, in which the heat from underground hot water is transferred to fluid hydrocarbon, whose vapors drive the turbine of a generator.

Tracking the *maximum power point* (MPP) constitutes an important part of power conversion and control. At given illumination, the MPP on the voltage-current characteristics of a solar cell marks the maximum product of the voltage and the current, that is, the maximum power drawn from the cell. Similarly, at a given wind speed, the power coefficient of a wind-turbine reaches maximum at a specific value of the tip speed ratio and, if applicable, the blade pitch angle. Control systems of PV or wind-power sources continuously track the MPP in order to draw the maximum power from the sun or wind.

The traditional power grid is based on large electric power plants located possibly close to fossil fuel sources, such as coal mines or natural gas fields. The energy is then delivered to population centers over high-voltage transmission lines. Recently, the power infrastructure has started to change. In addition to the gas and coal fired plants, distributed generation systems employ variety of renewable energy sources, reducing the demand for fossil fuels and environmental damage. In areas without an access to the grid, such as remote locations of North America or numerous underdeveloped countries, power microgrids are formed from renewable and nonrenewable sources and energy storage devices. Diesel generators and electric batteries are usually used to smooth the temporal variations of power supply. For the large number of sources of varying type, ratings, and intermittency to operate efficiently and harmoniously, the electric power generated by individual must undergo various kinds of conversion and control.

From the beginnings of modern power electronics, energy conservation has been an important factor supporting the growth of that discipline. Most of the electric power consumed in industry is spent in drives of such fluid-handling machinery as pumps, fans, blowers, and compressors. If, for example, a fixed-speed motor powers a pump in a hydraulic system, the flow intensity must be controlled by valves. If weak flow is required, the valve is almost closed and the pump basically mixes the stagnant fluid, wasting most of the energy drawn by the motor. Controlling the flow by adjusting the pump speed in a variable-speed drive is a much more efficient solution. Huge energy savings in industrial plants provided main motivation for the rapid growth of power electronics and electric drives in the 1970s and 1980s. The trend of replacing fixed-speed drives with variable-speed drives continues unabated. Recently, the variable-speed drives have entered the automobile world, appearing in electric and hybrid cars. Majority of variable-speed drives are of the ac type, because ac motors are distinctly less expensive and more robust than their dc counterparts. Therefore, as already mentioned in Chapter 7, inverters constitute one of the most popular types of power electronic converters.

As seen from the presented overview, power electronics constitute a vital component of most clean energy systems. More details of applications of power converters in those systems are described in the subsequent sections.

9.2   SOLAR AND WIND RENEWABLE ENERGY SYSTEMS

The role of power electronic converters in solar and wind renewable energy systems depends on the type and ratings of the source, as well as the type of load. Most often, the load is the electric power grid, which collects and distributes electrical energy supplied from a number of sources. Sometimes, though, a renewable energy source directly feeds a specific load, such as a battery pack supplying an off-grid household. In each case, the source voltage varies randomly, while the output of the system must provide fixed voltage and frequency required by the load. Maximum power tracking must constitute a part of the control strategy.

In a distributed generation system, where a variety of "small" sources (PV, wind, small hydro, etc.) provide electrical power along large traditional generators, care must be taken to detect and prevent the so-called islanding. If a fault occurs on the grid, all those small sources must disconnect from the grid, leaving only the large, centrally controlled power plants. Otherwise, the small sources would back-feed into a fault and worsen the situation. Also, safety of people, especially the maintenance crews, could be endangered. It is also required that the small sources cease to operate in an unintentional "island" separated from the grid. In order to supply certain loads when the grid has failed, intentional islands are allowed after taking steps preventing the grid from back-energizing. Islanding can be detected by sensing sudden changes in the system frequency, voltage, and real and reactive power output.

9.2.1   Solar Energy Systems

The quantum-mechanical PV effect is utilized in solar cells, most commonly based on silicon semiconductor material, for converting sunlight to electricity. The open-circuit voltage, V_{oc}, of a solar cell is typically 0.6–0.7 V and the short-circuit current, I_{sc}, is 20–40 mA/cm^2. To increase the output voltage, a number of cells, usually 36 or 72, are connected in series to form a solar, or PV, module. An assembly of modules forms a PV panel, which constitutes a mechanical and electrical entity. A PV array is a collection of several panels. For generality, in the subsequent considerations, the term "array" is used rather loosely, meaning any set of PV modules connected in series and parallel to form a source of dc voltage.

Voltage and power as functions of the current drawn from a solar array are illustrated in Figure 9.1. It can be seen that the power curve has a sharp peak, so that even a small deviation from the MPP significantly reduces the power yield. Note that a specific voltage-current characteristic depends on the irradiation and temperature of the cells. Therefore, MPP tracking usually involves some type of perturb-and-observe technique. Other approaches have also been proposed. Once the MPP is located, the power electronic interface must maintain the current close to the I_{MMP} level.

PV arrays come in various sizes, the small ones often mounted on buildings, including houses. Typically, the owner of the array sells the energy to the local utility by interfacing the array with the grid through a single-phase utility power line. Large arrays can form PV power plants connected to the grid via three-phase power lines.

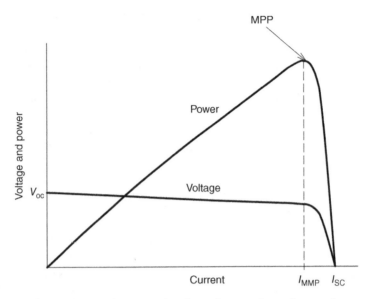

Figure 9.1   Voltage and power as functions of current drawn from a solar array.

Thus, the grid-side ac-output converter can be of the single- or three-phase variety. Without a loss of generality, a three-phase connection to the grid is shown in the subsequent diagrams of the interfaces.

The simplest solution for a PV array interface with the grid is shown in Figure 9.2. The array is connected to the grid through a dc link, voltage-source inverter, and a filter that smoothes the current ripple. If the array's voltage is insufficiently high to match the grid voltage, a transformer is needed, as in the system shown previously in Figure 7.70. A high-frequency transformer can also be embedded in an isolated dc-to-dc converter (see Section 8.3) at the output of the array. If galvanic isolation is not required, the voltage of the PV array can be increased in a transformer-less

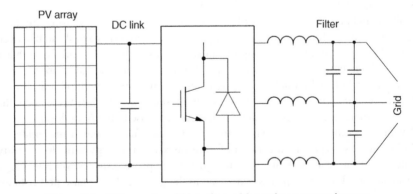

Figure 9.2   PV array-to-grid interface with a voltage-source inverter.

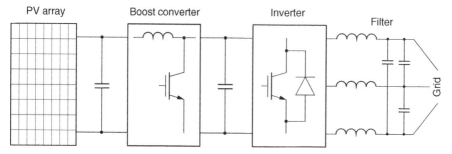

Figure 9.3   PV array-to-grid interface with a dc-to-dc boost converter and voltage-source inverter.

dc-to-dc boost converter, as illustrated in Figure 9.3. Another solution, shown in Figure 9.4, involves a current-source inverter. Thanks to the inductive dc link, the inverter provides boost to the output voltage. The filter must include capacitors to avoid the dangerous connection of two inductors carrying different currents.

Connecting a PV array to a single converter creates an inflexible structure, making an expansion (or reduction) of the array difficult without retrofitting the whole system. Therefore, the recent tendency in PV systems is to use "strings" of series-connected PV modules, with each string feeding its own inverter. The number of modules in the string is large enough to avoid the necessity of voltage boosting. More radical solutions are:

(a) multi-string interface, in which each string of modules is equipped with its own dc-to-dc converter interfaced to a single inverter;

(b) ac-module system, in which a power electronic interface is integrated into each module; and

(c) single-cell system with an inverter interface, employing a large PV cell, for example, of the photo electro-chemical type, which can be made arbitrarily large.

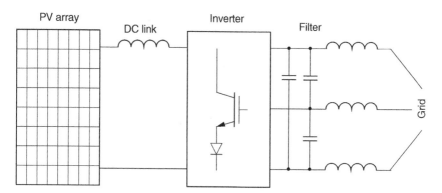

Figure 9.4   PV array-to-grid interface with a boost current-source inverter.

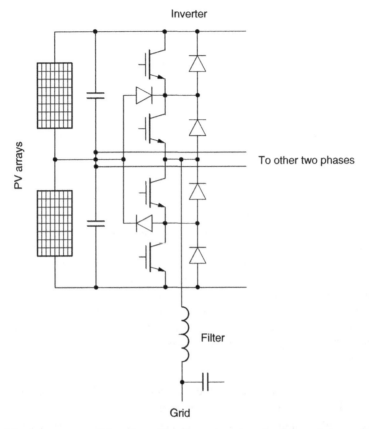

Figure 9.5 PV array-to-grid interface with a three-phase three-level neutral-clamped inverter (one leg shown only).

Multilevel inverters are well suited as PV array-to-grid interfaces. Various levels of the dc input voltage can easily be set up using an appropriate interconnection of PV modules. The higher is the number of levels, the higher quality of output voltage is obtained, either in the low-loss square-wave mode or a PWM mode with low switching frequency. One leg of a three-phase three-level neutral-clamped inverter supplied from two PV arrays is shown in Figure 9.5, while. Figure 9.6 depicts a single-phase cascaded H-bridge inverter with each bridge supplied from an array. Three such inverters can be connected as a three-phase inverter if a three-phase input to the grid is needed (see Figure 7.59a).

Various types of inverters and, especially, dc-to-dc converters, some of them developed specifically for this application, can be used in the interfaces. Therefore the presented review is far from complete. Interested readers are referred to the many existing specialized publications—alternative energy is one of the hottest topics of today's engineering.

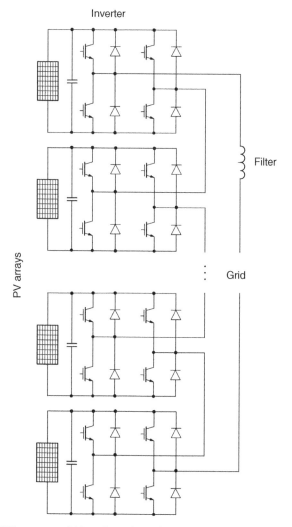

Figure 9.6   PV array-to-grid interface with a single-phase cascaded H-bridges inverter.

9.2.2   Wind Energy Systems

Wind energy systems enjoy rapid growth and, as of now, produce more power than any other renewable energy source apart from hydroelectric plants. Wind is less intermittent than sunlight, and a wind-turbine occupies less space than a solar array of comparable power. Large three-blade horizontal-axis turbines, with the rated power of several MW each, are usually grouped into commercial "wind farms." Low-power turbines of various designs are slowly making their way into residential market of renewable energy, but they are still relatively rare in practice.

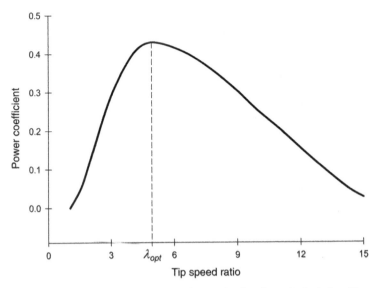

Figure 9.7   Power coefficient versus tip speed ratio of a typical wind-turbine.

The aerodynamic power, P_a, of a wind-turbine, that is the amount of power extracted from wind, is given by

$$P_a = \frac{1}{2}\rho C_p A v_w^3,\tag{9.1}$$

where ρ denotes air density, C_p is the power coefficient, A is the area swept by the blades, and v_w is the wind speed. The power coefficient strongly depends on the tip speed ratio, λ, defined as

$$\lambda = \frac{v_t}{v_w},\tag{9.2}$$

where v_t denotes the linear speed of tip of the blade. In pitch-controlled wind-turbines, C_p also depends on the blade pitch angle. A typical relation between the power coefficient and tip speed ratio is shown in Figure 9.7. Clearly, to best utilize the turbine, its speed should be varied with the wind speed to maintain the optimum value, λ_{opt}, of that ratio.

The output power versus wind speed relation for a typical variable-speed wind-turbine system is shown in Figure 9.8. With the increase in wind speed, the optimum tip speed ratio is maintained until the turbine reaches its rated speed, and the generator driven by the turbine produces the rated amount of electrical power. Further increase of turbine speed must be prevented to protect the generator from overloading. At this point, active control of the pitch of blades is employed to reduce the energy capture capability of the turbine. In turbines without pitch control, aerodynamic stall produces a similar result. The blades are so shaped that the smooth flow of air around the blades

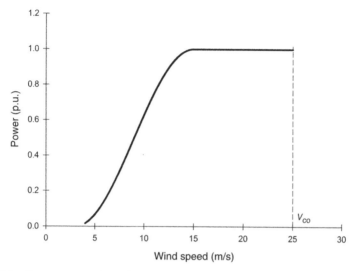

Figure 9.8 Output power of a typical wind-turbine system as a function of the wind speed.

becomes turbulent when the wind speed exceeds a specified threshold, and the forces driving the blades decrease. If speed of the wind exceeds the so-called cut-off value, v_{co}, the turbine is stopped to avoid structural damage.

Fixed-Speed Systems. If an ac squirrel-cage induction machine is employed as a generator, the speed of the turbine is practically constant, varying only within the limits of slip of the machine, that is, a few percentage points. The generator is driven by the turbine via a gearbox and, typically, connected to the grid through a transformer. Frequency of the grid voltage dictates speeds of the generator and turbine. To improve the power factor at the point of connection with the grid, a capacitor bank is employed as a reactive power compensator. The pitch control or aerodynamic stall limits the power output at high speeds. To avoid an overcurrent at the start-up of the system, a soft-starter is employed. A block diagram of the described system is shown in Figure 9.9. The speed range can be somewhat increased by replacing the squirrel-cage generator with a wound-rotor one and adding controlled resistances to rotor windings.

Variable-Speed Systems. Conditioning the power produced by the generator allows significant improvements of operation of wind energy systems in terms of MPP tracking and control of the real and reactive powers. The MPP tracking requires a variable speed of the turbine, which means that the frequency of the generator voltage is floating. Therefore, a frequency changer between the generator and transformer is needed. As shown in Figure 9.10, the frequency changer consists of a rectifier, a dc link, and an inverter (see Figure 7.75b). Control of real power is performed in the rectifier and that of reactive power in the inverter. A matrix converter can be used in place of the rectifier–inverter cascade.

In high-power systems, an electrically excited ac synchronous generator can be employed. Both the real and reactive powers are controlled in the inverter. As seen in

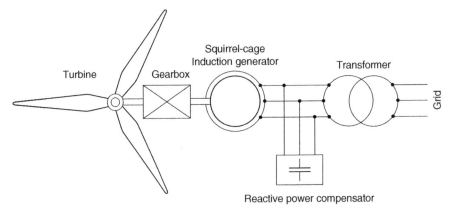

Figure 9.9   Wind-turbine system with induction generator (soft-starter not shown).

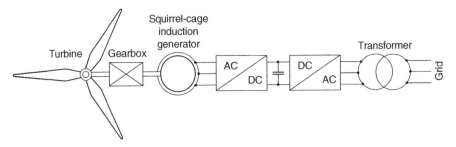

Figure 9.10   Wind-turbine system with an induction generator and frequency changer.

Figure 9.11, dc voltage must be provided to field winding to produce magnetic field in the generator. The rated power of the rectifier employed for that purpose is much lower than that of the frequency changer in the system in Figure 9.10.

Using a low-speed generator, that is, one with a large number of magnetic poles makes it possible to dispose of the gearbox and improve on the cost and reliability of the system. Employed in low- and medium-power systems, permanent-magnet synchronous generators do not require the electrical excitation circuit.

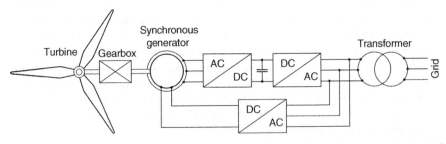

Figure 9.11   Wind-turbine system with an electrically excited synchronous generator and frequency changer.

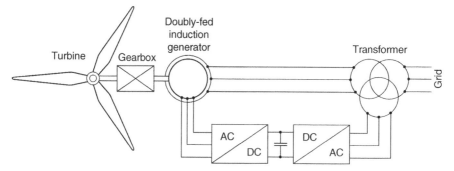

Figure 9.12   Wind-turbine system with a doubly fed induction generator.

In a scheme depicted in Figure 9.12, use of a doubly fed wound-rotor induction generator allows reduction of the rated power of power electronic converters employed. Recall that the ac-to-dc-to-ac converter cascade can pass power in both directions. Indeed, if the generator runs with a super-synchronous speed, that is, the speed of the rotor is higher than that of the stator field, the electrical power is delivered, via the transformer, to the grid from both the stator and rotor. However, a sub-synchronous speed results in electric power flowing into the rotor. As a result, a wide range of turbine speed is achieved, while the converter can be rated at a fraction (25–30%) of the rated power of the generator. The real and reactive powers are controlled in the converter connected to the rotor. This scheme is nowadays most common in the high-power wind-turbine systems.

Recently, matrix converters started to appear in simple low-power residential wind-energy systems as an interface between a permanent-magnet synchronous generator and the grid. An example of such system is shown in Figure 9.13. It is

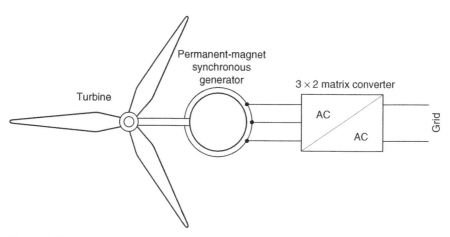

Figure 9.13   Wind-turbine system with a permanent-magnet synchronous generator and a matrix converter.

assumed that in spite of the lack of gearbox the generator voltage is high enough for transformer-less connection to the 120-V single-phase grid.

Wind-turbine systems produce an increasing share of the total electric power in many countries, with Denmark leading the way. It is expected that in the near future as much as half of the total power in that country will be generated in wind farms, mostly offshore ones. Wind farms, which constitute self-contained power plants, must satisfy strict tolerances on the frequency and voltage levels, which require precise real and reactive power control. Also, quick response to transients is necessary to ensure stability of the grid. Different configurations of wind farms are used, such as a local ac network or a local dc network. Various arrangements of control and delivery systems, such as high-voltage dc transmission lines for distant offshore farms, are employed depending on the specifics of the farm.

9.3 FUEL CELL ENERGY SYSTEMS

A fuel cell (FC) directly converts the chemical energy of a fuel to electrical energy. It is a clean source of power as it mostly emits water vapor, with only small amount of pollutants. The principle of energy conversion is similar to that of the electric battery, with the fuel oxidized at the anode and an oxidant reduced at the cathode. The produced ions are exchanged through an electrolyte, and electrons through the load circuit.

At the rated current, a single cell generates about 0.6–0.7 V. Therefore, a practical FC constitutes as stack of cells. Various types of FCs have been developed, the important distinction being the type of fuel used, that is, liquid or gaseous. Liquid fuels include various hydrocarbons, mostly alcohols, with chlorine and chlorine dioxide as oxidants. If gaseous hydrogen is fed to the FC, oxygen serves as the oxidant.

Increasingly, FCs find applications in transportation and portable and distributed power systems, the type of the FC depending on the application requirements, such as specific weight, power density, operating temperature, or start-up speed. The output power must be conditioned. The conditioning requirements include the allowed power and current ranges, as well as the change rate, polarity (negative current is forbidden), and ripple of the current.

An example power system for an FC-powered vehicle is shown in Figure 9.14. The motor driving wheels of the vehicle is of the ac type (permanent-magnet synchronous motor or induction motor), and the normally-charged battery allows bursts of power at acceleration. The converters allow for regenerative braking of the vehicle, with the recovered power charging the battery. Consequently, the dc-to-dc converter connected to the battery must be of the two-quadrant type. In place of the battery a supercapacitor pack can also be employed. In another version of the system, two or more wheels of the vehicle can be driven separately. In 2015, a hydrogen-powered SUV, the Hyundai Tucson Fuel Cell, will appear on the automobile market (initially limited to Los Angeles). With the range of 265 miles, it can be refueled in just 10 minutes. Other major car manufacturers are expected to follow suit. FC drives also appear in unmanned aerial vehicles.

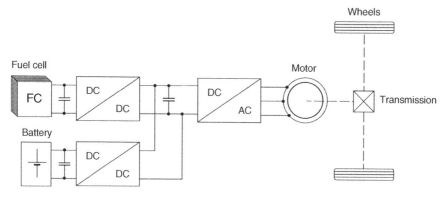

Figure 9.14   FC-powered drive system of a vehicle.

FCs are very well suited for application in distributed generation systems because of their non-polluting operation. In one version of the clean energy policy of the future, hydrogen is employed as a medium for energy storage and transport. It can be produced from water using electrolysis fed by excess energy from PV or wind sources. When power from renewable energy sources decreases, for example, after sunset, the same hydrogen in an FC is oxidized back into water and the energy produced is transferred to the grid. If FC voltage is not much lower than that of the grid, the system shown in Figure 9.15 and consisting of a boost dc-to-dc converter, dc link, and voltage-source inverter can be used. The boost converter provides stable and suitably increased dc voltage to the inverter, which converts it to ac voltage of the magnitude and frequency required by the grid. Notice the similarity of systems in Figures 9.3 and 9.14, which is obvious as both the PV arrays and the FCs are sources of low dc voltage.

If the grid voltage is much higher than the voltage of FC, and/or the transformer isolation is desired, a system shown in Figure 9.16 can be used. The inverter produces a high-frequency voltage to minimize the transformer size. The secondary voltage of the transformer is then converted to a grid-frequency voltage using a cycloconverter. If the FC is to supply power to a dc network, the system in Figure 9.15 can be modified by elimination of the load-side inverter (and dc link). Similarly, the cycloconverter in the system in Figure 9.16 can be replaced with a rectifier.

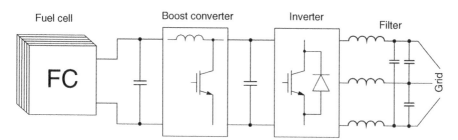

Figure 9.15   FC power system with boost converter for distributed generation.

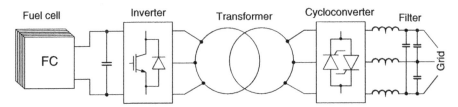

Figure 9.16   FC power system with transformer isolation for distributed generation.

9.4   ELECTRIC CARS

In spite of the limited amount of energy that today's batteries are capable to store, purely electric cars are gaining a foothold on the automobile market. Electric cars are quiet, nonpolluting, and capable of rapid acceleration from standstill thanks to the high starting torque of electric motors. The disadvantages include the need of frequent recharging, scarce charging-stations infrastructure, and prices still higher from those of comparable traditional automobiles.

In general, apart from the battery issues, electric vehicles represent an attractive proposition, mostly due to the fact that the average efficiency of electric machines is some three times higher than that of internal combustion engines. In an electric car, the engine is replaced with an electric motor fed from a battery-supplied inverter as shown in Figure 9.17. The lithium-ion battery is protected from overcharging, for example, when the vehicle coasts down a long slope, by a switched braking resistance (see Figure 6.25). Permanent-magnet synchronous motors are most often employed, although induction motors can also be encountered. When braking or driving down-hill, the motor acts as a generator, recharging the battery. Coaxial motors in certain cars simplify the transmission structure. Specifications of example commercial electric cars are listed in Table 9.1.

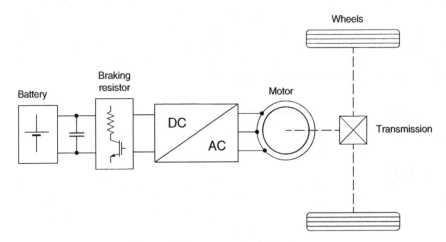

Figure 9.17   Powertrain of electric car.

Table 9.1   Example Electric Cars

Make/Model	Power (hp)	Torque (lb-ft)	0–60 mph (s)	Range (miles)	Battery (kWh)	Charging (kW)
BMW i3	170	184	7.2	81	22	6.6
Chevrolet Spark EV	130	327	7.2	82	21	3.3
Fiat 500e	111	147	8.4	87	24	6.6
Ford Focus Electric	143	184	9.4	76	23	6.6
Mercedes B-Class						
Electric Drive	177	310	7.9	85	28	10.0
Nissan Leaf	107	187	10.2	84	24	6.6
Tesla Model S	362	317	5.4	265	85	10.0
Toyota RAV4 EV	154	218	7.0	100	42	10.0
Volkswagen E-Golf	114	199	10.1	85	24	7.2

In-wheel motors allow disposing of the transmission. A simplified block diagram of such an EV powertrain is shown in Figure 9.18. Each wheel is separately driven by a low-speed high-torque ac motor fed from an inverter. The inverters are supplied from a battery through a capacitive dc link. The decentralized structure of the power scheme allows independent control of the torque and speed of each wheel, eliminating the need for mechanical transmission, differential, and gears. All sophisticated drive functions, such as stability control, non-skid regenerative braking, and optimal distribution of the torque between wheels on slippery roads, can be implemented in

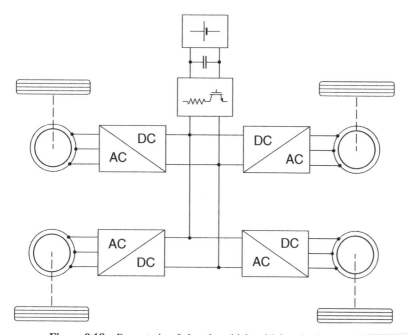

Figure 9.18   Powertrain of electric vehicle with in-wheel motors.

a microcomputer-based digital control system. As of now, such systems have not yet reached the commercial market due to several technical barriers. They include the limited power and vulnerability of in-wheel motors to dirt, heat, and vibration. The existing in-wheel motor drives are employed in certain low-power vehicles, such as electric wheelchairs, scooters, or golf carts.

It is hoped that further progress in technologies of batteries and FCs will in the foreseeable future allow production of long-range electric vehicles. Note that Tesla Model S by Tesla Motors has already attained a 265 miles range, as seen in Table 9.1. However, comparison with other cars displayed in that table shows uniqueness of Tesla automobile with respect to that feature.

9.5   HYBRID CARS

According to the International Electrotechnical Commission, a hybrid electric vehicle (HEV) is "one in which propulsion energy (…) is available from two or more kinds or types of energy stores, sources, or converters (…)." In practice, today's hybrid cars combine a battery, an electric motor, and an internal combustion engine to provide driving power. The battery is often supplemented with a supercapacitor bank, which allows short bursts of extra energy for acceleration.

In a hybrid electric drive system, advantage is being taken from the difference between mechanical characteristics of electric motors and internal combustion engines. For instance, an electric motor can generate full torque at any speed, even at standstill, while the engines require a considerable speed to achieve the maximum torque ability. The efficiency of an electric machine is high under all normal operating conditions, while internal combustion engines operate most efficiently at mid-range speeds and high torque levels. The driving modes include acceleration, cruising, coasting, braking, and downhill or uphill driving. Varying conditions include the amount of load, state of the road surface, wind speed and direction, or ambient temperature. A micropocessor-based control system, equipped with a number of sensors and governed by intelligent operating algorithms, can take advantage of the flexibility of the hybrid drivetrain to ensure best dynamic performance while minimizing fuel consumption.

Three basic architectures of the HEV powertrain are *series*, *parallel*, and *series-parallel*. The series powertrain, illustrated in Figure 9.19, uses an electric motor to drive the wheels. A battery and an electric generator driven by an internal combustion engine are two independent sources of electrical energy. Depending on operating conditions, the motor is fed from the battery (with the engine stopped), from the generator, or from the both sources. Also, if needed, the battery can be charged from the generator, or, vice-versa, the battery can supply power to the generator when it operates as a starter motor for the engine. This flexibility helps saving significant amounts of fuel, especially during city driving. There, frequent starts and stops allow the best use of the torque-speed characteristic of the motor, and recovery of the kinetic energy of the vehicle during regenerative braking. Also, the engine can operate most of the time under the most fuel-efficient conditions.

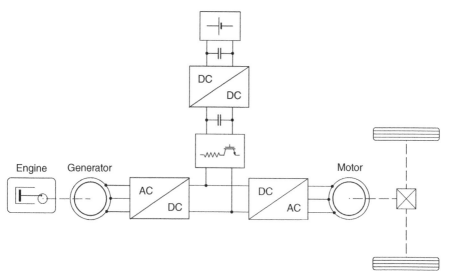

Figure 9.19   Series hybrid powertrain.

In comparison with internal combustion engines, electric machines have lower power density, that is, they are heavier than their internal combustion counterparts of comparable power. Therefore, series hybrid powertrains employing two electric machines are mostly used in heavy commercial and military vehicles, buses, and small locomotives. Still, Chevrolet engineers have decided to employ a series powertrain in the compact Chevy Volt sedan.

The parallel hybrid drive system shown in Figure 9.20 fits in the existing cars better because it does not require a generator. The engine and electric motor combine their torques through a mechanical coupling (gearbox, pulley or chain unit, or a common axle). The torque is then supplied to the wheels via a mechanical transmission. This powertrain configuration is common for several hybrid passenger cars.

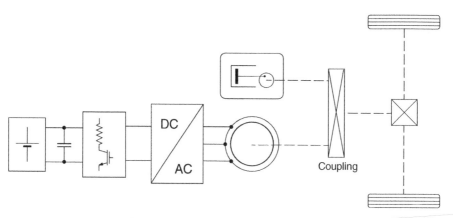

Figure 9.20   Parallel hybrid powertrain.

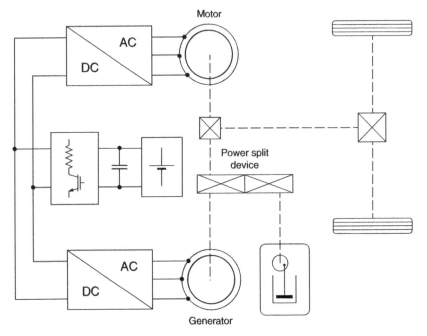

Figure 9.21 Series–parallel hybrid powertrain.

The series–parallel hybrid powertrain, depicted in Figure 9.21 and employed in the popular Toyota Prius, has higher operational flexibility than the described series and parallel drive systems. Its planetary *power split device* divides the engine power between the wheels and the generator, whose main function is to charge the battery. The same generator serves as a starter motor. The driving motor, fed from the dc bus, provides additional torque to the wheels. With the battery fully charged and the generator mechanically disconnected from the engine, this is the regular operating mode. The generator is activated if the battery needs adding some charge. Other modes are also feasible, with the microprocessor based control system selecting the one that is most appropriate for given driving conditions and requirements. For example, while braking, the motor acts as a generator, opposing the motion of wheels and charging the battery. It is even possible to have the generator, driven by the engine, to supply the motor, which then provides the whole needed power to the wheels with no direct help from the engine.

The powertrain of Chevy Volt, the popular US hybrid car, can operate both in the series and series–parallel configurations. Operation modes of that car are illustrated in Figure 9.22.

Hybrid electric cars, examples of which are listed in Table 9.2, have entered the automobile market for good, the more so that gasoline prices do not seem to be set for dropping significantly in any foreseeable future. The HEVs are expected to represent a 15–20% share of that market, and power electronics have found there a large application area.

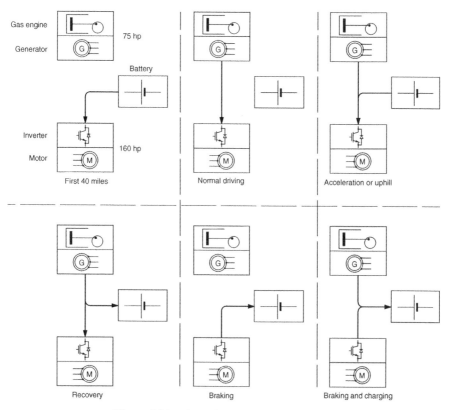

Figure 9.22   Operation modes of Chevy Volt.

Table 9.2   Example Hybrid Cars

Make/Model	Power (hp)	Torque[a] (lb-ft)	0–60 mph (s)	Range[b] (miles)	Mileage[c] (mpge)	Battery (kWh)	Charging (kW)
Cadillac ELR	181	86/295	8.1	35	33	16.5	3.3
Chevy Volt	149	93/273	9.0	38	37	16.0	3.3
Ford Fusion Energi	188	129/117	7.9	21	43	7.0	3.3
Honda Accord Hybrid	196	122/226	7.2	13	46	6.7	6.6
Porsche Panamera S E-Hybrid	416	325/229	5.2	22	50	9.0	3.0
Toyota Prius	134	105/153	10.7	11	50	4.4	3.3

[a]Gas engine/electric motor.

[b]On battery charge.

[c]EPA equivalent mileage calculated as mpge $= \dfrac{\text{miles driven}}{\text{gallons consumed} + \frac{\text{kWh used}}{33.7}}$.

9.6 POWER ELECTRONICS AND ENERGY CONSERVATION

In comparison with other methods of power conversion and control, the use of power electronic converters invariably saves energy. The motor–generator sets of the past were significantly less efficient than the static converters of today, and many processes involving flow control by means of hydraulic, pneumatic, mechanical, or electrical devices employed some sort of "choking." For instance, as explained in Section 9.1, flow of a fluid was traditionally controlled by manipulating a valve, as that in a faucet, and flow of current was controlled by a rheostat (see Section 1.4). Such approach can be compared to driving a car with the transmission and gas pedal stuck in fixed positions, using only the brakes to slow or accelerate the vehicle. In so controlled systems, the efficiency decreases with the load, as the choking is particularly severe at low flow intensities.

The use of power electronic converters allows wide-range torque and speed control in adjustable speed drive systems, leading to huge energy savings in a variety of commercial enterprises, such as the manufacturing and food processing industries, or electric transportation. On the domestic market, many countries impose high efficiency standards on appliances, forcing the manufacturers to install adjustable speed drives in "white goods," such as refrigerators, washers, and dryers. Majority of the electric drives use ac motors, mostly the squirrel-cage induction and permanent-magnet synchronous ones, supplied from inverters. However, rectifier-fed dc drives, easier to control than the ac drives, still find certain applications, such as high-performance positioning systems.

Power electronic converters in the national grid improve the power factor and facilitate control of the real and reactive power, raising the efficiency of transmission and distribution of electric energy. The so-called FACTS (flexible ac transmission systems) devices, such as STATCOMs (static synchronous compensators), SVRs (static var compensators), TCPARs (thyristor-controlled phase angle regulators), TCRs (thyristor-controlled reactors), or TSCs (thyristor-switched capacitors), are increasingly common in the grid.

An example STATCOM, based on a cascaded multilevel inverter (see Figure 7.59), is shown in Figure 9.23a, and a constituent H-bridge in Figure 9.23b. Inductor L represents the combination of both a physical reactor and a leakage inductance of the coupling transformer, or just the latter inductance if the former one is not needed. The STATCOM draws reactive power in the inductive mode of operation when the converter voltage, V_{con} is made lower than the bus voltage, V_{bus}. Vice-versa, the STATCOM delivers reactive power in the capacitive mode when V_{con} is made higher than V_{bus}. Capacitors C in the H-bridges are protected from overvoltage by switched resistors R, which are also employed when the capacitors must be discharged for maintenance or repair of the STATCOM.

Triac-based ac voltage controllers are increasingly used in homes and other buildings for lighting control. Switching power supplies in computer equipment, radios and TV sets, cell phones and portable media players efficiently provide high-quality power. The high power factor, obtained by employing pre-regulators and sophisticated control schemes, minimizes the current drawn from the supply source and diminishes the associated ohmic losses.

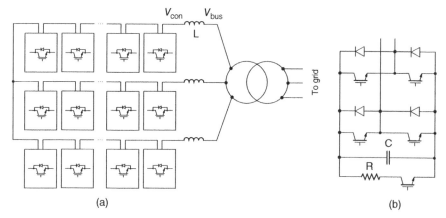

Figure 9.23 STATCOM: (a) block diagram, (b) constituent H-bridge.

The important role of power electronics in modern, clean-energy conscious societies cannot be underrated, and most engineers sooner or later encounter them in their everyday practice. Developing countries also take advantage of the technological progress in power electronics and renewable energy sources. Many communities in the world lack an access to an electric grid. To make things worse, fuel used in the commonly used diesel generators tends to be more expensive there than in the affluent nations. Therefore, solar panels and wind-turbines spring in many places, mostly to provide light and power for basic appliances. The hope for a better future brought about by simple means of electricity production is arguably as important as the tangible results of availability of cheap energy.

SUMMARY

Power electronics and clean energy are closely tied. Alternative energy sources, such as solar arrays, wind-turbines, or FCs, require interfacing with the electric grid or other type of load. Power electronic converters provide the necessary power conditioning to match the source with the load. Various configurations of power electronic interfaces have been developed for the solar, wind, and FC systems. Bidirectional ac-to-dc and dc-to-dc converters, with possible transformer isolation, are the most common components of the interfaces.

Public acceptance of purely electric cars still suffers from the limited energy storage capability and inconvenient re-charging process of today's batteries. However, the market of electric and hybrid automobiles is steadily growing. In electric cars, the internal combustion engine is replaced with an electric motor. In hybrid vehicles, the engine is supplemented with an electric supply of driving power to ensure optimal use of fuel under varying driving conditions. Batteries, often complemented with supercapacitors, constitute the source of electric energy, and a generator driven by the engine provides charging power, especially during regenerative braking. In the advanced hybrids, the power of the engine and the motor are combined in mechanical power split devices which, in addition to two electric machines, allows high flexibility of operation of the powertrain.

Power electronic converters appear in modern electric grids, electric drives, buildings, appliances, computer and communication equipment, and other applications, where electrical power is transmitted or converted. Enormous amounts of energy are saved thanks to the efficiency-optimized operation of those systems. Both the wealthy and poor societies enjoy the benefits of clean electric energy helped by power electronics.

FURTHER READING

[1] Blaabjerg, F., Chen, Z., and Kjaer, S. B., Power electronics as efficient interface in dispersed power generation systems, *IEEE Transactions on Power Electronics*, vol. 19, no. 5, pp. 1184–1194, 2004.

[2] Blaabjerg, F., Consoli, A., Ferreira, J. A., and van Wyk, J. D., The future of electronic power processing and conversion, *IEEE Transactions on Power Electronics*, vol. 20, no. 3, pp. 715–720, 2005.

[3] Ehsani, M., Gao, Y., Gay, S. E., and Emadi, A., *Modern Electric, Hybrid Electric, and Fuel Cell Vehicles: Fundamentals, Theory, and Design*, CRC Press, Boca Raton, FL, 2005.

[4] Ehsani, M, Gao, Y., and Miller, J. M., Hybrid electric vehicles: architecture and motor drives, *Proceedings of the IEEE*, vol. 95, no. 4, pp. 719–728, 2007.

[5] Hingorani, N. S. and Gyugui, L., *Understanding FACTS: Concepts and Technology of Flexible AC Transmission Systems*, IEEE Press, New York, 2000.

[6] Special section on renewable energy systems—Part I, *IEEE Transactions on Industrial Electronics*, vol. 58, no. 1, pp. 2–212, 2011.

[7] Special section on renewable energy systems—Part II, *IEEE Transactions on Industrial Electronics*, vol. 58, no. 4, pp. 1074–1293, 2011.

[8] Teodorescu, R., Liserre, M., and Rodriguez, P., *Grid Converters for Photovoltaic and Wind Power Systems*, John Wiley & Sons, Ltd., Chichester, 2011.

APPENDIX A
Spice Simulations

For better understanding of operating principles of power electronic converters, 46 Spice circuit files have been developed to be used with this book. They are available at http://www.wiley.com/go/modernpowerelectronics3e. The name of a file clearly identifies the topic, for example, the *Boost_Conv.cir* file models the dc-to-dc boost converter. An asterisk after the sequential number of a computer assignment indicated existence of an accompanying circuit file. For example, assignment CA8.2*, which involves simulation of the boost converter, requires use of the already mentioned *Boost_Conv.cir* file.

LTspice from Linear Technology or PSpice Lite from Cadence is recommended for simulations. Go to http://www.linear.com/designtools/software/ltspice.jsp for a free download of the LTspice. For a manual, go to http://denethor.wlu.ca/ltspice. Alternatively, the PSpice Lite software can be downloaded from https://www .cadence.com/products/orcad/Pages/orcaddownloads.aspx. You will need to create an account to get it. It is almost 700 MB, so give it a while to download. You can also order a free CD.

Both software packages are easy to use. The average and rms values in LTspice are obtained by holding down the control key and clicking the label of the trace of interest. The PSpice Lite allows plotting avg(…) and rms(…) traces. They represent running computations, thus the most accurate values should be read at the right-hand end of a trace. Many universities post Spice primers on the Internet. For instance, the University of Pennsylvania's one is available at http://www.seas.upenn.edu/~jan/spice /PSpicePrimer.pdf.

The generic voltage-controlled switches available in Spice are employed in all converter models. Circuit diagrams of four such switches appearing in the circuit files, specifically (a) single-gate switch, SGSW, (b) two-gate switch, TGSW, (c) unidirectional switch, UDSW, and (d) generic SCR, are shown in Figure A.1. In some circuit files, additional components, such as a series RC snubber circuit, are connected to the switches to improve operating conditions and secure the convergence of computations. The hysteresis current control in the three-phase voltage-source inverter in the *Hyster_Curr_Contr.cir* file is realized by subcircuit HCC depicted in Figure A.2. The

Introduction to Modern Power Electronics, Third Edition. Andrzej M. Trzynadlowski.
© 2016 John Wiley & Sons, Inc. Published 2016 by John Wiley & Sons, Inc.
Companion website: www.wiley.com/go/modernpowerelectronics3e

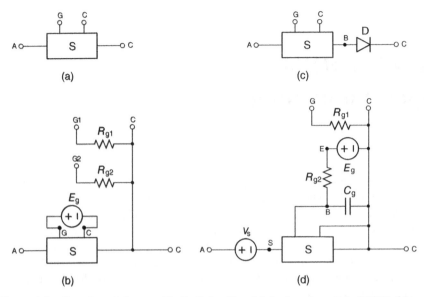

Figure A.1   Generic switches used in the Spice files: (a) single-gate switch, SGSW, (b) two-gate switch, TGSW, (c) unidirectional switch, UDSW, (d) generic SCR.

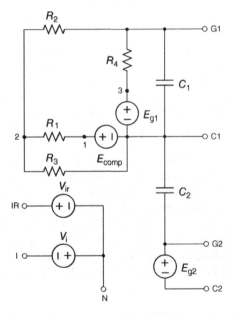

Figure A.2   Hysteresis current controller, HCC, for one leg of an inverter.

inductor-coupling Spice device, K, is used to model transformers in isolated dc-to-dc converters.

The Spice was originally developed for the simulation of analog electronic circuits, and switched networks are sometimes poorly handled. Therefore, certain files contain the message "RUN AS IS!" to indicate that any parameter changes will endanger the convergence. The reader can also notice that the RON and ROFF resistances of the voltage-controlled switches, Sgen, vary from file to file. The same applies to the .OPTIONS statement. Those variations have resulted from attempts to get the simulations run smoothly. Generally, the convergence can be improved by changing (usually relaxing) the tolerances in the .OPTIONS statement and reducing the ROFF/RON ratio in the Sgen model. Regrettably, some less important waveforms are still marred by "spikes" resulting mainly from the idealized nature of the assumed switches. To view such waveforms, the amplitude (Y-axis) scale should be manually reduced.

The manual setting of ranges of the axes is also needed in the case of rippled dc waveforms, for example, the output voltage of the voltage-type PWM rectifier. The automatic setting results in a misleading display of only the ripple.

No effort was spared to make the simulations user friendly. In particular, text strings, not numbers, are used to designate nodes, for example, "OUT" for an output terminal. This facilitates selection of waveforms to be displayed using the mock-oscilloscope Probe program. Also, each circuit file contains information about the textbook figure depicting the simulated circuit. Each assignment involving the Spice programs should begin with copying the corresponding figure and marking the nodes in accordance with the circuit file. The work involved will pay off by faster memorization of power electronic circuits.

For measurement of waveform components and determination of figures of merit, the $avg(...)$ and $rms(...)$ operators should be used. To determine the fundamental of a waveform, employ the "FFT" or "Fourier" options. Remember that the harmonic spectra show peak, not rms, values of the waveform harmonics.

The provided Spice circuit files represent, in a sense, a virtual laboratory of power electronics. Performing the simulations, observing the waveforms and power spectra of voltages and currents, and determining waveform components and figures of merit should give the reader better feel and appreciation of power electronic converters. Wherever possible, compare the simulation results with theoretical ones given by equations in the book. By no means should the learning exercises be limited to the suggested assignments. Students (and instructors) are encouraged to "tinker" with the modeled converters and expand the set of "virtual experiments" by developing new files.

List of the Spice Circuit Files

1.	*AC_Chopp.cir*	Single-phase ac chopper
2.	*AC_Volt_Contr_1ph.cir*	Single-phase ac voltage controller
3.	*AC_Volt_Contr_3ph.cir*	Three-phase ac voltage controller
4.	*Boost_Conv.cir*	Boost converter
5.	*Buck_Aver_Model.cir*	Averaged model of buck converter
6.	*Buck_Conv.cir*	Buck converter

APPENDIX B
Fourier Series

A function $\psi(t)$ is called *periodic* if

$$\psi(t + T) = \psi(t), \tag{B.1}$$

where T is the *period* of $\psi(t)$. The *fundamental frequency*, f, is defined as

$$f \equiv \frac{1}{T}\text{Hz}, \tag{B.2}$$

and the fundamental radian frequency, ω_1, as

$$\omega \equiv \frac{2\pi}{T} = 2\pi f \text{ rad/s}. \tag{B.3}$$

A periodic function can be represented by an infinite *Fourier series* as

$$\psi(t) = a_0 + \sum_{k=1}^{\infty} [a_k \cos(k\omega t) + b_k \sin(k\omega t)], \tag{B.4}$$

where

$$a_0 = \frac{1}{T} \int_0^T \psi(t)\,dt \tag{B.5}$$

$$a_k = \frac{2}{T} \int_0^T \psi(t) \cos(k\omega t)\,dt \tag{B.6}$$

$$b_k = \frac{2}{T} \int_0^T \psi(t) \sin(k\omega t)\,dt. \tag{B.7}$$

Introduction to Modern Power Electronics, Third Edition. Andrzej M. Trzynadlowski.
© 2016 John Wiley & Sons, Inc. Published 2016 by John Wiley & Sons, Inc.
Companion website: www.wiley.com/go/modernpowerelectronics3e

Alternately, $\psi(t)$ can be expressed as

$$\psi(t) = c_0 + \sum_{k=1}^{\infty} c_k \cos(k\omega t + \theta_k)dt, \tag{B.8}$$

where

$$c_0 = a_0 \tag{B.9}$$

$$c_k = \sqrt{a_k^2 + b_k^2} \tag{B.10}$$

$$\theta_k = \begin{cases} -\tan^{-1}\left(\dfrac{b_k}{a_k}\right) & \text{if} \quad a_k \geq 0 \\[3mm] -\tan^{-1}\left(\dfrac{b_k}{a_k}\right) \pm \pi & \text{if} \quad a_k < 0 \end{cases}. \tag{B.11}$$

Coefficient c_0 constitutes the *average value* of $\psi(t)$, and c_1 is the *peak value of the fundamental* of $\psi(t)$. If $\psi(t)$ represents a voltage or current, terms "dc component" for c_0 and "peak value of fundamental ac component" for c_1 are also in use. Coefficients $c_2, c_3, \ldots$, are peak values of *higher harmonics* of $\psi(t)$, the subscript denoting the *harmonic number*. Thus, the kth harmonic is a sinusoidal component of $\psi(t)$ with the peak value of c_k and radian frequency of $k\omega_1$. The harmonic angle, θ_k, is of minor importance in power electronics, except for the fundamental angle, θ_1.

Many practical waveforms are characterized by certain symmetries, which allow reducing the computational effort required for determination of Fourier series. To take advantage of those symmetries, the average value, if any, should first to be subtracted from the analyzed function $\psi(t)$, leaving a periodic function $\vartheta(t)$ given by

$$\vartheta(t) = \psi(t) - a_0 \tag{B.12}$$

and whose average value is zero. Clearly, coefficients a_k and b_k are the same for both $\psi(t)$ and $\vartheta(t)$. Thus, the latter function can be used in place of $\psi(t)$ in Eqs. (B.6) and (B.7) to determine those coefficients.

The most common symmetries are:

(1) *Even* symmetry, when

$$\vartheta(-t) = \vartheta(t). \tag{B.13}$$

Then,

$$b_k = 0 \tag{B.14}$$

$$c_k = a_k \tag{B.15}$$

for all values of k from 1 to ∞. In particular, the cosine function has the even symmetry, and that is why the sine coefficients, b_k, of the Fourier series are nulled.

(2) *Odd symmetry*, when

$$\vartheta(-t) = -\vartheta(t). \tag{B.16}$$

Then,

$$a_k = 0 \tag{B.17}$$

$$c_k = b_k \tag{B.18}$$

for all values of k from 1 to ∞. In particular, the sine function has the odd symmetry, and that is why the cosine coefficients, a_k, of the Fourier series are nulled.

(3) *Half-wave symmetry*, when

$$\vartheta\left(t + \frac{T}{2}\right) = -\vartheta(t). \tag{B.19}$$

Then $a_k = b_k = c_k = 0$ for $k = 2, 4, \ldots$, and the remaining coefficients of Fourier series can be calculated as

$$a_k = \frac{4}{T} \int_0^{T/2} \vartheta(t) \cos(k\omega t)\, dt \tag{B.20}$$

$$b_k = \frac{4}{T} \int_0^{T/2} \vartheta(t) \sin(k\omega t)\, dt \tag{B.21}$$

for odd values of k. Both the sine and cosine functions have the half-wave symmetry.

If $\vartheta(t)$ has both the even symmetry and half-wave symmetry, coefficients c_k can directly be found as

$$c_k = \begin{cases} \dfrac{8}{T} \displaystyle\int_0^{T/2} \vartheta(t) \cos(k\omega t)\, dt & \text{for} \quad k = 1, 3, \ldots \\ 0 & \text{for} \quad k = 2, 4, \ldots \end{cases} \tag{B.22}$$

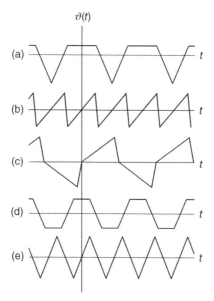

Figure B.1 Examples of various symmetries: (a) even, (b) odd, (c) half-wave, (d) even and half-wave, (e) odd and half-wave.

Analogously, if $\vartheta(t)$ has both the odd symmetry and half-wave symmetry, then

$$
c_k = \begin{cases} \dfrac{8}{T} \displaystyle\int_0^{T/2} \vartheta(t) \sin(k\omega t)\, dt & \text{for} \quad k = 1, 3, \ldots \\[4pt] 0 & \text{for} \quad k = 2, 4, \ldots \end{cases}
\tag{B.23}
$$

Various symmetries are illustrated in Figure B.1. The equations presented can easily be adapted to functions expressed in the angle domain, that is, $\psi(\omega t)$ and $\vartheta(\omega t)$ instead of $\psi(t)$ and $\vartheta(t)$. The period T is then replaced with 2π.

Certain waveforms, such as the optimal switching pattern in Figure 7.26, possess the so-called *quarter-wave symmetry*, which means that

$$
\vartheta\left(t + \frac{T}{4}\right) = \vartheta\left(\frac{T}{4} - t\right)
\tag{B.24}
$$

and

$$
\vartheta\left(t + \frac{3T}{4}\right) = \vartheta\left(\frac{3T}{4} - t\right).
\tag{B.25}
$$

APPENDIX C
Three-Phase Systems

Three-phase systems consist of three-phase sources and three-phase loads connected by three-wire or four-wire lines. A three-phase source is composed of three interconnected ac sources, and three interconnected loads comprise a three-phase load. Ideal sinusoidal voltage sources are assumed in the subsequent considerations. It is also assumed that the sources and loads are balanced. It means that the individual source voltages have the same value, differ from each other by the phase shift of 120°, and the load consists of three identical impedances with (if any) EMFs satisfying the same balance conditions as the source voltages.

A three-phase source can be connected either in wye (Y) or in delta (Δ), as shown in Figure C.1. These two connections also apply to three-phase loads, depicted in Figure C.2. Clearly, there are four possible arrangements of a three-phase source supplying a three-phase load: Y–Y, Y–Δ, Δ–Y, and Δ–Δ. As illustrated in Figure C.3a, three- or four-wire lines can be used between the source and load in the Y–Y system. The four-wire connection is not feasible in the other three systems, such as the Δ–Y system in Figure C.3b.

Two types of voltage and two types of current, all indicated in Figure C.3, appear in three-phase systems. These are:

(1) Line-to-neutral voltages, v_{AN}, v_{BN}, and v_{CN}, given by

$$v_{AN} = \sqrt{2}V_{LN}\cos(\omega t)$$
$$v_{AN} = \sqrt{2}V_{LN}\cos\left(\omega t - \frac{2}{3}\right),$$
$$v_{AN} = \sqrt{2}V_{LN}\cos\left(\omega t - \frac{4}{3}\right)$$

$$(C.1)$$

where V_{LN} denotes the rms value of these voltages. Somewhat imprecisely, the line-to-neutral voltages are also called *phase voltages*.

Introduction to Modern Power Electronics, Third Edition. Andrzej M. Trzynadlowski.
© 2016 John Wiley & Sons, Inc. Published 2016 by John Wiley & Sons, Inc.
Companion website: www.wiley.com/go/modernpowerelectronics3e

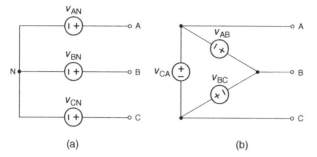

Figure C.1   Three-phase sources: (a) wye-connected, (b) delta-connected.

(2) *Line-to-line voltages*, v_{AB}, v_{BC}, and v_{CA}, given by

$$v_{AB} = \sqrt{2}V_{LL} \cos\left(\omega t + \frac{1}{6}\pi\right)$$

$$v_{AN} = \sqrt{2}V_{LL} \cos\left(\omega t - \frac{1}{2}\pi\right), \qquad (C.2)$$

$$v_{AN} = \sqrt{2}V_{LL} \cos\left(\omega t - \frac{7}{6}\pi\right)$$

where V_{LL}, which equals $\sqrt{3}V_{LN}$, denotes the rms value of these voltages. It is worth stressing that the V_{LL} value is universally used as the rated voltage of three-phase lines and apparatus. Alternately, the line-to-line voltages are called *line voltages*[a].

(3) Line currents, i_A, i_B, and i_C, given by

$$i_A = \sqrt{2}I_L \cos(\omega t - \varphi)$$

$$i_B = \sqrt{2}I_L \cos\left(\omega t - \varphi - \frac{2}{3}\pi\right), \qquad (C.3)$$

$$i_C = \sqrt{2}I_L \cos\left(\omega t - \varphi - \frac{4}{3}\pi\right)$$

where I_L denotes the rms value of these currents and φ is the *load angle*.

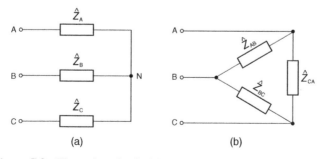

Figure C.2   Three-phase loads: (a) wye-connected, (b) delta-connected.

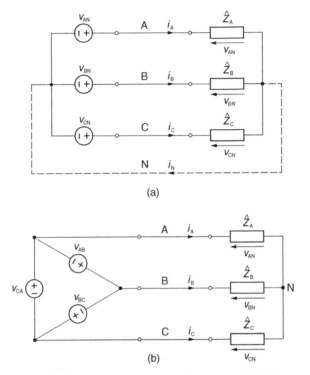

(a)

(b)

Figure C.3   Three-phase source-load systems: (a) Y–Y, (b) Δ–Y.

(4) *Line-to-line currents*, i_{AB}, i_{BC}, and i_{CA}, also known as *phase currents* and given by

$$i_{AB} = \sqrt{2}I_{LL}\cos\left(\omega t - \varphi + \frac{1}{6}\pi\right)$$
$$i_{BC} = \sqrt{2}I_{LL}\cos\left(\omega t - \varphi - \frac{1}{2}\pi\right). \qquad \text{(C.4)}$$
$$i_{CA} = \sqrt{2}I_{LL}\cos\left(\omega t - \varphi - \frac{7}{6}\pi\right)$$

Note that in delta-connected loads these voltages appear across individual load impedances, analogously to line-to-neutral voltages in wye-connected loads. Thus, logically, they are "phase voltages", too.

where the rms value, I_{LL}, of these currents equals $I_L/\sqrt{3}$.

Phasor diagram of the voltages and currents is shown in Figure C.4. The *apparent power*, S, expressed in volt-amperes (VA) and defined as

$$S \equiv V_{AN}I_A + V_{BN}I_B + V_{CN}I_C = V_{AB}I_{AB} + V_{BC}I_{BC} + V_{CA}I_{CA}, \qquad \text{(C.5)}$$

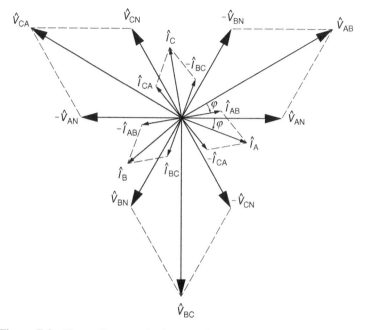

Figure C.4   Phasor diagram of voltages and currents in a three-phase system.

where the right-hand side quantities denote rms values of the respective voltages and currents, equals

$$S = 3V_{LN}I_L = 3V_{LL}I_{LL} = \sqrt{3}V_{LL}I_L \qquad (C.6)$$

in a balanced system. The right-hand side expression is most commonly used because the line-to-line voltage and line current are easily available for measurement in both the three- and four-wire systems.

The *real power*, *P*, expressed in watts (W), transmitted from the source to the load, is given by

$$P = S\cos(\varphi) = \sqrt{3}V_{LL}I_L\cos(\varphi) \qquad (C.7)$$

and the *reactive power*, *Q*, expressed in volt-amperes reactive (VAr), by

$$Q = S\sin(\varphi) = \sqrt{3}V_{LL}I_L\sin(\varphi). \qquad (C.8)$$

Equations, (C.6) through (C.8) are valid for all four source-load systems shown in Figure C.3. The ratio of the real power to the apparent power is defined as the *power factor*, PF:

$$PF \equiv \frac{P}{S} = \cos(\varphi).$$

(C.9)

It must be stressed that the power factor equals the cosine of load angle only in the case of sinusoidal voltages and currents, and identical load angles of all the three phase loads.

Considering Eq. (C.7), if P and V_{LL} are constant, then the low value of $\cos(\varphi)$ must be compensated by a high value of I_L. Thus, low power factor causes extra losses in the resistances of the power system. To recoup these loses, utility companies charge users for the amount of reactive energy consumed (a time integral of reactive power). Indeed, according to Eqs. (C.8) and (C.9), a low power factor corresponds to a large power angle and high reactive power.

INDEX

Introduction to Modern Power Electronics, Third Edition. Andrzej M. Trzynadlowski.
© 2016 John Wiley & Sons, Inc. Published 2016 by John Wiley & Sons, Inc.
Companion website: www.wiley.com/go/modernpowerelectronics3e

Printed and bound by CPI Group (UK) Ltd, Croydon, CR0 4YY

16/04/2025

14658583-0005